Citrus pests
and their
natural enemies

Authors

Dan Smith
Andrew Beattie
Mali Malipatil
David James
Dan Papacek
James Altmann
Andrew Green
Bill Woods
Geoff Furness
Chris Freebairn
Mark Stevens
Greg Buchanan
David Madge
Craig Feutrill
John Kennedy
Brian Gallagher
Paul Jones
Roger Broadley
Megan Edwards
Greg Baker
Scott Dix
Peter Burne
Bill Frost
Jeff Watson
Miriam Pywell

INFORMATION SERIES
QI97030

Citrus pests
and their
natural enemies

INTEGRATED PEST MANAGEMENT IN AUSTRALIA

Edited by Dan Smith,
GAC Beattie & Roger Broadley

ISSN 0727–6273
Agdex 220/610

First published 1997

This book was produced by the following organisations, working in partnership, and with contributions and support from the citrus growers of Australia, and from many people involved in integrated pest management.

National Library of Australia cataloguing-in-publication data:

Citrus pests and their natural enemies: integrated pest management in Australia

Bibliography.
Includes index.
ISBN 0 7242 6695 X.

1. Citrus fruits—Diseases and pests—Biological control—Australia. 2. Citrus fruits—Diseases and pests—Integrated control—Australia. I. Smith, Dan. II. Beattie, G. A. C. III. Broadley, R. H. (Roger H.). IV. Queensland. Dept of Primary Industries. (Series: Information series (Queensland. Dept of Primary Industries); QI97030).

634.30490994

Typeset in Goudy Old Style 10/12 point, and Optima

Editing and production by
Ruth Ridgway and Jim Arthur
Desert Oak Publishing Services

Designed by Anne-Maree Althaus

Cover photographs
front cover: spined citrus bug adult, nymphs and eggs
back cover: top right, Valencia oranges damaged by citrus rust mite; top left, aphytis wasp laying an egg on red scale; middle left, brown lacewing, a predator of scales and aphids; bottom left, soft brown scale; right, citrus gall wasp

Copyright © State of Queensland, Department of Primary Industries, and Horticultural Research and Development Corporation, 1997

Copyright protects this publication. Except for purposes permitted by the Copyright Act, reproduction by whatever means is prohibited without prior written permission. Copyright inquiries should be addressed to:
Manager Publishing Services
Department of Primary Industries
GPO Box 46
Brisbane Q 4001
AUSTRALIA

For sales and orders, contact DPI Publications at the above postal address, or by
email: books@dpi.qld.gov.au

Disclaimer
The information in this book was the best available at the time of publication. However, integrated pest management is a dynamic and growing field, and readers should seek professional advice on developments since publication, particularly in the areas of monitoring and pest management.

Contents

	Foreword	ix
	Acknowledgements	xi
	About the authors	xii
	Glossary	xiv
1	*The Australian citrus industry*	1
	Citrus species and related plants	1
	Origins of the Australian citrus industry	1
	Varieties of citrus grown in Australia	1
	Rootstocks	3
	Crop cycle	3
	Tree and orchard management	5
	Production and climate	5
	Diseases	9
2	*Concepts of IPM*	14
	Biological control	14
	Cultural control	14
	Chemical control	14
	Advantages of IPM	14
	Components of IPM	15

CITRUS PESTS AND THEIR NATURAL ENEMIES

	Mites	17
3	*Plant-feeding mites and other mites*	17
	Brown citrus rust mite	17
	Citrus rust mite	21
	Citrus bud mite	24
	Broad mite	25
	Two-spotted mite	28
	Oriental spider mite	31
	Citrus red mite	32
	Citrus flat mite	34
	Other mites	35
	Insects	37
4	Ants	37

	Scales	39
5	*Soft scales*	39
	Citricola scale	39
	Green coffee scale	41
	Soft brown scale	43
	Long soft scale	46
	Black scale	47
	Hemispherical scale	49
	Nigra scale	50
	Cottony citrus scale (pulvinaria scale)	52
	Pink wax scale	53
	Florida wax scale	56
	White wax scale	57
	Hard wax scale (Chinese wax scale)	59
	Cottony cushion scale	61
6	*Hard or armoured scales*	64
	Red scale	64
	Yellow scale	68
	Circular black scale (Florida red scale)	69
	Citrus snow scale (white louse scale)	70
	Mussel scale (purple scale)	73
	Glover's scale	74
	Chaff scale	76
7	*Mealybugs*	78
	Citrophilous mealybug	78
	Longtailed mealybug	80
	Citrus mealybug	82
	Spherical mealybug	85
	Rastrococcus mealybug	87
8	**Aphids**	**89**
	Citrus aphids	89
	Melon aphid and spiraea aphid	92
9	*Planthoppers and leafhoppers*	95
	Planthoppers (flatids)	95
	Passionvine hopper	98
	Citrus leafhopper (citrus jassid)	99
10	*Cicadas*	102
	Bladder cicada	102
11	*Whiteflies*	104
	Australian citrus whitefly	104
	Aleurocanthus whitefly	106

12	*True bugs*	108
	Spined citrus bug	108
	Bronze orange bug	111
	Green vegetable bug	113
	Citrus blossom bug	115
	Crusader bug	116
13	*Moths and butterflies*	118
	Citrus leafminer	118
	Lightbrown apple moth	122
	Orange fruitborer	124
	Citrus rindborer	125
	Other fruitborers	126
	Fruitpiercing moths	129
	Lemon bud moth and citrus flower moth	131
	Citrus leafroller	133
	Corn earworm and native budworm	134
	Banana fruit caterpillar	137
	Citrus butterflies	138
14	*Beetles*	141
	Fuller's rose weevil	141
	Citrus leafeating weevil	142
	Elephant weevil	144
	Citrus fruit weevil	145
	Spinelegged citrus weevil	146
	Apple weevil	147
	Citrus branch borer	148
	Fig, pittosporum and citrus longicorns	149
	Speckled longicorn	151
	Monolepta beetle and rhyparida beetle	153
15	*Katydids, crickets and grasshoppers*	155
	Katydids	155
	Citrus leafeating cricket	157
	Giant grasshopper	159
16	*Flies and midges*	160
	Queensland fruit fly	160
	Mediterranean fruit fly (medfly)	163
	Papaya fruit fly	165
	Citrus blossom midge	166
17	*Wasps*	168
	Citrus gall wasp	168
	Paper wasps	170

18	*Thrips*	172
	Scirtothrips	172
	Megalurothrips	174
	Citrus rust thrips (orchid thrips)	175
	Greenhouse thrips	176
	Plague thrips	179
19	*Termites*	181
	Giant northern termite	181
20	*Other insects and mites*	183
	Insects	183
	Mites	184
21	*Spiders*	185
	Brown house spider (webbing spider)	185
22	*Nematodes*	186
	Citrus nematode	186
	Root lesion nematodes	188
	Stubby root nematodes	189
23	*Snails*	190
	Common garden snail	190
24	*Natural enemies of citrus pests*	192
	Parasites	192
	Predators	195
	Pathogens	199

INTEGRATED PEST MANAGEMENT (IPM)

25	*Practical integrated pest management (IPM)*	201
	A template for IPM in citrus	201
	Monitoring guides	215
26	*Pesticide application*	226
	Sprayer types	226
27	*Petroleum spray oils*	231
	Equipment and coverage	231
	Timing of PSO sprays	232
28	*Pesticide toxicity to natural enemies*	234

APPENDIXES

Appendix 1	*Exotic citrus pests*	238
Appendix 2	*Keys for identifying common wasp parasites of scales and mealybugs*	240
Appendix 3	*A brief history of IPM in Australian citrus*	252

BIBLIOGRAPHY 256

INDEX 263

Foreword

The publication of *Citrus pests and their natural enemies*, the first book of its kind in Australia, is most timely. The citrus industry in Australia, like many other horticultural industries, is going through a period of substantial growth and change, due to the development and expansion of Australian and export markets. Consumers want clean, attractive, high-quality, good-tasting fruit that is safe to eat.

One of the challenges for growers and distributors in competitively producing and marketing such high-quality fruit is to minimise the use of pesticides and other chemicals in controlling pests and diseases. It is imperative that growers and others involved in pest management develop a thorough knowledge and understanding of pests in the orchard and how to control them effectively and economically.

A wide range of pests attack citrus in Australia, and there are many others throughout the world that would be a significant threat to the citrus industry if introduced into this country. Responsible adherence to quarantine requirements for interstate, export and import trade is critically important for the continued prosperity of the industry.

Integrated pest management (IPM) and biological control methods have been well established in Australian citrus for many years. However, increased consumer demand for 'clean and green' products, greater understanding of occupational health and safety on farms, and higher chemical costs have all combined to highlight the need for universal adoption of IPM by citrus growers in their quest for international competitiveness and sustainability.

Citrus pests and their natural enemies deals with over one hundred pests. Some cause problems every season, but most cause significant damage in some areas in some seasons. An amazingly high number of natural enemies of pests are active in the citrus orchard. By encouraging these natural enemies to control pests, growers can reduce pesticide use to a light spray program of low-toxicity chemicals, and produce the desired 'clean and green' fruit. An essential part of such integrated pest management is the systematic and regular monitoring or orchards to identify pests, their natural enemies, and action levels (when pest control is needed).

The citrus industry will find this outstanding book invaluable in meeting the challenges of increasing economic and environmental pressures. I congratulate the authors and editors, and thank citrus growers, and State government research agencies for funding and supporting the book's production in partnership with the Horticultural Research and Development Corporation.

Michael Keenan
Director
Horticultural Research and
Development Corporation

Acknowledgements

Funds to produce this book were provided by the Horticultural Research and Development Corporation (HRDC), and Australian citrus growers. Without this support, the book could not have been published.

Many people besides the authors and editors made valuable contributions to the book. We thank everyone, and would like to particularly acknowledge the following people and their work.

The technical assistance and specimens provided by citrus pest scouts Gary Artlett, Malcolm Wallace, Margot Gallagher, Neil Renfrey, Stuart Pettigrew and entomologist Graham Baker are greatly appreciated.

We thank Hugh Cope, John Forsyth and John Owen-Turner for commenting on chapter 1, 'The Australian citrus industry', Sharyn Foulis for assistance with information on mealybugs, Greg Walker for reviewing the nematode section, and John Dick for his input on the Western Australian perspective. Valuable advice was given by plant pathologists Pat Barkley, Peter Mayers and Bob Emmett on diseases, and by David Holdom and Bob Teakle on pathogens. We are grateful to John Donaldson, Ian Naumann, Murray Fletcher, Mary Carver, Gordon Gordh, Jan Green, David Rentz, Gerard Prinsloo and Max Moulds for answers to many taxonomic queries. Deanna Chin, Stuart Smith and Graham Young kindly supplied information on the status of citrus pests in the Northern Territory.

We thank Lorraine Chapman, who produced initial computer graphics used in the section on pest monitoring. Michael Thomas drew the colour distribution maps for individual pests, and assisted with computing.

Thank you, too, to the staff of the Maroochy Horticultural Research Station, particularly to Sonia Silcock, Susan Butler and Lisa Langton who helped wordprocess the manuscript, and to Paul McCarthy for excellent administrative support.

We gratefully acknowledge Lisa Forster, Bob Luck, and Elizabeth Grafton-Cardwell, University of California, Riverside, for permission to use the life-cycle drawing of red scale (from *Life stages of California red scale and its parasitoids*, University of California Publication #21529).

Life cycle drawings were produced specifically for this book by Susan Phillips, Catherine Symington, and Chris Lambkin. These beautiful drawings enhance the book's value as an aid in identifying pests and their natural enemies. In addition, some life cycle drawings have been reproduced from *The GOOD BUG book*, published by Australasian Biological Control Inc. We thank this group for permission to use these drawings, and also the tables on pages 234–236, which have been adapted from tables in *The GOOD BUG book*.

The following people made valuable contributions to the keys for identifying common wasp parasites of scales and mealybugs in Australia. Dr M. Hayat, of Aligarh University, India, identified some parasites included in the keys, critically read an early draft of the key, and kindly checked key characters against identified specimens in his collection. Dr S. V. Triapitsyn, of the University of California, Riverside, arranged the loan of identified specimens, and read an early draft of the keys. Dr Ian Naumann, of the CSIRO Division of Entomology, provided general comments on the draft keys. Ms C. Symington prepared the illustrations for the keys.

We pay tribute to the quality of the photographs, which are a major feature of the book. There are over 400 colour plates illustrating insect and mite pests, and their natural enemies. Of these photographs, Chris Freebairn contributed 200, Dan Smith 40, Dan Papacek 30, and David Ironside 30. The remainder were contributed by Miriam Pywell, John Kennedy, Andrew Beattie, Andrew Green, Geoff Furness, David James, James Altmann, Roger Broadley, Harry Fay, Jeff Watson, Bill Woods, Mark Stevens, Dick Drew, Murray Fletcher and Harold Browning. Plant pathologists Pat Barclay, Bob Emmett and Peter Mayers contributed some of the disease photographs, John Bagshaw over 30 photographs of mechanical damage for the monitoring guides, and John Forsyth and Slade Lee photographs of citrus cultivars.

In addition to contributing a large number of the photographs, Chris Freebairn helped with the final selection of photographs, and with writing the captions. He also clarified information on several pests and their natural enemies.

We would also like to thank Ruth Ridgway and Jim Arthur, Desert Oak Publishing Services, for editing and producing the book in digital form, and Anne-Maree Althaus for design. These three people have contributed greatly to realising our vision of a beautiful, and at the same time functional, book.

Finally, a special acknowledgement. Our families have been a tremendous help to us in many ways during our labours on the book. Thank you all for your support.

In particular, Dan Smith would like to dedicate his contribution to the book to his wife, Helen, and his parents, Marion and Thomas Smith.

Dan Smith

Andrew Beattie

Roger Broadley

and the author team

About the authors

Dan Smith is a Senior Principal Entomologist with the Department of Primary Industries, Queensland, based at the Maroochy Horticultural Research Station, Nambour. Dan has worked in citrus entomology for 28 years, achieving the successful introduction of over a dozen biocontrol agents, and making a major contribution to the implementation of integrated pest management (IPM) in Queensland citrus. He has also researched IPM in bananas, custard apples, passionfruit and papaws.

Andrew Beattie was New South Wales Agriculture's citrus entomologist from 1975 to 1996 at the Biological and Chemical Research Institute at Rydalmere in Sydney. He is now Associate Professor, School of Horticulture, University of Western Sydney, Hawkesbury, at Richmond in New South Wales.

Roger Broadley is a Senior Principal Horticulturist (Integrated Crop Protection) who has worked for the Department of Primary Industries, Queensland, for 25 years, and is based at the Maroochy Horticultural Research Station, Nambour. One of his major interests is developing and testing integrated crop protection systems for Queensland horticultural crops. Roger is the editor and co-author of 12 crop protection books, and several professional videos and wall charts. Recently he has worked on the development and adoption of sustainable cropping systems for strawberries, and in extension with custard apple, subtropical banana and ginger growers.

Mali Malipatil is Senior Insect Taxonomist with Agriculture Victoria, based at the Institute for Horticultural Development, Knoxfield. He is curator of the Victorian Agricultural Insect Collection and has published extensively on the taxonomy of heteropteran bugs. His other interests include the taxonomy and biology of thrips, whiteflies, other pests, and beneficial insects associated with fruit and other crops.

David James is a Senior Research Scientist at the Yanco Agricultural Institute of NSW Agriculture in the Riverina district, with extensive experience in entomological research. He has developed IPM and biological control systems for insect and mite pests of citrus, grapes, stonefruit and pastures. With a background in insect biology and ecology, his interests include acarology and the chemical ecology of insects and mites.

Dan Papacek is an entomologist, and Director of Integrated Pest Management Pty Ltd (Bugs for Bugs), based in Mundubbera, Queensland. He has been directly involved with the implementation of a working IPM program for the Queensland citrus industry since 1978. During that time he has undertaken cooperative research projects, especially with Dan Smith of the Department of Primary Industries, Queensland. He has a keen interest in classical biological control of citrus pests (especially mealybugs and scales), and has been involved in the search for suitable biological control organisms in overseas countries. More recently Dan has been engaged as a consultant to assist with the implementation of IPM programs for tree fruit crops in Thailand.

James Altmann is Manager of Biological Services at Loxton in South Australia. Biological Services commercially rears and sells beneficial organisms for use in citrus, pome, stonefruit and greenhouse crops. James is a consultant with Fruit Doctors in the Riverland and Sunraysia districts. He is particularly interested in IPM and biological control programs, but also advises on general horticultural practices in citrus, stonefruit, pome fruit, avocados and almonds.

Andrew Green has over eight years experience in the field of citrus entomology. He began work with Yandilla Park Ltd as an IPM consultant, and is currently employed by the Citrus Board of South Australia to advise the industry on a broad range of technical issues relating to citriculture. He is currently working on insect quarantine issues relating to citrus exports, and developing IPM courses for growers.

Bill Woods is a Senior Entomologist with Agriculture Western Australia, and is based in Midland. He is involved in developing pest management strategies for a wide range of horticultural crops including citrus.

Geoff Furness is a Senior Research Scientist with the South Australian Research and Development Institute (SARDI), Loxton Research Centre. He has 25 years research and extension experience in IPM and oil sprays in citrus. In the last 15 years he has also conducted research and development projects on air-assisted spray application equipment in a range of horticultural and agricultural crops.

Chris Freebairn is an Entomologist with the Department of Primary Industries, Queensland, at the Maroochy Horticultural Research Station, Nambour. He has been active in IPM research in citrus for two years, working on citrus leafeating weevil, red scale and citrus leafhopper. During the preparation of this book, Chris has developed considerable skills as an insect photographer, and many of his photographs are included in the book.

Mark Stevens is a Research Scientist with NSW Agriculture, based at Yanco Agricultural Institute in the Riverina district of New South Wales. He has worked on the ecology and control of ants and soft scales in citrus for four years. His other research interests include rice entomology and leafhopper taxonomy.

Greg Buchanan has worked on pests of grape (grape phylloxera, mites and lightbrown apple moth) and citrus (red scale and soft scales) in the Sunraysia for over 20 years, during which he was awarded MSc and PhD degrees. Greg is also leader of the Plant Protection group at the Sunraysia Horticultural Centre at Mildura, part of Agriculture Victoria. In 1994, the group was awarded the Daniel McAlpine medal for contributions to the implementation of IPM in the Sunraysia.

David Madge has been working with Agriculture Victoria at the Sunraysia Horticultural Centre, Mildura, since 1983. In 1988, David joined the Plant Protection group at that centre. Since then he has worked on IPM in citrus and grapevines, including pest and disease computer modelling and automatic weather data acquisition.

Craig Feutrill was formerly a Consultant Entomologist for Yandilla Park, a large grower–packer in South Australia's Riverland. In 1992 he started Horticultural Pest Management Services, an IPM-based consultancy which now services the Riverland, Sunraysia and Riverina districts.

John Kennedy, after working with the South Australian Department of Agriculture, joined the Plant Protection group with Agriculture Victoria at the Sunraysia Horticultural Centre, Mildura, in 1992. Since then John has worked in the field of citrus entomology, investigating biological control agents of soft scales and their interactions with ants. He is currently involved in IPM projects in citrus and viticulture.

Brian Gallagher was employed for ten years at Integrated Pest Management Pty Ltd (Bugs for Bugs) in Mundubbera, until setting up his own consultancy, Citrus Monitoring Services Pty Ltd, in 1993. This company provides Gayndah growers with a form of IPM focusing on sustained profitability.

Paul Jones is an IPM specialist located in south-east Queensland. He was originally employed by MacFarms as a pest scout in 1989–1990, and has since developed his own consulting business, Horticultural Crop Pest Monitoring. He pioneered the commercial implementation of IPM programs in strawberries, and specialises in IPM in coastal citrus and a range of other crops, e.g. cut flowers, low-chill stonefruit and custard apples. Paul is also involved in mass field-rearing of the predatory mite *Phytoseiulus persimilis*.

Megan Edwards is a Research Scientist with Agriculture Victoria at the Sunraysia Horticultural Centre, Mildura. She has extensive experience in nematology of horticultural crops, and has recently been studying the biology and control of Fuller's rose weevil and thrips on citrus.

Greg Baker is a Senior Entomologist with the South Australian Research and Development Institute (SARDI), Adelaide. He has broad research experience with both agricultural and horticultural pests, and is presently engaged in an IPM study of citrophilous mealybug in Riverland citrus in South Australia.

Scott Dix is an Entomology Manager with Yandilla Park Services, and IPM consultant to their citrus and grape growers. He is part of a team of horticultural consultants, providing a wide range of technical services, and has eight years experience in the Riverland and Sunraysia districts.

Peter Burne is the Horticultural Adviser for the Riverland Fruit Cooperative, which has 1000 members, and is based at Renmark in South Australia. He has 25 years experience in general horticulture, including 12 years with Primary Industries, South Australia. Peter currently provides a general horticultural advisory service to the cooperative's citrus, stonefruit, pomefruit and grape growers. In addition, he monitors pests for 70 citrus growers, and has a major input into the cooperative's citrus packing division.

Bill Frost is an entomologist with the South Australian Research and Development Institute (SARDI). He has been involved in the assessment and control of mites in a variety of crops, including citrus.

Jeff Watson is an entomologist working part-time for the Department of Primary Industries, Queensland, and as a private IPM consultant. He is based at Mareeba. Before moving to north Queensland, he worked as an IPM scout in Mundubbera citrus orchards. Since 1992, he has concentrated on citrus and tropical tree crops, such as papaws.

Miriam Pywell works on pest and disease projects in grapevines, citrus and vegetables at the Sunraysia Horticultural Centre, Mildura, part of Agriculture Victoria. She has worked on citrus IPM since 1990, during which time she has made collections of honeydew-producing insects and their natural enemies, and developed a photographic collection to assist other researchers.

Glossary

abdomen	the posterior end of an insect body, containing the intestines, and reproductive organs.
abscission	formation of a weakened area of tissue between plant parts, leading to fruit or leaf drop.
action level	the level of a pest population at which control measures should be implemented (*see also* economic injury level).
albedo	the white part of the pith inside the citrus rind.
antenna	(*pl.* antennae); elongate sensory structures ('feelers') projecting from the heads of insects.
asynchronous reproduction/generations	reproduction where eggs are laid continuously or at intervals of less than one generation, so the populations consist of individuals at different developmental stages (*see also* synchronous reproduction/generations).
augmentative release	release of beneficial species to supplement existing populations and to achieve the desired level of pest control after the beneficials reproduce further in the orchard (*see also* inundative release).
axil	the upper hollow where a leaf or flower attaches to a stem.
beneficial organisms	natural enemies (parasites, predators and pathogens) of pests, and other organisms, such as bees, that are useful in orchard management.
biological control	control of pests by beneficial organisms (parasites, predators and pathogens).
broad-spectrum pesticide	a chemical active against a broad range of organisms, usually both pests and beneficials.
BT	*Bacillus thuringiensis*, a bacterial insecticide with a high level of activity against caterpillars.
budline/bud union	the junction between the rootstock and the scion (budwood).
budwood	the upper part of the citrus plant which is grafted onto the rootstock.
calyx	whorl of residual flower parts, the sepals, at the stem end of a fruit.
canopy	leaves, stem and branches of a plant.
chemical control	control of pests through the use of chemicals.
cm	symbol for 'centimetre'.
chlorosis	yellowing of foliage or green fruit.
crawler	mobile form of first instar scale insects.
cultural control	control of pests by modifying management practices or environmental conditions, e.g. skirting, using physical barriers; part of integrated pest management (IPM) and often used in conjunction with biological control.
diapause	temporary cessation of development caused by a physiological response to unfavourable conditions.
economic injury level (EIL)	the level of crop damage at which control procedures become cost-effective (*see also* action level).
ectoparasite, ectoparasitoid	a parasite that develops on the external surface of its host (*see also* endoparasite).
EIL	*see* economic injury level.
endoparasite, endoparasitoid	a parasite that develops inside the body of its host (*see also* ectoparasite).
epidermal cells	layer of plant cells immediately beneath the waxy covering on the surface (the cuticle).
exudate	fluid released naturally from the body openings of insects and mites, or from wounds in insects, mites, or plant tissues.
fecundity	the number of young produced by a species or an individual.
femur	the upper part of the leg.
flush	period of rapid growth in the terminal shoots of citrus trees, which may be associated with flowering or which may be totally vegetative.
foliar	relating to foliage (leaves).

frass	pellets of excreta produced by some insects.
fungicide	a chemical used to control a fungal disease.
g	symbol for 'gram'.
gall	a distorted area of stem or leaf tissue, arising from a plant's response to attack by certain pests and pathogens.
generation	time from any given stage in a life cycle to the same stage in the offspring.
girdling	injury to the stem or trunk where the bark is destroyed around the entire circumference of the affected part; often associated with borers and fungal disease.
gumming, gummosis	form of damage in which sap exudes from injured plant parts.
ha	symbol for 'hectare'.
halo mark	a circular blemish on fruit caused by pest damage to the rind around the calyx during early fruit development.
honeydew	a sugar-rich solution excreted by soft scales, aphids, whiteflies and other sap-feeding insects; sooty mould grows on honeydew on plant surfaces.
hedging	pruning of tops and sides of tree canopy.
inundative release	large-scale releases of beneficial organisms designed to provide effective pest control straight away (i.e. without having to wait for the beneficials to reproduce in the orchard) (*see also* augmentative release).
inoculum	infective material of a pathogen (or sometimes low-level populations of a pest) that act as sources of subsequent outbreaks.
insecticide	a chemical used to control an insect pest.
instar	individual larval or nymphal stages of an insect or mite.
integrated pest management (IPM)	a method of pest control that incorporates aspects of separate control strategies: biological, cultural, varietal and chemical. The objective is to provide cost-effective pest management with minimal disruption to the environment.
IPM	*see* integrated pest management.
L	symbol for 'litre'.
larva	(*pl.* larvae, *adj.* larval); the immature stage of insects such as moths and beetles, which pass through a pupal stage before emerging as an adult that is very different from the immature stage.
m	symbol for 'metre'.
mandibles	the jaws of insects with chewing mouthparts, e.g. caterpillars.
miticide	a chemical used to control mites.
mL	symbol for 'millilitre'.
mm	symbol for 'millimetre'.
mobile	able to move ('motile' can also be used).
moult	the process in which an insect nymph or larva sheds its external skeleton when changing to the next instar.
mycelium	vegetative fungal growth.
natural enemy	an organism (a parasite, predator or pathogen) that has an adverse effect on another organism.
necrotic	describes damage caused by cell death (necrosis) in either plants or other organisms.
nymph	(*pl.* nymphs, *adj.* nymphal); the immature stages of insects such as true bugs, planthoppers, and katydids, which look similar to the adult, and develop to the adult stage through a series of incremental changes, without passing through a pupal stage (*see also* larva).
oleocellosis	a form of fruit damage in which oil cells within the rind collapse; may occur as a consequence of feeding by sucking insects.
oviposition	egg laying.
ovisac	a specialised structure produced by some scale insects (e.g. cottony citrus scale) in which the eggs are deposited.
parasite	term commonly used instead of 'parasitoid' (*see below*).

parasitoid	an organism which develops in its host (endoparasitoid), or on its host (ectoparasitoid), and causes the death of the host; although commonly called a 'parasite', the correct term is 'parasitoid'.
parthenogenesis	(*adj.* parthenogenetic); a form of reproduction where females produce young without fertilisation by a male; parthenogenesis is common in scale insects, particularly soft scales.
pathogens	organisms that causes disease, e.g. viruses, bacteria, microsporidia etc.
pesticide	a chemical used to control some form of pest; includes insecticides, miticides, fungicides, herbicides, nematicides, molluscicides etc.
phenology	the study of insect, mite and plant development in relation to seasons.
pheromones	chemical substances secreted by an animal which influence the behaviour of another animal, e.g. mating behaviour or aggregation. Synthetic pheromones are important management tools for controlling some pests.
predator	an animal that feeds on other animals, killing two or more prey during its lifetime.
pre-pupa	(*pl.* pre-pupae, *adj.* pre-pupal); the non-feeding last instar larva of moths, beetles, flies and wasps.
prey	(*noun*) an animal eaten by another; (*verb*) to seek for and seize prey.
propupa	(*pl.* propupae, *adj.* propupal); a non-feeding stage in the development of thrips and male scales immediately before formation of the pupa.
pulp	the internal part of the citrus fruit, beneath the rind and albedo layers.
pupa	(*pl.* pupae, *adj.* pupal); the non-feeding developmental stage of some insects (such as beetles, flies and moths) during which the larva develops into an adult very different from the larva.
resistance, pesticide	the tendency of pest populations repeatedly exposed to pesticides to become more difficult to control with pesticides. This resistance develops because the most vulnerable individuals in a population are more severely affected by a pesticide, leaving the least vulnerable individuals to breed the next generation.
resistance, varietal	tendency of some citrus varieties to be able to withstand attack by pests and pathogens, or to not attract pests and pathogens
ring scar	(*see* halo mark.)
rosette	term used to describe first and second instar wax scales, e.g. white wax scale.
russeting	a form of damage characterised by red or brown patches on the fruit.
scion	(*see* budwood.)
sedentary	non-mobile.
selective chemical	a chemical that is effective against a certain pest, but less harmful to other pests and beneficial organisms; often preferred in IPM programs because of the lower impact on beneficial organisms (*see also* broad-spectrum pesticide).
skirting	pruning to remove branches that reach ground level.
stippling	damage consisting of minute marks on the fruit rind and leaves.
susceptible	vulnerable, e.g. to a pesticide (*see also* resistance).
susceptible variety	a variety, e.g. a citrus variety, that is vulnerable to attack by a pest or pathogen.
synchronous reproduction/generations	reproduction in which all females lay their eggs at approximately the same time, resulting in populations consisting of individuals all roughly at the same stage of development (*see also* asynchronous reproduction/generations).
systemic insecticide	insecticide which enters the plant through one kind of tissue or in one region and is conducted through the plant to all the other parts of the plant.
thorax	the middle section of the insect body, to which the legs and wings are attached.
varietal control	control of pests or pathogens by selecting plant varieties resistant to attack.
vegetative growth	growth of non-reproductive parts of a plant, e.g. stems and leaves.
vmd	short for 'volume median diameter', a measure of droplet size in a spray; 50% of the droplets have a larger diameter than the vmd, and 50% of the droplets have a smaller diameter than the vmd.
webbing	silken threads produced by some mites, spiders and caterpillars, often to construct shelters.

1 The Australian citrus industry

Citrus species and related plants

Citrus belong to the plant family Rutaceae. Plants in the genus *Citrus* have large, fragrant flowers, and fruits with juice sacs.

Virtually all of the citrus species with edible fruits originated in the area extending from eastern India through South-East Asia to China. The most common species with edible fruits are sweet orange (*Citrus sinensis*), mandarin (*Citrus reticulata*), lemon (*Citrus limon*), grapefruit (*Citrus paradisi*) (now regarded by some as a variety of pummelo), lime (*Citrus aurantifolia*), citron (*Citrus medica*), pummelo (*Citrus maxima*), and sour orange (*Citrus aurantium*). Kumquats belong to a related genus, *Fortunella*.

In Australia, six native species of the genus *Microcitrus* occur in coastal rainforest areas of northern New South Wales and Queensland. Some of these trees grow to 10 m high in the rainforest. The fruits of all six species are unpalatable. Another distant citrus relative is the 2–3 m high native kumquat or desert lime (*Eremocitrus glauca*). These indigenous relatives of citrus support a range of native insects and mites (some harmful to citrus, and some beneficial), which have spread to the cultivated, introduced species of citrus.

Plate 1.1 Valencia oranges, Berri, South Australia.

Origins of the Australian citrus industry

Orange, lime and lemon seeds and plants were first brought to Australia with the First Fleet and planted on the day of arrival at Port Jackson (Sydney) on 26 January 1788. These seeds and plants were obtained from Rio de Janeiro and the Cape of Good Hope. The colony's chaplain planted a small orange grove in Sydney, and these became the first orange trees to bear fruit in Australia. Plants and seeds from China and South-East Asia were probably introduced shortly after 1788. By the 1820s all of the common species of citrus, except grapefruit, were growing in and around Sydney. Many pests and diseases were introduced on imported plants before quarantine regulations were enforced in the 1920s.

In 1887, the Chaffey brothers came to Australia from California. With assistance from the governments of South Australia and Victoria, they established Renmark and Mildura as irrigation settlements. They imported Washington navel and Valencia orange trees from California, and these trees became the forerunners of the citrus industry along the Murray Valley.

Varieties of citrus grown in Australia

Oranges

▶ **Valencia orange**

Valencia orange trees make up 46% of all citrus trees cultivated commercially in Australia. The Valencia orange is the main variety used for processing. It is generally harvested in Queensland from July to October, and in southern states from September to April, or later for the fresh fruit market.

▶ **Navel oranges**

Navel orange trees make up 32% of all citrus trees cultivated commercially in Australia. The main commercial variety, Washington, is harvested from April to August. Different selections of navel oranges, including early—Thompson,

Plate 1.2 Navelina navel oranges.

Fulwood, Pasin, Navelina, Leng, Atwood, Fischer—and late or summer navels—Barnfield, Chislett, Powell, Summer Gold, Lane, Witten, Rhode, Wilson—are now extending the marketing period.

▶ Other oranges

The Joppa is grown in Queensland mostly for processing. It is harvested from May to July. Other mid-season varieties are Parramatta, Mediterranean Sweet, Hamlin, Pineapple and Siletta. In southern states, these all mature in late winter to early spring.

Mandarins

Mandarin trees make up 12.6% of all citrus trees cultivated commercially in Australia.

▶ Imperial mandarin

The Imperial mandarin is harvested from late March to June in Queensland, but in southern states can be harvested until August.

▶ Ellendale mandarin

The Ellendale mandarin is actually a tangor (a mandarin × sweet orange hybrid), which is harvested in Queensland from June to August and in southern states from late July to late September.

▶ Murcott mandarin

The Murcott mandarin, or tangerine, is probably a tangor. It is harvested in Queensland from July to September, and in southern states from mid-August to early October.

Plate 1.3 Murcott mandarins, ready to pick.

▶ Minor varieties of mandarin and mandarin hybrids

Minor varieties of mandarin and mandarin hybrids include Clementine (various selections), Emperor, Hickson, Glen Retreat, Dancy, Kara, Satsuma, Fewtrell and Wallent, the newer varieties Amigo, Nova, Ellenor, Sunburst and Fremont, and the tangelos (mandarin × grapefruit hybrids) Orlando, Minneola and Seminole.

Lemons and limes

Lemon and lime trees make up 5.2% of all citrus trees cultivated commercially in Australia.

▶ Lisbon, Eureka, and Villafranca lemons

The lemon crop is harvested from February to October in Queensland, and from June in southern states. Lemons produce a small, valuable summer crop in southern states, particularly in coastal areas.

Plate 1.4 Eureka lemons.

▶ Meyer lemon

The Meyer lemon is thought to be a hybrid between a lemon and an orange. It is a popular, heavy-bearing variety in southern coastal Queensland, where fruit are harvested from December to May.

Plate 1.5 Meyer lemon orchard, Glasshouse Mountains, south-east Queensland.

▶ Limes

The main commercial variety of lime is the Tahitian, which bears a heavy summer crop. The West Indian lime is of lesser importance.

Grapefruit

Grapefruit trees make up 2.6% of all citrus trees cultivated commercially in Australia.

In Queensland, Marsh seedless grapefruit are harvested from March to August. Grapefruit grown in southern areas mature from August, but can be held on the tree until October–November, or into January–February with stop-drop hormone sprays.

Plantings of new red-fleshed grapefruit, e.g. Star Ruby, Ray Ruby, Henderson, Rio Red, and Flame, are increasing. Thompson, a pink-fleshed variety, is mainly grown in New South Wales, while Ruby is grown in Western Australia and the Northern Territory.

Miscellaneous citrus

Other kinds of citrus grown commercially in Australia include rough and smooth Seville oranges (for marmalade production), and the pummelo.

Plate 1.6 New red-fleshed grapefruit varieties. Left, Ray Ruby; right, Henderson.

Rootstocks

All citrus trees are propagated by budding or grafting onto selected seedling rootstocks. Rootstocks affect the tree size and health, resistance to disease and pests, productivity and quality of fruit. See table 1.1 for varieties of rootstocks recommended for Australian citrus.

Crop cycle

Citrus grows in a wide range of climates from tropical to temperate. Cool night temperatures near harvest produce well-coloured fruit.

In the tropics, citrus will flower and fruit continually. Under subtropical and temperate conditions, citrus species other than lemons flower once a year, except when stressed (e.g. by lack of water). Lemons flower more than once a year.

Flushing

Shoot growth of most varieties occurs in distinct flushes, but may be almost continual for limes and lemons in warmer coastal areas.

The major flush is in the spring. It produces most shoots and carries the flowers for the main crop.

The autumn flush produces vigorous shoots with large, wide leaves. It does not carry flowers except on lemons, or when it results in an out-of-season crop, which is of no economic value. Management practices such as irrigation and fertiliser application can be used to control autumn vegetative growth. The aim generally is to minimise it.

Flowering and fruit set

All citrus varieties flower in the spring. Flowering continues for 4–6 weeks.

Pollination requirements vary widely for different citrus species. Often seediness in mandarins is increased by unnecessary cross-pollination, although some cross-pollination may improve crop load and fruit size.

Flower induction can be inhibited by petroleum oil sprays (but the risk with modern oils is very low) or by late picking. More than 30% of flower buds drop off before reaching full bloom. Most of the remaining flowers set fruit, but only a few of these reach maturity.

Fruit drop

There are three types of fruit drop:
- initial flower and fruitlet drop
- physiological fruit drop
- pre-harvest fruit drop.

Excess flowers and fruitlets fall during flowering and initial fruit setting. Losses caused by insects such as citrus blossom bug may not be significant, as the flowers or fruitlets they destroy may have fallen anyway.

Physiological fruit drop (of fruit 5–20 mm in diameter), which lasts from mid-November to late December, is caused by high temperatures, moisture stress, growth regulators (if applied for thinning) and lack of nutrients.

Pre-harvest fruit drop results from the formation of an abscission layer between the fruit and the stem. Low levels of such fruit drop occur naturally, but can be prevented by growth regulator sprays. However, the levels can be increased by wind and damage caused by some insects.

Fruit development

From flowering until mid-December, increase in fruit size is mainly due to the growth of the rind. From mid-December, the juice sacs enlarge and the juice content increases. Sugar content increases and acidity decreases; the rind changes from a dark to a light green, then finally shows traces of the colour of ripe fruit as maturity approaches.

Once the rind has changed to its final 'ripe' colour, it begins to soften and eventually deteriorates. Growth regulators (e.g. gibberellic acid) can be applied to fruit on the tree to delay the onset of rind colouring.

Table 1.1 Rootstock varieties recommended for Australian citrus

ROOTSTOCK	PREFERRED SOIL	ADVANTAGES	DISADVANTAGES	SUITABLE VARIETIES
sweet orange (*Citrus sinensis*)	deep sandy	• good yield and quality • long-lasting large trees • can hold fruit late • tolerant of calcareous soils • tolerant to tristeza and exocortis	• susceptible to phytophthora root and collar rots • sensitive to dry conditions • susceptible to the stem-pitting strain of tristeza • susceptible to nematodes • intolerant of excess soil moisture	oranges mandarins
Carrizo and Troyer citrange (*Citrus sinensis* × *Poncirus trifoliata*)	wide range	• good yield and quality • tolerant of phytophthora root rot • performs best in replant sites • cold hardy • tolerant of tristeza • medium to large trees	• incompatible with Eureka lemon • compatibility with some minor varieties unknown • prone to micronutrient deficiencies • sometimes rootstock overgrows the scion (especially in Imperial mandarin) • requires exocortis-free budwood • moderately susceptible to citrus nematode	oranges grapefruit limes mandarins
trifoliate orange (*Poncirus trifoliata*)	wide range (prefers loams)	• good fruit quality • highly resistant to phytophthora root rot, tristeza and citrus root nematode • cold hardy • small trees	• not suitable on light sandy soil • intolerant of exocortis • sensitive to dry conditions • intolerant of salt • intolerant of highly acid or lime soil • incompatible with Eureka lemon	oranges Ellendale mandarin
rough lemon (*Citrus jambhiri*)	deep virgin sand	• good yield • tolerant of dry conditions • tolerant of tristeza and exocortis • large trees • easy to propagate • slightly earlier maturity	• susceptible to phytophthora root and collar rots • intolerant of excess soil moisture • susceptible to citrus nematodes • short life • unsuitable for most mandarins, including Ellendale • poor internal fruit quality and coarser rind	lemons grapefruit limes oranges Murcott mandarin
Cleopatra mandarin (*Citrus reticulata*)	wide range (prefers loams)	• good fruit quality for mandarins • salt-tolerant • tolerant of tristeza and exocortis	• smaller fruit • susceptible to phytophthora root and collar rots • susceptible to citrus nematode	mandarins
Benton citrange (*Citrus sinensis* × *Poncirus trifoliata*)	wide range	• resistant to phytophthora root and collar rots • compatible with Eureka lemon • tolerant of tristeza	• requires exocortis-free budwood • tolerance of nematodes not known	Eureka lemon
Swingle citrumelo (*Citrus paradisi* × *Poncirus trifoliata*)	wide range	• resistant to phytophthora root and collar rots • tolerant of tristeza • resistant to citrus nematode • cold hardy • vigorous	• incompatible with Eureka lemon	grapefruit

Tree and orchard management

Pruning, hedging and skirting

Citrus trees are pruned to eliminate dead and weak wood. Young trees are shaped to produce a framework of three or four well-spaced main limbs, 450–600 mm above the ground.

Pruning inside canopies of mature trees, particularly of dead wood, will reduce the incidence of diseases like melanose, pink disease, septoria, anthracnose, stem end rot, root and collar rot. Pruning improves spray and light penetration.

Major pruning cuts should be as close as possible to the trunk or main branch to allow the wound to heal smoothly and to prevent attack by diseases and wood borers. Large cuts should be painted with an anti-fungal waterproofing sealant.

Mechanical hedging is commonly practised to correct overcrowding and improve access to trees in older, high-density plantings. Hedging can also be used to thin the crop, and to encourage new growth during a heavy cropping year.

Skirting is important to encourage fruit development well above ground, and to minimise diseases (brown rot) and pests (snails, Fuller's rose weevil and ants). It improves distribution of water by under-tree sprinklers, fertiliser spread and herbicide coverage. Skirting usually involves mechanical removal of foliage and minor branches up to 900 mm above ground level, and is done before flowering.

Nutrition

The annual nitrogen requirement is usually applied in late winter – early spring to be available for the spring growth flush, flowering and fruit set. On light sandy soils, the application is split, with an application in early spring and one in early summer. The timing of phosphorous and potassium application is not as critical, and these can be applied with the nitrogen in mixed fertilisers for convenience, or applied separately.

Foliar fertiliser sprays are most effective in correcting zinc, manganese and magnesium deficiencies. Iron deficiency is corrected with soil applications.

Nitrogen levels can influence tree growth patterns, and the size and fecundity of pests, particularly soft scales. Deficiencies can lead to delayed development of the pests, smaller scales and fewer eggs. Excessive levels may have the opposite effects. The recommended range for levels of nitrogen in mature leaves in February–March is 2.4–2.8% nitrogen.

Irrigation

About 1000–1250 mm of water (including rainfall) is needed annually by citrus trees. Methods of irrigation include the use of under-tree microsprinklers, fixed overhead sprinklers, drippers, or furrow irrigation. Most newer plantings have low-level, under-tree sprinklers or microjets, or drippers.

Weed control and inter-row cover crops

Cultivation has been a common method of weed control, especially in southern inland orchards. Herbicide usage is now more common because of the increasing use of microsprinklers, closer tree plantings and increased labour costs. Total elimination of weeds within the rows by either cultivation or herbicides can have a detrimental effect on pest management (see the sections on dust and heat below).

Between the rows, a combination of strip herbicide use and sod culture (cultivation of selected grasses and/or legumes) is generally practised. Mowing between rows is used in winter to reduce the risk of frost damage. This system results in a more favourable environment for the natural enemies of pests, and provides supplementary food (pollens and nectars) for them.

Mulching beneath trees increases soil organic matter, improves soil structure, reduces root temperature fluctuations, and increases water retention.

Wind protection

Wind is a major cause of rind blemish, and natural windbreaks should be retained or established for all new plantings. Some windbreak species, e.g. *Eucalyptus torelliana*, act as good reservoirs for beneficial organisms such as predatory mites.

Dust

Dust from access roads or cultivated inter-rows is hazardous to beneficial insects, especially small parasitic wasps. Minimise dust by sealing or watering roads and keeping vehicle speeds to a minimum.

Heat

Very hot conditions caused by heat waves and inappropriate use of cultivation and herbicides can increase the number of annual generations of pests (e.g. of red scale), and reduce the effectiveness of natural enemies. High temperatures may also kill young stages of pests like black scale and lightbrown apple moth.

Hygiene

Fruit left to rot on the ground are a significant source of infection by the diseases blue mould and brown rot. Fallen fruit are also a breeding place for Queensland fruit fly. If practicable, remove or slash decaying fruit, particularly where fruit fly or fruit borers are a problem.

Production and climate

In Australia in 1994, 697 kilotonnes of fruit were produced from 10.2 million trees on 34 000 hectares. The crop was valued at $321 million. About 15% of fruit was exported in 1994, and this proportion is increasing.

The main areas of citrus production are the Riverina district (including the Murrumbidgee Irrigation Area (MIA)) of New South Wales, the Riverland district of South Australia, and the Sunraysia district of New South Wales and Victoria, with the remainder produced in other areas of New South Wales, Victoria, Queensland, Western Australia and the Northern Territory. (See table 1.3 for state production statistics.)

The climates of the various citrus regions vary considerably, ranging from humid coastal conditions requiring some supplementary irrigation, to drier inland regions which depend on irrigation. Representative long-term weather data for the major districts are presented in table 1.2. (However, note that the climates of Perth and Darwin are milder than those of their nearby citrus districts.)

Table 1.2 Annual climatological data for Australian citrus districts

The data in the table is based on long-term records and analysis over periods ranging from 15 to 137 years.

PLACE	LOCATION AND ELEVATION	TEMPERATURES (°C)		
		MEAN DAILY MAXIMUM	MEAN DAILY MIMIMUM	HIGHEST RECORDED
New South Wales				
Coastal				
Coffs Harbour	30°19′S 153°07′E, 5 m	23.2	13.9	43.3
Somersby plateau	33°18′S 151°14′E, 300 m	21.4	11.3	40.8
Richmond	33°36′S 150°47′E, 19 m	23.7	11.0	47.8
Inland Riverina (including MIA)				
Narromine	32°13′S 148°14′E, 237 m	23.8	10.7	43.4
Hillston	33°29′S 145°31′E, 122 m	24.1	10.9	46.0
Griffith	34°19′S 146°04′E, 126 m	23.1	9.6	43.9
Leeton	34°34′S 146°24′E, 140 m	23.0	10.3	44.3
South Australia				
Riverland				
Loxton	34°26′S 140°36′E, 66 m	23.4	9.2	46.6
Renmark	34°10′S 140°45′E, 20 m	24.2	10.9	47.4
New South Wales and Victoria				
Upper and mid- Murray River				
Tocumwal	35°48′S 145°35′E, 114 m	22.9	9.6	45.8
Echuca	36°09′S 144°45′E, 99 m	22.2	9.4	45.3
Kerang	35°44′S 143°55′E, 78 m	22.7	9.3	45.6
Sunraysia				
Wentworth	34°06′S 141°54′E, 37 m	24.3	10.9	48.1
Mildura	34°15′S 142°05′E, 50 m	23.6	10.3	46.9
Swan Hill	35°20′S 143°33′E, 70 m	23.0	9.7	46.1
Queensland				
Coastal				
Mareeba	17°04′S 145°25′E, 473 m	28.9	17.7	40.7
Nambour	26°38′S 152°56′E, 25 m	25.7	13.8	40.8
Inland				
Emerald	23°31′S 148°09′E, 179 m	29.5	15.2	44.6
Gayndah	28°37′S 151°36′E, 106 m	28.1	13.5	43.7
Western Australia				
Perth	31°57′S 115°52′E, 19 m	23.3	13.3	46.2
Carnarvon	24°53S 113°40E, 4 m	27.0	17.2	47.7
Northern Territory				
Darwin	12°25′S 130°54′E, 31 m	31.9	23.2	38.9

Source: Bureau of Meteorology

		MEAN RELATIVE HUMIDITIES (%)		MEAN ANNUAL RAINFALL (mm)	MEAN NUMBER OF DAYS WITH RAIN	MEAN NUMBER OF CLOUDY DAYS	MEAN NUMBER OF FROSTS
LOWEST RECORDED	MEAN NUMBER OF DAYS >35°C	0900 h	1500 h				
−3.2	1.4	67.3	62.8	1698	142	100	6.7
−1.9	5.4	71.7	57.1	1254	135	116	15.9
−8.3	16.5	73.6	46.9	808	110	106	27.6
−5.4	21.6	67.8	47.3	654	71	47	38.0
−4.4	36.1	62.4	39.8	365	61	90	27.0
−5.4	21.4	62.7	41.5	406	81	55	59.6
−3.9	21.8	62.1	42.9	436	80	92	37.4
−5.7	26.7	64.4	41.9	273.5	84	104	58.0
−3.8	33.8	62.8	38.4	261	64	95	9.1
−4.3	26.8	74.4	47.4	454	76	105	25.2
−5.5	19.3	66.9	44.6	433	88	109	25.6
−4.6	22.9	68.4	44.0	377	73	91	6.3
−2.8	38.4	61.1	37.6	289	60	77	13.8
−4.0	29.3	66.1	38.5	298	74	59	27.8
−3.6	24.8	68.2	42.5	349	71	97	18.9
7.8	18.3	70.5	50.2	718	97	87	0.0
−2.9	4.5	66.7	58.9	1737	145	83	9.3
−2.3	59.1	62.1	37.8	640	60	68	4.8
−3.2	22.0	66.5	43.3	773	72	78	13.5
1.6	19.3	63.0	49.4	869	120	60	1.0
2.4	26.5	58.7	54.3	224	41	39	0.1
10.4	8.8	72.9	53.7	1670	109	110	0.0

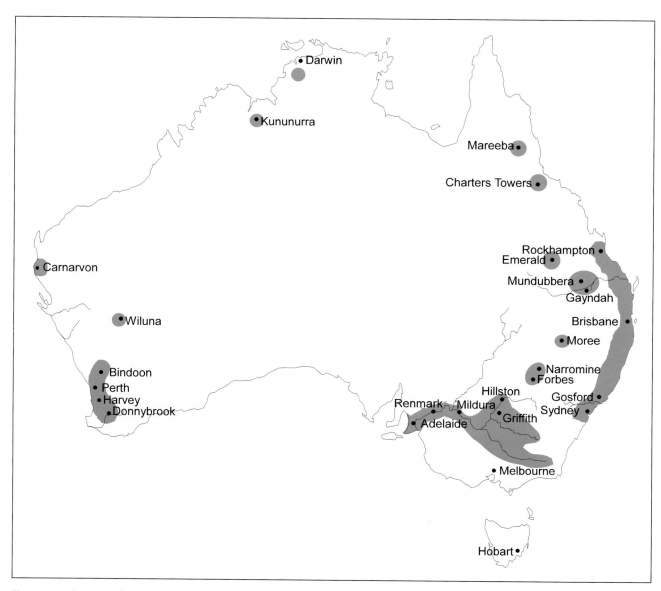

Figure 1.1 Citrus production areas in Australia.

New South Wales

The inland Riverina district is the largest citrus growing area of New South Wales (64% of the state's total citrus growing area). Production is centred around Griffith, Leeton, Yanco, Coleambally, Darlington Point and Narrandera near the Murrumbidgee River, Hillston on the Lachlan River, and Narromine on the Macquarie River. The area also produces grapes, canning fruits and vegetables.

There are also significant plantings within the Sunraysia district of New South Wales (23% of the state's total citrus growing area) at Coomealla, Monak, and Curlwaa, and smaller plantings along the mid-Murray, from Goodnight to Tocumwal, with the largest concentration at Barham (near Kerang in Victoria).

The central coastal citrus districts are near Sydney at Windsor, Richmond, Somersby, Peats Ridge, Kulnura and Mangrove Mountain, and further north at Maitland near Newcastle.

There are small plantings on the north coast, particularly at Grafton and Coffs Harbour.

Table 1.3 Citrus production statistics for different states of Australia

State	Percent of total trees[1]	Percent of total production[2]	Value of production ($ million)
New South Wales	43	38	95
South Australia	26	36	101
Victoria	17	16	65
Queensland	11	9	54[3]
Western Australia	3	1	6
Northern Territory	<0.1	<0.1	0.25

Source: Australian Bureau of Statistics 1994

[1]Total 10.2 million trees or about 34 000 hectares

[2]Total fruit production 697 kilotonnes

[3]Increased in 1996 to $75 million.

South Australia

The main citrus growing areas in South Australia are concentrated along the Murray River at Waikerie, Berri, Loxton and Renmark, with smaller plantings at Cadell, Swan Reach and Myopolonga. The area is also an important grape, dried fruit, canning fruit and vegetable area. Navel and Valencia oranges are the main varieties grown, but production of mandarins has greatly increased since 1990.

Victoria

In Victoria, production is centred around the Sunraysia districts of Nangiloc, Merbein, Boundary Bend, Mildura and Robinvale. Citrus is also grown along the Murray River from Swan Hill to Wangaratta. Navel and Valencia oranges are the main varieties grown, then mandarins, grapefruit and lemons. Citrus is often produced in conjunction with grapes.

Queensland

The main citrus producing area of Queensland is the Central Burnett, around Gayndah and Mundubbera. Other subcoastal citrus areas are near Gatton (Lockyer Valley), Charters Towers and Mareeba, while about 400 hectares of citrus have recently been established at Emerald in central Queensland.

Citrus is grown in coastal districts from Caboolture to Bundaberg, and at Byfield near Rockhampton.

Mandarins (Imperial, Murcott and Ellendale) make up over 50% of commercial citrus production in Queensland. The state produces about 60% of Australia's mandarins. Oranges (Washington navel and late Valencia) are also grown, and small numbers of lemon, lime and grapefruit.

Western Australia

The main production areas in Western Australia include Harvey, Gingin, Bindoon, Chittering, Capel, Donnybrook and Carnarvon. Apart from orchards at Carnarvon, Wiluna and Kununurra, most citrus orchards are within 200 km of Perth.

Diseases

About a dozen serious diseases, affecting most parts of the tree, occur on citrus in Australia. Post-harvest disease control is particularly important to prevent serious fruit losses.

It is also important to obtain healthy trees from nurseries using selected rootstock and approved scion material to avoid subsequent problems with phytophthora root rot and viral diseases. Growers should specify the variety, budline and rootstock they require. Approved buds should be obtained from the Australian Citrus Budwood Scheme.

Southern inland growers should obtain nursery trees from inland nurseries to avoid introducing diseases and pests such as citrus red mite from coastal orchards. Introduction of budwood or nursery trees from Queensland to southern states is prohibited to prevent introduction of the orange stem-pitting strain of tristeza virus.

Root and trunk diseases

▶ Phytophthora

The fungi *Phytophthora citrophthora* and *Phytophthora parasitica* cause serious diseases giving rise to collar and root rots. Spores of these water moulds are soil-borne and spread rapidly after heavy rain or where soil remains moist.

Trees suffering from root rot become unthrifty, with a sparse canopy. The leaves turn yellow, and twig dieback follows.

Symptoms of collar rot in susceptible rootstocks include dark, water-soaked areas, often with gum exuded on the trunk near ground level. Cutting through the bark reveals the bark to be soft and damp, with tan to darker brown discoloured wood underneath. Later the dead bark breaks away from the wood in strips.

On resistant rootstocks, trunk rot is usually confined to the area above the bud union. In persistently wet conditions, phytophthora-infested soil splashed onto trunk, branches, leaves and fruit can cause gummosis and brown rot on aerial parts of the tree.

The risk of disease can be reduced by: using phytophthora-tolerant rootstocks; ensuring adequate drainage; applying systemic fungicides; keeping bud unions well above ground level; and skirting trees to improve air circulation.

▶ Armillaria root rot

Armillaria root rot occurs more in coastal orchards in New South Wales, and is caused mainly by the fungus *Armillaria luteobubalina*.

Symptoms include slow decline of trees, with leaf yellowing and leaf drop. White, fan-shaped mycelial mats with a strong mushroom odour occur under the bark of affected roots and trunks.

This 'shoe-string' fungus spreads from old stumps, e.g stumps from eucalypt and other tree crops, from roots and from tree to tree using its black rhizomorphs (bootlace-like fungal growths). The fungus then grows on the outer bark of the roots. In autumn, honey-coloured toadstools appear in clumps around the bark of the tree.

▶ Sudden death syndrome

Trees of all varieties on trifoliate orange or citrange rootstocks can succumb to sudden death syndrome with losses as high as 80%. Temporary waterlogging is a major contributing factor. Losses are greatest when trees are 9–15 years old.

Limb, leaf and fruit diseases

▶ Pink disease

This is a minor disease caused by the fungus *Corticium salmonicolor*.

Symptoms of pink disease include wilting caused by branch girdling, splitting and gummosis of individual branches. A pink fungal growth that later turns white or grey is visible on the bark, particularly in wet weather. The bark splits and lifts off and eventually the branch is girdled and dies. The disease is more common in moist tropical areas next to rainforest.

▶ Scab

Scab is caused by the fungus *Sphaceloma fawcettii* var. *scabiosa*. Scab can be a serious problem, particularly in lemons and Kara mandarins grown in moist coastal conditions.

Only young tissues are affected. Symptoms on green twigs, leaves and fruit are raised, grey to light-brown, corky scabs 1 mm in diameter.

Spores from old scabs, spread by wind or rain, are the main source of infection. Watering with overhead sprinklers when susceptible young growth or fruit are present increases the risk of scab infection.

❯ Melanose

Melanose is a serious disease caused by the fungus *Diaporthe citri* (also known as *Phomopsis citri* in the conidial, or asexual, state).

Only young leaves, twigs and fruit (up to 3 months old) are susceptible to infection. Symptoms include reddish to black, raised, corky pustules 1 mm in diameter. Dense brown concentrations of infection on the fruit are known as 'mudcake'.

Plate 1.7 Melanose on Marsh seedless grapefruit.

Dead twigs are a source of spores which are dispersed by splashing and dripping water. As a result, some infections show a 'tear stain' pattern, and are known as 'tear stain' melanose. Citrus trees grown in areas with wet summers are more prone to melanose infection.

The disease is not often a problem in young trees, as they have less dead wood and twigs carrying spores. Pruning dead wood from older trees helps control the disease.

The fungus that causes melanose also causes stem end rot of citrus fruit, and crotch rot of Hickson mandarins. Stem end rot is characterised by a dark-tan rot of fruit beginning at the stem end. Crotch rot, a serious disease in coastal areas, develops in the limb base of Hickson mandarins, where water retained in the limb hollow provides the ideal conditions for infection. The infected bark cracks and fungal decay progresses through the trunk. Girdling of the trunk or branch is followed by loss of leaves and the death of the tree.

❯ Sooty mould

Any of several species of *Capnodium*, *Cladosporium* or *Meliola* fungi can cause the minor disease sooty mould. Twigs, leaves and fruit become enveloped in a dark fungal mat. The fungi grow on honeydew produced by aphids, mealybugs, planthoppers and soft scales.

❯ Greasy spot

Greasy spot is a minor disease caused by the fungi *Colletotrichum gloeosporioides*, *Septoria citri* and *Mycosphaerella* sp., and is confined to inland citrus growing areas. It infects leaves 4–6 months old, initially showing as a black spot with a yellowish halo on the undersurface of the leaf. At a later stage, greyish blisters up to 3 mm in diameter are formed on the undersurface of the leaves. The blisters become slightly raised and have a greasy appearance. The damage extends through the leaf to the upper surface, with the spots visible on both upper and lower leaf surfaces. Leaf drop and fruit lesions can occur.

❯ Shell bark of lemons

In Queensland shell bark is common in Lisbon, Villafranca and Eureka lemons. The bark of the trunk and main limbs cracks and peels, accompanied by some gumming. The cause of the disease is unknown, but it can be avoided by using approved budwood.

Virus and virus-like diseases

❯ Citrus tristeza virus

The tristeza closterovirus blocks food-conducting vessels at the bud union of certain stock–scion combinations. The roots are starved and deficiency patterns develop in the leaves, followed by dieback of twigs and stunting of the tree.

Tristeza is the cause of quick decline of oranges and mandarins on sour orange rootstocks. Most rootstocks are

Plate 1.8 Stem pitting of grapefruit trunk, caused by citrus tristeza virus.

tolerant, but sour orange must be avoided. A serious orange stem-pitting strain of tristeza affecting Washington navel and Ortanique tangor occurred in 1990 in the Central Burnett area of Queensland.

All varieties of citrus can be infected by the tristeza virus, but in Australia, normally only grapefruit show obvious symptoms. A pitting and grooving of the wood surface is visible when the bark is removed. The virus is spread by black citrus aphid and by using budwood from infected trees.

In grapefruit, it can be avoided by using budwood carrying a mild closterovirus that protects against infection by the tristeza virus. Such budwood is available from the Australian Citrus Budwood Scheme.

▶ Exocortis or scaly butt

The citrus exocortis viroid affects citrus on susceptible rootstocks, such as *Poncirus trifoliata*, citrange and Rangpur lime. It causes cracking and scaling of the bark below the bud union, the bark often peeling off in strips. Symptoms appear in trees more than 4 years old. Affected trees are stunted. The disease is spread in infected budwood, and mechanically on pruning equipment.

Plate 1.9 Lisbon lemon affected by exocortis or scaly butt.

▶ Psorosis

Psorosis (the name given to a group of virus diseases which share common symptoms) is uncommon in Australia, thanks to the Australian Citrus Budwood Scheme. Symptoms are scaling and flaking of the bark on the trunk and limbs. Other forms occasionally seen in Australia are concave gum, and blind pocket.

The disease is graft transmitted (i.e. by using infected budwood, or susceptible rootstocks). Symptoms are rarely seen in trees younger than 10 years old.

Fruit diseases

▶ Melanose
See page 10.

Plate 1.10 Orange tree affected by psorosis.

▶ Brown rot

Brown rot is mainly caused by the fungus *Phytophthora citrophthora*. (See the information on phytophthora in the section on root and trunk diseases (page 9).)

Prolonged wet weather promotes the spread and development of brown rot. Fruit near the ground are commonly affected, developing a dull brown leathery rot. The disease can also affect leaves, causing death of young tip growth and serious leaf drop.

▶ Black spot

Black spot is a serious disease of all citrus in coastal and subcoastal areas, and is caused by the fungus *Guignardia citricarpa* (also known as *Phyllostictina citricarpa* in the asexual, or conidial, state).

Fruit symptoms are dark-brown or black speckling of the rind, or sunken spots with pale centres and red to black margins. In a virulent attack of black spot, the spots grow rapidly and coalesce into large black areas. Symptoms are rarely seen in green leaves, but the fungus develops its fruiting bodies abundantly in leaf litter after several cycles of wetting and drying.

Infection usually takes place when fungal spores are carried by the wind to young leaves and fruit. Although infection takes place in spring and early summer during the first 20 weeks after flowering, fruit first show symptoms only on ripening.

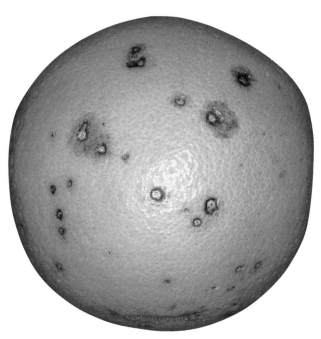

Plate 1.11 Washington navel orange affected by black spot.

▶ Alternaria brown spot

Alternaria brown spot is a serious disease caused by a toxin-producing strain of the fungus *Alternaria alternata*. It is a particular problem in Murcott and Emperor mandarins in Queensland, and in Minneola tangelo in coastal areas of Queensland and New South Wales.

Small brown spots and yellowing of veins can occur on leaves. Leaves often curl and fall. Young leaves and shoots are severely affected by the toxin produced by the fungus, and become totally blighted and blackened. Mature infected leaves develop dark, water-soaked spots.

Brown, depressed lesions up to 5 mm wide with slightly raised centres develop on the fruit which can ripen prematurely and fall.

The fungus survives on old infected twigs, producing large numbers of spores which are spread by wind and water. The lime and rough lemon strain of the fungus infects leaves and stems of rough lemon seedlings used as rootstock.

▶ Centre rot

Centre rot is caused by a non-toxin-producing strain of the fungus *Alternaria citri*. Fruit symptoms are a dark-brown or black stem end or stylar end (or flower end) rot and a black internal core rot. The disease is mainly a post-harvest problem, particularly in navel oranges and Ellendale mandarins.

▶ Blue and green moulds

Blue mould (*Penicillium italicum*) and green mould (*Penicillium digitatum*) affect mature fruit, and are mostly a post-harvest problem.

Early symptoms are small, soft, water-soaked areas, which expand rapidly over the whole fruit. Rotted areas become covered with a dense white fungal growth covered with masses of blue or green spores.

▶ Sour rot

Sour rot is caused by the fungus *Geotrichum candidum* which enters through rind wounds. Symptoms include a pale, soft decay of ripe fruit in storage. Sour rot can be a serious post-harvest problem, particularly during wet harvesting periods.

▶ Septoria spot

Septoria spot is caused by the fungus *Septoria citri*, and can be a serious disease, particularly on export fruit from inland areas. It causes light tan pits, 1–2 mm wide, on the fruit. These appear in winter. As the fruit mature, the pits may become reddish-brown then darken as they enlarge and merge.

Plate 1.12 Leaf symptoms of alternaria brown spot.

Plate 1.13 Septoria spot on lemon.

▸ **Anthracnose**

Anthracnose is the term given to disfiguring blotches on the fruit rind caused by the fungus *Colletotrichum gloeosporioides*. This fungus is commonly present on the fruit rind where it is a weak pathogen. It affects fruit that are damaged or stressed by poor picking practices or by sun or spray burn, or fruit that are overripe or held too long in storage.

The anthracnose occurring in Australia is not as severe as the anthracnose caused by another strain of *Colletotrichum gloeosporioides* in Florida.

▸ **Citrus blast**

Citrus blast is a minor disease caused by the bacterium *Pseudomonas syringae*. It occurs during wet weather, and is associated with wind-driven rain and whipping of the leaves and twigs. Infection follows mechanical injury of leaves and fruit. On leaves, black lesions appear in the leaf stalk which eventually lead to leaf drop. Sunken black pits appear on the rind of infected fruit.

▸ **Sooty blotch**

Sooty blotch is a minor disease in humid areas caused mostly by the fungus *Gloeodes pomigena*. The fungus grows on the surface of the fruit rind, forming blotches consisting of dark masses of fungal filaments. Sooty blotch looks like light sooty mould but can be more difficult to remove by washing.

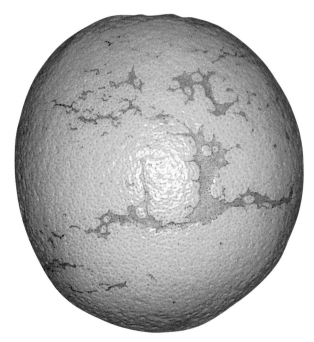

Plate 1.14 Wind rub on orange. Wind rub is a major cause of rind blemish on citrus fruit.

Physical blemishes and physiological disorders of fruit

There are several physical injuries and physiological disorders that spoil the appearance of citrus fruit and can be confused with insect or mite damage.

The most common physical injury is wind rub, caused by fruit rubbing against each other or against twigs or leaves in the wind. 'Netting' refers to a scattered network of scratches, possibly caused when small fruit are scratched by twigs or leaf edges.

There are several types of fruit burn: sunburn, spray burn and water burn. Water burn (usually on Imperial and Emperor mandarins) occurs during wet weather when the bottom of the fruit is wet for a prolonged period.

Fruit may also be damaged by hail, bruised or split during picking and packing, pierced by thorns or sometimes pecked by birds.

Physiological disorders include: oleocellosis, or damage to the rind by oil released from damaged oil glands (e.g. after insect attack, or mechanical injury); rind creasing (caused by collapse of the underlying albedo, which is sometimes associated with wet weather after extended dry periods); corrugations; chimeras; navel deformation; regreening of Valencia oranges held on the tree after maturity; and papillae (raised pimples) on the rind of Imperial mandarins.

Plate 1.15 Imperial mandarins damaged by sulphur-crested cockatoos. Ripening citrus fruit, particularly mandarins, are attractive to cockatoos and other birds, including crows, currawongs, lorikeets, and, in rainforest areas of north Queensland, even cassowaries.

2 Concepts of integrated pest management (IPM)

Integrated pest management (IPM) is a strategy which encourages the reduction of pesticide use by using a variety of controls in harmonious combination to contain or manage pests below their economic injury levels. The aim is to produce quality fruit at minimal cost by intelligently managing pests.

IPM is the complete opposite of relying solely on pesticides to control pests (and diseases). It aims to maximise the use of biological and cultural controls. Other measures, such as chemical control, must play a supportive rather than disruptive role.

Pesticides should be used strictly when needed, as determined by systematic monitoring of pests and their natural enemies. Selective, non-disruptive chemicals are preferred over broad-spectrum chemicals, such as organophosphates, carbamates and synthetic pyrethroids, which are very harmful to natural enemies of pests.

Biological control

Biological control is the use of natural enemies (parasites, predators or pathogens) of pests. (The natural enemies are also called 'beneficial organisms' or 'beneficials'.) The aim of biological control is to establish and maintain populations of natural enemies which will keep pest populations below economically damaging levels.

Natural enemies may be mass-reared and released into orchards. Augmentative releases aim to ensure that the natural enemy is present in sufficient numbers in time to control the pest. Before mass-rearing and augmentative release programs are initiated, it is necessary to know the true value of the various beneficials in controlling the pests.

Cultural control

Cultural practices should help to conserve existing natural enemies of pests. In citrus, these practices include the management of other plants growing in the orchard, tree skirting, trunk banding and good orchard hygiene.

Maintaining trees in general good health is also important for both pest and disease control. Healthy trees are usually better able to withstand attack by pests than stressed trees. The microhabitat for beneficials is also of better quality in healthy orchards than in unthrifty orchards.

In general, some weed growth can be of benefit in an IPM orchard, although this must be balanced against the need for weed control to minimise competition for nutrients and water, and to avoid frost damage. Certain pollen-producing plants, for example, encourage high populations of predatory mites, and nectar-producing plants may provide supplementary nourishment for tiny wasp parasites.

In Queensland citrus orchards, inter-row sod culture of Rhodes grass (*Chloris gayana*) is practised to encourage high populations of predatory phytoseiid mites. The numbers of these mites can increase three- or four-fold if good pollen-producing stands of the grass are present. Growers are encouraged to slash every second row of Rhodes grass on an alternating cycle, so that weed management is balanced against predator enhancement.

The use of windbreak trees, such as *Eucalyptus torelliana* mixed with other types of eucalypt, also increases predatory mite populations.

Chemical control

Pesticides are powerful tools to use against pests, and can play a role in IPM programs. Note, however, that the use of most broad-spectrum pesticides, even only once or twice in a season, can seriously disrupt an IPM program. Spray drift between blocks may compound these problems.

Pest control programs that rely mainly on chemicals can seem attractive, simple and low-risk. They do not require detailed knowledge of the pests and their natural enemies, and appear to be an easy solution to the problems of quickly reducing pest numbers, and guaranteeing production of quality fruit.

However, continual use of broad-spectrum pesticides on a regular calendar basis leads to a number of serious problems. These include:

- resistance of the pests, first to one pesticide, then to a whole chemical group and then across groups, which ultimately leads to poor control of the pests and increasing losses in production and quality
- increased costs for developing and producing pesticides to control resistant pests
- secondary pest problems because the pesticides kill natural enemies which normally suppress a wide range of minor pests
- environmental contamination
- pesticide residues in or on the fruit
- potentially detrimental effects on the health of orchardists, their families and their staff.

Advantages of IPM

The adoption of IPM can result in production of high-quality fruit at less cost to the grower than the cost of chemical control programs. A recent economic survey in Queensland citrus showed that IPM resulted in savings of 37–53% compared to the costs of chemical control. There are also potential marketing advantages for 'clean and green' citrus produced with minimal pesticide use.

Other advantages of IPM are:

- the development of pest resistance to pesticides is delayed or avoided due to the less frequent use of chemicals
- growers develop a thorough knowledge of pests and beneficials, and can use this to improve orchard management

- long-term control of citrus pests is improved through the increased abundance and diversity of natural enemies
- as a result of reduced pesticide use, safety is improved for people working in orchards
- as a result of reduced pesticide use, environmental contamination is reduced
- pesticide residues in or on fruit are minimised, enhancing consumer acceptance of the produce.

Components of IPM

IPM must be based on a sound knowledge of the pests and their natural enemies, action levels and management strategies. Building up to a high-level IPM program is a gradual process, needing careful attention. This can lead ultimately to minimal or no use of pesticides, with pest control achieved mostly by natural enemies and/or cultural control.

Identification of pests and natural enemies

Specialist training in pest and disease diagnosis (e.g. by working alongside a pest scout or entomologist), and understanding of IPM techniques are required.

Monitoring of pests and natural enemies

Monitoring of pests and their natural enemies is a vital component of IPM. Monitoring can be carried out by a commercial pest scout or by the orchard manager. In either case, it must be done regularly in a systematic manner.

Action levels

Action must be taken to control pests before the point at which they cause economic loss. Action levels are usually estimates based on research and experience. Pest management plans should be closely linked with disease management decisions.

Appropriate action

When action levels of a pest are reached, the most appropriate control strategy must be selected, e.g. release of parasites, or use of a selective pesticide. Such decisions are critical to IPM success.

Avoid using broad-spectrum pesticides. If a pesticide has to be used, its effect on natural enemies must be considered first. Choose a 'soft' pesticide if possible, e.g. petroleum spray oil.

Citrus pests
and their
natural enemies

This section gives detailed information on the insect, mite, spider, nematode and snail pests which attack citrus. The pests are arranged in their natural groupings, not alphabetically. Where possible, life cycle illustrations of different groups are provided to assist in understanding their habits, and how damage is caused. (Note that most of these illustrations are not life size. Actual sizes of the organisms are given in the text.)

Descriptions are given of each pest, of the damage it causes, of its natural enemies, and its management. For each pest, a distribution map is given to show the main areas where it occurs on citrus in Australia.

The importance of each pest in each area is graded as 'major', 'occasionally important', or 'minor'. Major pests are those that occur most frequently, and can seriously affect fruit yield or quality, or tree health. A grading of 'occasionally important' means that the pest occurs more sporadically, or causes less damage to fruit or tree health. A pest of minor importance has the potential to cause problems, but is usually present only in small numbers. Some minor pests, e.g. white wax scale and circular black scale, were major problems 25 years ago, but are now effectively controlled by specially introduced natural enemies. Nearly every pest in this book, however, even when graded as 'minor' can cause damage if not properly managed.

Regular monitoring of the levels of pests and their natural enemies is vital for good pest management. Monitoring procedures are given for each pest.

The term 'action level' is used to define the point at which some action should be taken to avoid economically significant crop or tree damage by the pest. The action level is an estimate only, and will vary according to citrus variety, district, market prices and demand. Growers and pest scouts can adjust the action level as necessary, and according to their experience of the particular orchard and environment. To achieve good IPM, action must be taken only when necessary, and not before.

If an action level is reached, the decision must be made on what action to take to control the pest. The most appropriate control strategies are given for each pest in this book. (See also chapter 2 for information on general control strategies.)

Mites

Mites are minute arthropods, with the thorax and abdomen fused together. They are a separate group from insects. Mites mostly range in size from 0.1 mm to about 0.6 mm long. To view them closely, a hand lens or microscope is essential.

On citrus, a range of plant-feeding, predatory and other mites occurs. Feeding by pest mites causes surface scarring, russeting or deformation of fruit and leaves. Predatory mites feed on pest mites, and on other foods such as pollen and fungal spores. They cause no harm to citrus plants or fruit. Other mites feed on fungi, or on decaying plant and animal material on the bark, leaves and fruit. They also cause no harm to citrus plants or fruit.

Monitoring

The following points apply to the monitoring of all pest mites and their natural enemies:
- The number of trees from which samples are taken depends on block size (see chapter 25 for details).
- Additional care should be taken when monitoring blocks that have a history of economic damage.

Plant-feeding mites and other mites (Acarina) 3

Brown citrus rust mite

Tegolophus australis Keifer, Acarina: Eriophyidae

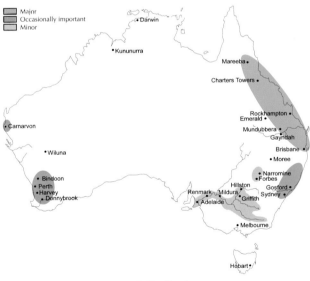

Figure 3.1 Importance and distribution of brown citrus rust mite.

Plate 3.1 Brown citrus rust mite. Shown are light-brown, wedge-shaped adults, smaller, paler nymphs, and eggs (arrowed).

Description

▶ **General appearance**

The brown citrus rust mite is light brown to dark brown and wedge-shaped, with the head at the broader end. It has two pairs of short legs at the head end of the body, and a pair of lobes or 'false feet' on the last body segment.

The adult female is 0.18 mm long and 0.07 mm wide across the head end. The male is smaller. The immature stages are lighter in colour and slimmer than the adults.

▶ **Distinguishing features**

Brown citrus rust mite can easily be confused with citrus rust mite. However, adults of brown citrus rust mite are light brown to dark brown, with a definite wedge shape, while adults of citrus rust mite are yellow to yellow-brown and cigar-shaped. A ×10 hand lens is necessary to identify and monitor rust mites.

Brown citrus rust mite is usually found on the side of the fruit facing the exterior of the tree, while citrus rust mite prefers the side of the fruit facing the interior of the tree.

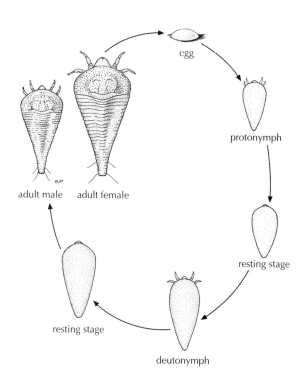

Figure 3.2 Life cycle of brown citrus rust mite.

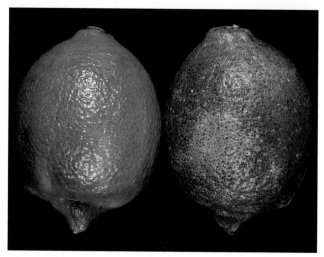

Plate 3.2 Rind damage on a young Lisbon lemon (right) caused by brown citrus rust mite. Compare with the undamaged fruit (left).

Plate 3.3 Rind damage on a Valencia orange caused by brown citrus rust mite.

▶ Life cycle

Eggs are translucent, shaped like a low dome, and 0.07 mm in diameter. They are laid in minute depressions in the fruit rind or upper leaf surface, and hatch into nymphs which are similar to the adult, but smaller and paler.

The female life cycle from egg to adult takes 10–30 days, depending on temperature. Adults live for 4–6 weeks and females lay an average of 30 eggs.

▶ Seasonal history

There are 20–30 generations of brown citrus rust mite per year.

▶ Habits

Mites infest young green fruit from early spring. Populations tend to peak on mature leaves up to one year old. Blocks of citrus can be infested with both brown and yellow citrus rust mites, but usually one species predominates.

Unlike citrus rust mite, brown citrus rust mite prefers exposed sites and warm conditions. Serious infestations of either species are more likely to occur in coastal districts. However, severe infestations can occur unexpectedly in the inland regions of the southern states, especially in warm humid years, along the major rivers, and where overhead sprinklers are used. Severe infestations can also result from the use of disruptive organophosphate insecticides. The mite's natural enemies are destroyed by the spray, and predatory mite populations take longer to recover than the pest mite's.

▶ Hosts

Only citrus is attacked.

▶ Origin and distribution

Brown citrus rust mite is native to Australia, and has not been recorded overseas.

Damage

Fruit
Fine background brown stippling to heavy, shiny brown russeting on the side facing the outside of the tree; reduced fruit size, and shrivelling.

Leaves
Stippled and blistered bronze areas on upper surfaces of leaves, leaf drop.

Twigs
Stippled and blistered bronze areas.

Brown citrus rust mite damage on fruit can be confused with melanose, chemical burn and weather staining. However, two key characteristics of severe damage by brown citrus rust mite are the presence of white cast skins, and extensive smooth, dark-brown areas with indistinct margins.

When mature, severely damaged fruit are also smaller than normal, and shrivelled. The blemish becomes more prominent as fruit colour.

Heavily infested fruit show damage after 3–4 weeks.

While brown citrus rust mite usually attacks fruit on the outside of the tree, or the side of fruit facing the outside of the tree, citrus rust mite usually attacks fruit on the inside of the tree, or the side of fruit facing the inside of the tree.

▶ Varieties attacked
All varieties of citrus are attacked.

Natural enemies

▶ Predators

Phytoseiid mites
Euseius victoriensis is an effective predator of brown citrus rust mite, particularly in inland citrus areas and in sub-coastal Queensland. In coastal areas of New South Wales, the phytoseiid mites *Euseius elinae*, *Amblyseius herbicolus* and *Amblyseius lentiginosus* are the most important predators.

If phytoseiids are present early in the season and in autumn, and not disrupted by pesticides, they give excellent biological control of brown citrus rust mite.

Adult phytoseiids are translucent, teardrop-shaped mites about 0.5 mm long, and just visible with the unaided eye. They have 4 pairs of legs on which they run rapidly. The first pair of legs project towards the front of the body, like an additional pair of antennae.

These predatory mites are most active at night. During the day they are commonly found in sheltered sites on the underside of leaves near the midrib, and under the calyx.

Adult females deposit relatively large spherical eggs, often attached to plant hairs or spider web on the underside of leaves.

Plate 3.5 Rhodes grass growing between rows of grapefruit trees. This grass is a good producer of pollen used as a supplementary food source by predatory mites.

Besides feeding on brown citrus rust mites, phytoseiid mites also eat citrus rust mite; some tarsonemid, tydeid, tetranychid and brevipalpid mites; pollen; plant sugars; and fungal spores. When they feed on brown citrus mite, their body contents take on the colour of their prey.

E. victoriensis and *E. elinae* feed readily on a range of pollens, and can be successfully reared in the laboratory. Certain plants in orchards can be very useful in providing pollen which blows onto the citrus to act as a supplementary food source. In Queensland, inter-row Rhodes grass (*Chloris gayana*), produces pollen for much of the year and significantly boosts predatory mite numbers, as do annual and perennial ryegrasses in southern Australia.

E. victoriensis and *E. elinae* occur commonly on a wide range of native and introduced shrubs and trees. *Eucalyptus torelliana*, for example, is an excellent host plant for *E. victoriensis* and, when used in windbreaks, acts as a reservoir of the predator. *E. elinae* is also common on brush box (*Lophostemon confertus*) and other plants.

Phytoseiids are susceptible to many broad-spectrum pesticides and some fungicides such as mancozeb and benomyl. Most other fungicides are relatively safe to *E. victoriensis*. Serious infestations of brown citrus rust mite often result from the use of broad-spectrum pesticides which destroy phytoseiid predators.

Stigmaeid mites
The yellow and carmine coloured stigmaeid mites commonly feed on brown citrus rust mite. Although they can reduce mite infestations, they are much less important than phytoseiids.

Cecidomyiid flies
Cecidomyiid fly larvae prey on brown citrus rust mite. These larvae are small, legless, cream to white maggots with no visible head. They suck the body fluids of the mites. They are of minor importance.

▶ Pathogens
Fungal pathogens of brown citrus rust mite have been found, but have not yet been identified. They are probably only important when prolonged humid conditions coincide with high populations of the rust mite.

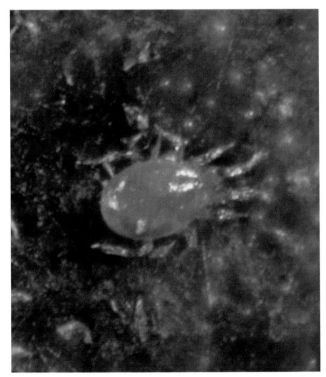

Plate 3.4 Predatory mite *Euseius victoriensis*. Note the brown colour of the gut, caused by feeding on brown citrus rust mite.

Management

▶ Monitoring

- Brown citrus rust mite occurs on all varieties of citrus.
- The pest and its predators should be monitored fortnightly, from the beginning of November to May harvest in Queensland and coastal New South Wales. Elsewhere monitor fortnightly to monthly on young fruit from January to April, but pay special attention to mature late Valencias in spring.
- Using a hand lens, check 5 randomly selected fruit per tree. It is advisable to take around 10–20% of fruit from the top centre of the tree. Examine 25% of the fruit surface that faces the outside of the tree for the presence or absence of mites. The predatory mites *Euseius victoriensis* and *Euseius elinae* should be monitored on the same trees by counting mite numbers on the undersides of 5 randomly selected mature leaves in the middle of each of 5 young branches.

▶ Action level

The action level depends on fruit size, percentage of fruit infested, and predatory mite activity.

Young fruit (October to December) can accommodate rind damage better than large fruit (January to harvest). Extra care with decision making is required later in the season.

When unaffected by pesticides, the predatory mites *E. victoriensis* and *E. elinae* will normally prevent brown citrus rust mite from causing damage.

The action level is 10% or more of young fruit infested with brown citrus rust mite. However, infestation of up to 80% or more of young fruit can be tolerated, provided there is no visible sign of damage, and there is clear evidence that predatory mites are abundant and their numbers are increasing rapidly. As a guide, 40 or more predators per 100 leaves is desirable (see figure 3.3).

In an infested block, a small number of exposed fruit (or fruit at the top of high trees) often have mite populations 2–3 weeks ahead of other fruit. The first signs of blemish on these fruit indicates that immediate action should be considered, provided live mites are present.

The action level for large fruit is 10% or more of fruit infested with brown citrus rust mite. The action level can be raised where there are good numbers of predators, but take extra care when considering this decision. Seek advice from an experienced pest scout.

▶ Appropriate action

Apply a miticide, preferably one that does not kill predatory mites. Alternatively, in New South Wales, spray thoroughly (5000–8000 L/ha for trees 4 m high) with petroleum spray oil (700 mL – 1 L in each 100 L of water) (see chapter 27). More than one spray may be necessary, particularly if warm humid conditions continue into late autumn.

The recommended volumes are for oscillating booms with outriggers, air-blast sprayers with towers, and rotary atomisers. The 8000 L/ha rate is for moderate to heavy infestations. Higher volumes may be necessary if low-profile air-blast sprayers are used.

▶ Additional management notes

Every effort should be made to encourage phytoseiid predators. Avoid or minimise the use of disruptive pesticides.

Encourage the growth of pollen-producing plants such as Rhodes grass or ryegrass in the inter-rows (sod culture). (This is also recommended for managing red scale and some other pests.) Discourage the growth of plants that are hosts of lightbrown apple moth, e.g. broadleaf weeds and legumes.

Avoid overzealous mowing. Mow alternate rows every 2–4 weeks. This will result in neat orchards while still permitting continual flowering and pollen production.

Establish peripheral windbreaks of predator-harbouring hosts, such as *Eucalyptus torelliana*, especially in new citrus blocks.

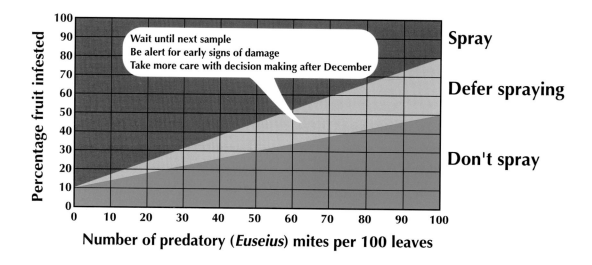

Figure 3.3 Action levels for brown citrus rust mite.

Citrus rust mite

Phyllocoptruta oleivora (Ashmead), Acarina: Eriophyidae

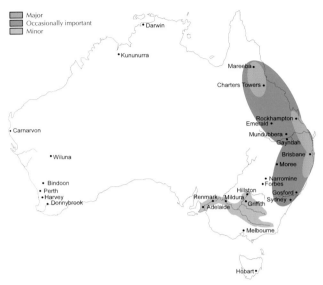

Figure 3.4 Importance and distribution of citrus rust mite.

Description

▶ General appearance
The citrus rust mite is lemon-yellow and cigar-shaped. The adult female is 0.15 mm long and 0.04 mm wide. The male is smaller.

The mites have two pairs of short legs located at the head end and a pair of lobes or false feet on the last body segment.

▶ Distinguishing features
Citrus rust mite adults are yellow and cigar-shaped, while brown citrus rust mite adults are light to dark brown, and wedge-shaped.

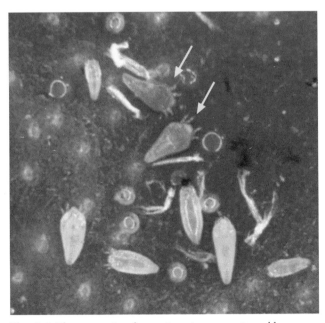

Plate 3.6 The two species of rust mites: citrus rust mite and brown citrus rust mite. Citrus rust mite adults are yellow and cigar-shaped, while brown citrus rust mite adults (arrowed) are more broad-shouldered or wedge-shaped, and brown.

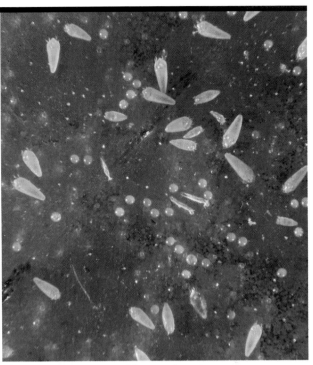

Plate 3.7 Citrus rust mite adults, nymphs and eggs.

▶ Life cycle
The eggs are white and translucent, spherical in shape, and 0.04 mm in diameter. They are laid singly or in groups in depressions in the fruit rind or in the leaf surface. They hatch into yellow, cigar-shaped nymphs, which moult twice before reaching adulthood.

The life cycle from egg to adult takes about 6 days at 30°C. The adult female lives 4–6 weeks and lays an average of 30 eggs.

▶ Seasonal history
There are 20–30 generations per year.

▶ Habits
Citrus rust mite infests leaves, green twigs, and developing fruit. Fruit infestation is the main problem.

The mite mainly infests the sheltered sides of young green fruit from late spring onwards. Attack on young fruit results in surface bronzing, while attack on mature green fruit often results in a darker, deeper scarring.

Citrus rust mite prefers less exposed sites than does brown citrus rust mite, although they occasionally occur together. The underside of leaves, the inside surface of the fruit, the lower and inside part of the tree, and the southern and eastern aspects are preferred.

Populations tend to peak on mature leaves up to a year old.

Citrus rust mite is most common in humid conditions, especially in coastal areas.

▶ Hosts
Citrus is the only known host of citrus rust mite.

▶ Origin and distribution
This mite originated from East and South-East Asia, and is now distributed throughout the world.

Damage

Fruit
On oranges, mandarins and grapefruit, a dusky grey-brown discolouration followed by thick brown russeting or silver-grey blemish, which turns black in heavy infestations.
On lemons, a thick, grey to dirty brown, 'sharkskin' appearance of the skin.

Leaves
Brown russeting on mature leaves.

Twigs
Brown russeting on green twigs.

The damage caused by citrus rust mite can be confused with that caused by brown citrus rust mite and broad mite. Damage caused by brown citrus rust mite is always shiny brown, usually on the exposed surfaces of fruit, and of a similar kind to damage caused by citrus rust mite, which is grey, brown, and/or black, and smooth, with indistinct margins.

Damage to fruit caused by broad mite appears on all varieties as a thin, silver-grey skin that can be readily scratched off. The dark blemish caused by citrus rust mite is coarser and cannot be scratched off.

Damage to fruit caused by citrus rust mite can also be confused with damage caused by melanose, chemical burn and weather staining. However two characteristics distinguish severe citrus rust mite damage: the presence of elongate white cast skins, shed at moulting; and the presence of damage on the inside-facing surfaces of the fruit.

Plate 3.9 Citrus rust mite damage on Valencia oranges on the tree. Note the range of colours of the damaged areas.

Varieties attacked
All citrus varieties are attacked.

Natural enemies

Predators

Phytoseiid mites
Euseius victoriensis is an effective predator of citrus rust mite, particularly in inland citrus areas and in sub-coastal Queensland. In coastal areas of New South Wales, the phytoseiid mites *Euseius elinae*, *Amblyseius herbicolus* and *Amblyseius lentiginosus* are the most important predators. Provided predatory mites are present early in the season, satisfactory control of citrus rust mite is likely. (See page 19 for more information on phytoseiid mites.)

Stigmaeid mites
Yellow and carmine coloured stigmaeid mites commonly feed on citrus rust mite. Although they can reduce mite infestations, they are much less important than phytoseiids.

Cecidomyiid flies
Cecidomyiid fly larvae are small, legless, cream to white maggots with no visible head. They feed by sucking the body fluids out of citrus rust mites. They are of minor importance in controlling these mites.

Plate 3.8 Citrus rust mite damage on the rind of Valencia oranges.

Pathogens

Fungal pathogens of citrus rust mite have been found, and include *Hirsutella* sp. They are probably only important when prolonged, very humid conditions coincide with high populations of the rust mite. Copper fungicides may reduce the efficacy of these fungal pathogens.

Management

Monitoring

- Citrus rust mite occurs on all varieties.
- The pest and its predators should be monitored fortnightly, from the beginning of November to May in Queensland and coastal New South Wales. Elsewhere monitor from January to April, but pay special attention to late Valencias in spring.
- Using a ×10 hand lens, check 5 randomly selected fruit per tree. It is advisable to take 10–20% of these fruit from the top centre of trees. Examine 25% of the fruit surface that faced the inside of the tree for the presence or absence of mites.

    The predatory mites *Euseius victoriensis* and *Euseius elinae*, should be assessed on the same trees by counting mite numbers on the undersides of 5 randomly selected mature leaves on each of 5 young branches.

Action level

The action level depends on fruit size, percentage of fruit infested, and predatory mite activity.

Young fruit (October to December) can tolerate rind damage better than large fruit (January to harvest). Extra care with decision making is required later in the season.

The predatory mite *E. victoriensis* is not as effective a predator on citrus rust mite as on brown citrus rust mite. (This may also be the case for *E. elinae*.) However, if predatory mites are present early in the season in good numbers, e.g. 40 predators per 100 leaves, they can achieve effective control of citrus rust mite.

The action level is 5–10% or more of young and old fruit infested with citrus rust mite. Great care is necessary in making the decision to defer spraying after this level of infestation has been reached (see figure 3.5).

Plate 3.10 Adult *Euseius elinae*, a predatory mite that feeds on citrus rust mite.

In an infested block, a small number of exposed fruit (or fruit from the top of high trees) often have mite populations 2–3 weeks ahead of other fruit. The first signs of blemish on these fruit indicate that immediate action should be considered, provided live mites are present.

Appropriate action

Spray thoroughly with a selective miticide, or with petroleum spray oil (700 mL/100L) (see brown citrus rust mite, page 20).

Additional management notes

Every effort should be made to encourage phytoseiid predators. Avoid or minimise the use of disruptive pesticides.

Encourage the growth of pollen-producing plants such as Rhodes grass or ryegrass in the inter-rows (sod culture). (This is also recommended for managing red scale and some other pests.) Discourage the growth of plants that are hosts of lightbrown apple moth, e.g. broadleaf weeds and legumes.

Avoid overzealous mowing. Mow alternate rows every 2–4 weeks. This will result in neat orchards while still permitting continual flowering and pollen production.

Establish peripheral windbreaks of predator-harbouring hosts, such as *Eucalyptus torelliana*, especially in new citrus blocks.

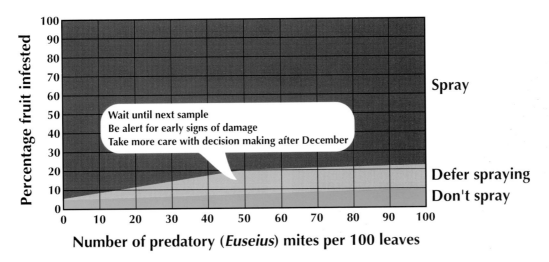

Figure 3.5 Action levels for citrus rust mite.

Citrus bud mite

Eriophyes sheldoni Ewing, Acarina:Eriophyidae

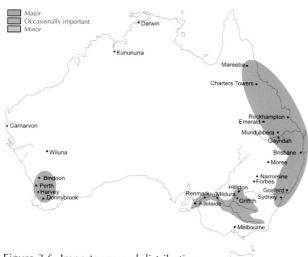

Figure 3.6 Importance and distribution of citrus bud mite.

Plate 3.11 Citrus bud mite adults, nymphs and eggs. These mites are very difficult to find without the aid of a microscope, as they are very small, white to cream, and hidden in buds.

Description

▶ General appearance
The adult citrus bud mite is a small, cream-coloured, cigar-shaped mite about 0.17 mm long, and very difficult to find without the aid of a microscope.

▶ Distinguishing features
The citrus bud mite is very similar in shape and size to citrus rust mite, but the bud mite is usually hidden in the buds of flowers and leaves, and is cream-coloured.

▶ Life cycle
The life cycle is similar to that of the brown citrus rust mite, taking about 10 days in the summer. The spherical, white eggs, 0.03 mm in diameter, are deposited singly or in groups in the young tissue of buds and flowers. The eggs hatch into small nymphs similar in shape to the adult mite. These undergo two moults before reaching adulthood.

▶ Seasonal history
Over 20 generations are produced in a year.

▶ Habits
The citrus bud mite feeds within leaf buds and blossoms. Up to 50 eggs can be deposited within a leaf bud. High populations cause bud death, but it is more common for small buds to develop in abnormal positions and for growth to be deformed and bunched.

▶ Hosts
Citrus is the only known host.

▶ Origin and distribution
Citrus bud mite occurs worldwide.

Damage

Flowers
Blackening (with heavy infestations) and distortion.

Fruit
Reduced fruit set; severe distortion resulting in bizarre shapes.

Leaves
Multiple budding, distortion and bunching of leaf growth.

Twigs
Shortening of twigs, sometimes flattening and distortion.

Plate 3.12 Flower distortion caused by citrus bud mite.

The twisting and distortion of fruit and leaves caused by citrus bud mite resemble symptoms caused by herbicides, physiological disorders or insects such as spherical mealybug and aphids.

▶ Varieties attacked

Most varieties of citrus are attacked. In Queensland, Lisbon lemons are attacked more frequently than other varieties, and in New South Wales, navel oranges.

Natural enemies

▶ Predators

The main natural enemies are phytoseiid, stigmaeid and cheyletid mites. The major phytoseiid predators are *Euseius victoriensis* and *Euseius elinae* (see page 19).

Management

▶ Monitoring

- Citrus bud mite occurs on all varieties, with Lisbon lemons, Marsh seedless grapefruit and Washington navel oranges being most commonly attacked.
- As this mite is hard to see with a hand lens, keep a close watch for the twisting of young shoots and distortion of young fruit from the beginning of November to May.
- If damage is observed, sample 5 young shoots per tree for the presence or absence of twisting and damage. Confirm the presence of citrus bud mite by carefully examining or dissecting young leaf buds under a binocular microscope.

▶ Action level

The action level is 10% or more of leaf buds infested.

▶ Appropriate action

A selective miticide should be applied when the action level is reached. Predatory mites (phytoseiid, stigmaeid and cheyletid) should be encouraged in the orchard (see brown citrus rust mite, page 20).

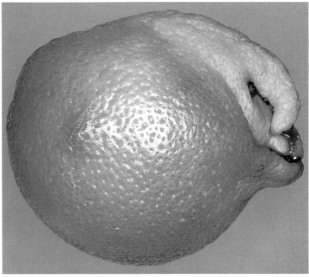

Plate 3.13 Distortion of a Lisbon lemon caused by citrus bud mite.

Plate 3.14 Distortion of grapefruit shoots caused by citrus bud mite.

Broad mite

Polyphagotarsonemus latus (Banks), Acarina: Tarsonemidae

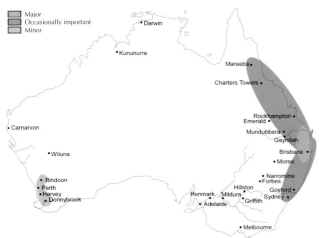

Figure 3.7 Importance and distribution of broad mite.

Plate 3.15 Broad mite adults, nymphs and eggs. The adults are yellow, the nymphs white, and the surface of the eggs covered with white 'studs' (bottom right).

Description

General appearance
The adult female mite is white or pale yellow, oval-shaped and 0.2 mm long. The male is smaller, with longer hindlegs modified into claw-like appendages, and is more active.

Distinguishing features
A distinguishing feature of broad mite is the male habit of carrying pre-adult females on their backs. Another identifying characteristic is the appearance of the eggs, which are dome-shaped, translucent, and covered with white 'studs' or tubercles.

Other pale or white mites which might at first be mistaken for broad mite are the harmless tydeid, acarid and bdellid mites. These are predominantly scavengers.

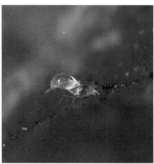

Plate 3.16 Bdellid mites (left) and broad mites (right). Bdellid mites are sometimes mistaken for broad mites. However, the bdellid male carries the pre-adult female lengthwise, while the male broad mite usually carries the pre-adult female crosswise.

Life cycle
Eggs are laid in depressions in fruit rind and on the lower side of the leaf. The female lives for about a fortnight producing 5 eggs per day. The total life cycle can be as short as 7 days in summer.

Seasonal history
There are 20–30 generations per year, and commonly more in coastal areas of warmer districts.

Habits
Broad mite feeds mainly on young foliage or fruit. Mites usually move onto the newly set fruit in spring and summer. Infestations develop rapidly and cause damage quickly compared to rust mite infestations (which take 3–4 weeks to cause damage).

Broad mite prefers humid situations and usually causes damage to citrus only in coastal districts. The mites are found mostly on the inward-facing side of the fruit (beginning at the stem end), in the lower inner part of the tree.

Hosts
A wide range of hosts are attacked, including many tropical and subtropical fruit and ornamentals. Broadleaf weeds are a significant source of infestations in orchards.

Origin and distribution
Broad mite has a worldwide distribution.

Damage

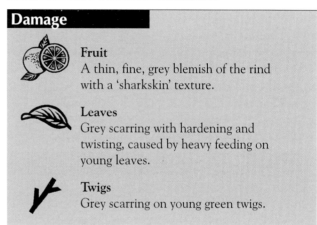

Fruit
A thin, fine, grey blemish of the rind with a 'sharkskin' texture.

Leaves
Grey scarring with hardening and twisting, caused by heavy feeding on young leaves.

Twigs
Grey scarring on young green twigs.

Broad mite damage to fruit can be similar in appearance to that caused by citrus rust mite. However, the broad mite damage on all varieties appears as a thin, silver-grey skin that can be readily scratched off. The dark skin damage caused by citrus rust mite is coarser and cannot be scratched off.

Lemons are less prone to broad mite damage after fruit are half-grown. Mandarins are susceptible until they begin to colour. Limes are susceptible until they are two-thirds mature.

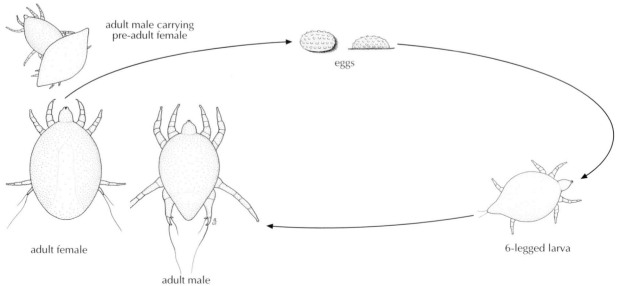

Figure 3.8 Life cycle of broad mite.

Plate 3.17 Rind damage on young Lisbon lemons caused by broad mite. The silvery-grey areas are on the surfaces of the fruit that faced the inside of the tree.

Plate 3.18 Rind damage on grapefruit caused by broad mite.

Most damage is done during spring – early summer, but second-crop fruit are often attacked later in the season. In New South Wales, damage is most frequent on valuable summer-crop fruit, which set in summer–autumn.

▸ Varieties attacked
Lemons and limes are the most consistently attacked. Young grapefruit, oranges and mandarins are also susceptible, particularly Hickson mandarins.

Plate 3.19 Rind damage on Hickson mandarin caused by broad mite.

Natural enemies

▸ Predators
Predatory phytoseiid mites (see page 19) are the main natural enemies of broad mite. In Queensland, *Euseius victoriensis* is an effective predator in subcoastal areas, but less effective in coastal orchards. *Euseius elinae* is another phytoseiid which feeds on broad mite, particularly in glasshouses in south-east Queensland.

In coastal New South Wales, *E. elinae* and *Amblyseius herbicolus* are common predators of broad mite, rust mites and citrus red mite.

A ladybird *Scymnus* sp. feeds on broad mites.

Management

▸ Monitoring
- Broad mite occurs on all citrus varieties, especially lemons, limes, grapefruit, and Hickson and Ellendale mandarins. Coastal navel oranges are also susceptible.
- Monitoring for broad mite and its predators should be done fortnightly, particularly on susceptible varieties, from the beginning of fruit set to green maturity. Monitoring for citrus rust mite will normally also detect broad mite.
- Check for the presence or absence of broad mites on 5 randomly selected fruit per tree. With a hand lens, examine the inward-facing surface of the fruit, near the calyx.
- Note that broad mite can be a problem in nursery plants.

▸ Action level
The action level is 5% or more of the fruit infested with live mites. Action must be taken quickly especially in coastal areas.

If good numbers of predatory phytoseiid mites are present (i.e. 40% or more of leaves with the predators visible), broad mite infestations will not usually develop.

▸ Appropriate action
Every effort should be taken to encourage good numbers of predatory mites (see brown citrus rust mite, page 20).

If the action level is reached, apply a selective miticide.

Two-spotted mite

Tetranychus urticae Koch, Acarina: Tetranychidae

Figure 3.9 Importance and distribution of two-spotted mite.

Plate 3.20 Two-spotted mite (right) and Chilean predatory mite (left).

Description

▶ General appearance
The adult female two-spotted mite is 0.5 mm long and greenish-yellow with a dark spot on each side of the body. The males are smaller and narrower than the females. Under stressful conditions, two-spotted mites turn a reddish colour.

▶ Distinguishing features
Adult females could possibly be confused with oriental spider mite. However, female two-spotted mites can be identified by the two spots on their bodies. Male two-spotted mites are not long-legged like male oriental spider mites, which have legs three times the length of their bodies.

▶ Life cycle
The eggs are spherical and translucent, changing to yellow as they age. They are laid on the leaf or fruit surface within spidery webbing produced by the mites. Each female lays about 70 eggs over a period of a fortnight. Larvae with 6 legs hatch from the eggs. After moulting to the first of 2 nymphal stages, they have 8 legs. The life cycle can take as little as 10 days in the summer.

▶ Seasonal history
There are 10–20 generations per year.

▶ Habits
Mites feed on mature leaves and on fruit. In heavy infestations, young shoots and young fruit are also infested and webbed by the spider mites.

▶ Hosts
Two-spotted mite has a wide range of hosts, including many commercially grown fruits, vegetables and ornamentals.

▶ Origin and distribution
Two-spotted mite has a worldwide distribution.

Damage

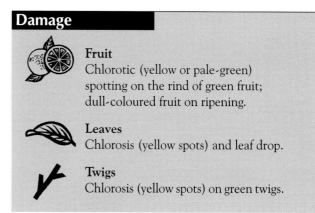

Fruit
Chlorotic (yellow or pale-green) spotting on the rind of green fruit; dull-coloured fruit on ripening.

Leaves
Chlorosis (yellow spots) and leaf drop.

Twigs
Chlorosis (yellow spots) on green twigs.

▶ Varieties attacked
All citrus varieties are attacked by two-spotted mite. However, in coastal areas, Meyer lemon is the most commonly infested variety.

Plate 3.21 Rind damage on Valencia orange caused by two-spotted mite. The rind is covered with pale spots.

3 Plant-feeding mites and other mites (Acarina)

Plate 3.22 Leaf (right) damaged by two-spotted mite. Compare with the healthy leaf (left).

Natural enemies

Predators

The major natural enemies of two-spotted mite are stethorus ladybirds (see information under oriental spider mite, pages 31–32), predatory mites, predatory thrips (*Scolothrips sexmaculatus*) and lacewing larvae.

The Chilean predatory mite (*Phytoseiulus persimilis*) is a voracious predator of two-spotted mite, particularly in coastal areas of Queensland. It is about 0.7 mm long, with a pear-shaped body, and a shiny, orange-red in colour. The adult female lives for about 4 weeks during which time she lays about 50 eggs. These hatch in 2–3 days and develop through two nymphal stages to the adult in about a week. The Chilean predatory mites multiply twice as fast as the two-spotted mites and each eat about 7 mites per day, so when present they usually completely control the pest within 6–8 weeks.

The predatory mite *Typhlodromus occidentalis* is also an important predator in hot, dry areas. Both *Typhlodromus occidentalis* and *Phytoseiulus persimilis* can be purchased from commercial suppliers for release into orchards to control two-spotted mite.

Native phytoseiid mites, e.g. *Euseius victoriensis* and *Euseius elinae*, also prey on two-spotted mite in Queensland and inland New South Wales citrus.

Lacewing larvae commonly occur on citrus, where they feed on scales, mealybugs, moth eggs, aphids, thrips and mites. Lacewing eggs are 0.5–1.0 mm long, and each one is attached by a long stalk to the laying site, e.g. the surface of a leaf. Groups of several eggs are laid close to each other. The larvae are cream to brown in colour and up to 8 mm long. They move actively amongst the prey, impaling their victims on their large sickle-shaped jaws and sucking out the body contents. The larvae attach the remains of their prey to their backs, probably for camouflage.

Pathogens

Some disease-causing fungi (e.g. *Neozygites* spp., *Hirsutella thompsonii*) may help control two-spotted mite populations, especially in coastal areas.

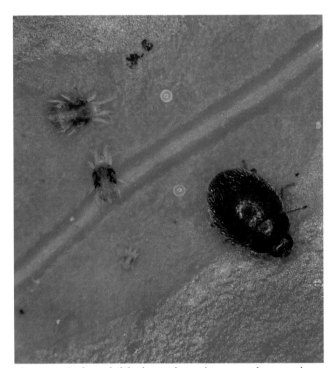

Plate 3.23 Stethorus ladybird, a predator of two-spotted mite, with some of its prey.

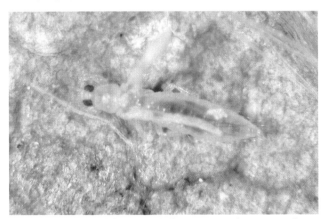

Plate 3.24 Six-spotted thrips (Scolothrips sexmaculatus), a predator of two-spotted mite.

Plate 3.25 A typhlodromid predatory mite, feeding on a two-spotted mite egg.

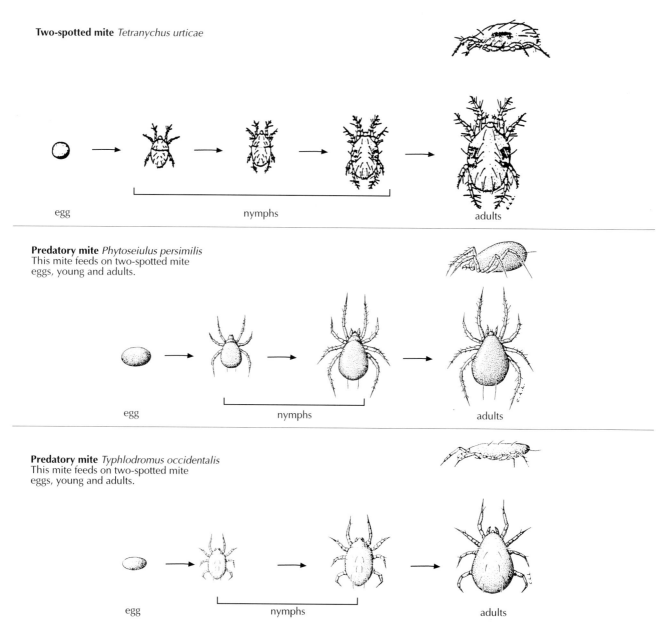

Figure 3.10 Life cycles of two-spotted mite, the Chilean predatory mite (*Phytoseiulus persimilis*) and the typhlodromid predatory mite (*Typhlodromus occidentalis*).

Management

▶ Monitoring
- Two-spotted mite occurs on all citrus varieties, and especially on Meyer lemons.
- The pest and its predators should be monitored fortnightly, from the beginning of November to May in Queensland and coastal New South Wales, if there is any evidence that the mite is present, or if its presence is suspected. Elsewhere monitor from January to April.
- Using a ×10 hand lens, check 5 randomly selected fruit or 5 leaves per tree. Check the upper leaf surfaces and exposed fruit surfaces.

▶ Action level
Action is required when more than 20% of fruit or leaves are infested.

▶ Appropriate action
Apply a selective miticide. If a persistent problem occurs (especially in nurseries), periodic releases should be made of the predatory mites, *Phytoseiulus persimilis* or *Typhlodromus occidentalis*.

▶ Additional management notes
Infestations of two-spotted mite are commonly the result of misusing broad-spectrum pesticides to control pests such as citrus leafminer, red scale or mealybugs, particularly in nurseries. Such misuse destroys the natural enemies of two-spotted mite. Avoid using such pesticides, and prevent spray drift.

Applying fruit-fly baits too high in the tree is also disruptive to predators and results in outbreaks of two-spotted mite, particularly in Meyer lemons. Bait sprays should be applied to the lower parts of the tree.

Oriental spider mite

Eutetranychus orientalis (Klein), Acarina: Tetranychidae

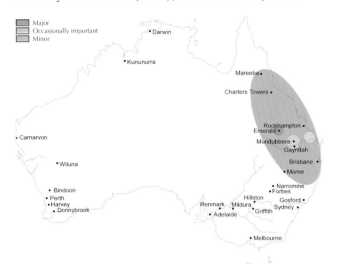

Figure 3.11 Importance and distribution of oriental spider mite.

Plate 3.26 Oriental spider mite. Left, male on top of darker female. Right, several males clustering around a pre-adult female. Note the long legs on the males. Unhatched eggs are brownish, while hatched eggs are colourless and translucent. The white objects are shed skins.

Description

General appearance
Oriental spider mite females are about 0.5 mm long, and brownish-green with lighter, orange-brown legs. Males have smaller bodies than the females, but much longer legs. The males are lighter in colour than the females, being bright orange-brown.

Distinguishing features
Female oriental spider mites may be confused with other tetranychid mites such as two-spotted mite. However, male oriental spider mites are distinguished by their legs, which are three times the length of the body. Males have a habit of clustering around pre-adult females.

Life cycle
The life cycle is very similar to that of two-spotted mite, and takes as little as 10 days. The eggs are orange-brown, and deposited singly on the upper leaf surface. They hatch into first-stage mites that have 3 pairs of legs. Two nymphal stages follow, before the adult stage is reached. Nymphs, like the adults, have 4 pairs of legs.

Seasonal history
These mites are mainly a problem in Queensland, where there are 10–20 generations per year.

Habits
Oriental red spider mites feed mostly on the upper surface of leaves, and on the outward-facing surfaces of fruit. Populations build up rapidly from February to May. Eggs are commonly laid on the upper surface of leaves near the midrib.

Hosts
This mite occurs on a wide range of host plants.

Origin and distribution
Oriental spider mite is also found in South-East Asia and Africa.

Damage

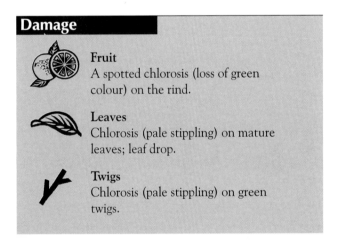

Fruit
A spotted chlorosis (loss of green colour) on the rind.

Leaves
Chlorosis (pale stippling) on mature leaves; leaf drop.

Twigs
Chlorosis (pale stippling) on green twigs.

Fruit is damaged only by heavy populations of the mite, which cause chlorosis on large areas of the rind, and a dull appearance at ripening.

Leaf drop is worse if trees are also stressed by lack of moisture.

Damage to leaves and fruit is similar to that caused by two-spotted mite and citrus red mite.

Varieties attacked
All citrus varieties are attacked, but particularly Murcott and Imperial mandarins.

Natural enemies

Predators
The major natural enemies of oriental spider mite are stethorus ladybirds (mainly *Stethorus histrio* and *Stethorus fenestralis*) and phytoseiid mites. The ladybirds are small black beetles 3 mm long. Adults and larvae both feed on the mites, particularly on the eggs and moulting stages.

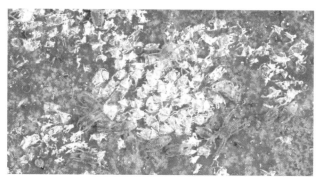

Plate 3.27 Oriental spider mites and white shed skins on damaged leaf tissue. The damage shows as spotted, pale, discoloured areas.

Ladybird eggs, 1 mm long, are laid amongst the mites and hatch into small, cream-coloured larvae. There are four larval instars before a black pupal case is formed, attached to the surface on which the mites are feeding. The adult emerges from this pupal case. The whole life cycle takes about 4 weeks during the summer.

The main phytoseiid mite preying on oriental spider mite is *Euseius victoriensis* (see brown citrus rust mite on page 19 for more information).

▶ **Pathogens**

Fungal pathogens probably attack oriental spider mite.

Management

▶ **Monitoring**
- Oriental spider mite occurs on all citrus varieties, especially in drier regions.
- The pest and its predators should be monitored fortnightly, from the beginning of November to May, when the presence of mites is evident or suspected (e.g. if some leaves show yellow stippling).

Plate 3.28 Leaves damaged by oriental spider mite. Note the pale, spotted appearance of the leaves.

- Using a ×10 hand lens, check 5 randomly selected fruit or leaves per tree. Check the upper leaf surfaces and exposed fruit surfaces for oriental spider mite. Repeat sampling fortnightly when the mite poses a threat.

▶ **Action level**

Action is required when 20% or more of fruit or leaves are infested and predators are absent.

▶ **Appropriate action**

Apply a selective miticide.

▶ **Additional management notes**

Oriental spider mite is almost always an induced pest, with infestation resulting from the destruction of its natural enemies by excessive spraying or spray drift. In particular, avoid spraying trees with synthetic pyrethroids.

Citrus red mite

Panonychus citri (McGregor), Acarina: Tetranychidae

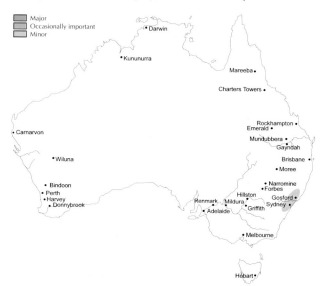

Figure 3.12 Importance and distribution of citrus red mite.

Plate 3.29 Citrus red mite. Note the purple-red colour of the body, the pale legs, and the long white body bristles.

Description

▶ **General appearance**

The adult female citrus red mite has an oval body, and is 0.5 mm long. The body is purple-red in colour with long white bristles on the back and sides. The legs are light yellow. Males are smaller, long-legged, and with a flatter body abruptly narrowed at the posterior end.

▶ **Distinguishing features**

The dark-red or purple-red colour of citrus red mite distinguishes it from other plant-feeding tetranychid mites on citrus. The egg is bright red and spherical, 0.13 mm in diameter, with a vertical stalk on the top.

▶ **Life cycle**

The adult lays 20–40 eggs at a rate of 2–3 per day. Eggs are commonly deposited along the midribs of the leaves. Newly hatched larvae are red and have 6 legs. They develop through two further nymphal stages with 8 legs before reaching adulthood.

In summer, the development time from egg to adult can take as little as 2 weeks. During winter the life cycle takes about 8 weeks. The adult females live about 18 days.

▶ **Seasonal history**

There are 8–10 generations per year.

▶ **Habits**

Citrus red mite feeds on citrus fruit, leaves and green twigs. The maturing leaf is preferred, and most damage occurs on the upper surface. In light infestations, feeding marks are most common at the base of the upper leaf surfaces.

Extremely hot, dry weather and periods of prolonged high humidity reduce the numbers of citrus red mite.

▶ **Hosts**

Other hosts include mulberry, hawthorn and legumes.

▶ **Origin and distribution**

Citrus red mite is native to Asia, and is widely distributed throughout the world.

Damage

Fruit
Pale stippling of the rind, and the fruit may appear dull.

Leaves
Pale stippling; with heavy infestations, leaf drop and twig dieback.

Twigs
Pale stippling.

Damaged mature oranges and lemons turn a pale, straw-yellow. Severe infestations results in leaf drop, fruit drop and twig dieback. Leaf drop is worse if trees are also stressed by lack of moisture.

Damage to foliage and fruit is similar to that caused by two-spotted and oriental spider mites.

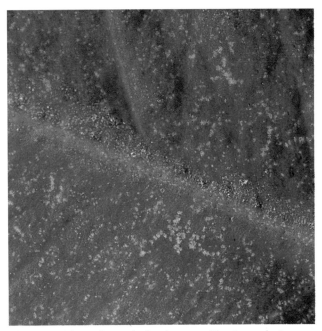

Plate 3.30 *Pale stippling on leaf surface caused by citrus red mite.*

▶ **Varieties attacked**

Most citrus varieties are susceptible to citrus red mite. Oranges are most affected on the central coast of New South Wales.

Natural enemies

▶ **Predators**

The major natural enemies of citrus red mite are the predatory phytoseiid mites *Euseius elinae*, *Amblyseius herbicolus* and *Amblyseius lentiginosus*, lacewing larvae, and larvae of the ladybirds *Stethorus nigripes*, *Serangium bicolor*, and the steel-blue ladybird (*Halmus chalybeus*).

Plate 3.31 *Larva of stethorus ladybird* (Stethorus nigripes) *feeding on citrus red mite egg.*

▶ **Pathogens**

Viruses are known to occur, but have not been identified. They are most commonly associated with high-density populations.

Management

▶ **Monitoring**

- Citrus red mite occurs on all varieties of citrus, and especially oranges.
- The pest and its predators should be monitored fortnightly when conditions are warm and dry, particularly in late winter, early spring and autumn.

- Using a ×10 hand lens, check 5 young mature leaves from each of 5 branches per tree. Examine both sides of the leaves for the presence of citrus red mite eggs and adults, and stippling.

▶ **Action level**

Action should be taken if there are live adults present, and stippling on 50% or more of the leaves.

▶ **Appropriate action**

Spray thoroughly with petroleum spray oil (700 mL/100L) (see chapter 27).

▶ **Additional management notes**

Citrus red mite in Australia is restricted to the central coast (Sydney and Gosford) of New South Wales, where it is subject to quarantine restrictions. Extreme care must be taken to ensure that it is not transferred to citrus areas outside the counties of Northumberland and Cumberland (contact NSW Agriculture for more details).

Citrus red mite is not a serious pest of citrus on the New South Wales central coast due to effective biological control.

Predators should be encouraged (see notes on brown citrus rust mite, page 20), and their disruption by pesticides avoided. Disruption of predators by broad-spectrum pesticides can lead to rapid defoliation of trees by citrus red mite.

Selective miticides should not be used to control this pest. They are costly, and overuse will lead to resistance.

The use of fungicide sprays containing 250–500 mL of petroleum spray oil per 100 L of water as a spreader has the side-benefit of suppressing citrus red mite.

Citrus flat mite

Brevipalpus lewisi McGregor, and *Brevipalpus californicus* (Banks), Acarina: Tenuipalpidae

Figure 3.13 Importance and distribution of citrus flat mite.

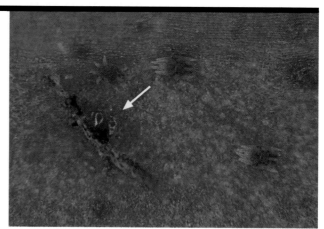

Plate 3.32 Citrus flat mite. Note the red eggs (arrowed).

Description

▶ **General appearance**

Citrus flat mites of both species are 0.3 mm long, flat, shield-shaped and bright red to reddish-brown. The eggs are elliptical, and orange to bright red.

▶ **Distinguishing features**

Mites of both species are slow-moving, and very flat in profile compared to other pest mites.

▶ **Life cycle**

The spherical eggs are found singly in depressions or crevices in the foliage, twigs or fruit. A 6-legged larva hatches from the egg, and develops through two 8-legged nymphal stages to become an adult. The life cycle takes 2–3 weeks in summer.

▶ **Seasonal history**

The mites survive winter as non-reproducing adults. There are probably 6–8 generations between spring and autumn.

▶ **Habits**

Citrus flat mites commonly feed on the exposed surface of the fruit, near the stem end or under the calyx, and are also found on the undersides of leaves and green twigs. They prefer green to colouring fruit. Movement onto the fruit occurs later in the season (from February to May).

▶ **Hosts**

Other hosts include grapevines and ornamentals.

▶ **Origin and distribution**

Flat mites are widely distributed throughout the world.

Damage

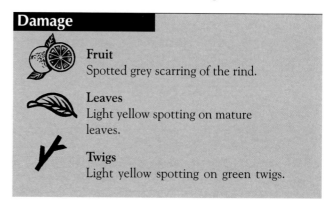

Fruit
Spotted grey scarring of the rind.

Leaves
Light yellow spotting on mature leaves.

Twigs
Light yellow spotting on green twigs.

When flat mites are present in high numbers, blemish on the fruit is often associated with depressions on the rind. The spotted, grey blemish is similar to that caused by broad mite, but rougher, and the spots do not join up. It differs from two-spotted mite damage in that it is a scarring of the rind, rather than chlorosis.

Plate 3.33 Spotted grey scarring of the rind of a Valencia orange caused by citrus flat mite.

▶ Varieties attacked

Most varieties of citrus are susceptible to flat mites, but oranges are most commonly affected in Queensland.

Natural enemies

▶ Predators

The major natural enemies of citrus flat mite are phytoseiid mites, e.g. *Euseius victoriensis*, and stigmaeid mites. For information on the commonly occurring *E. victoriensis*, see brown citrus rust mite (page 19).

Management

▶ Monitoring

- Citrus flat mites occur on all citrus varieties, especially oranges.
- When the presence of mites is evident or suspected, the mites and their predators should be monitored fortnightly from the beginning of November to May in Queensland and coastal New South Wales.
- Using a ×10 hand lens, check 5 randomly selected fruit per tree.

▶ Action level

Action is required if 20% or more of fruit are infested.

This is a very minor pest, and if predatory mites are present, action is not usually required.

▶ Appropriate action

Apply a selective miticide.

Predatory mites should be encouraged in the orchard (see brown citrus rust mite, page 20).

Other mites

Many species of non-pest mites are found in citrus, where they feed on fungi, or decaying plant and animal material on the bark, leaves and fruit. These harmless mites contribute to nutrient cycling, helping to break down organic debris, and may be an alternative food source for predatory mites.

Fungi-feeding and scavenging mites are very commonly found in association with sooty mould on the fruit and leaves. Their numbers are often closely related to the levels of mealybugs and other sap-feeding insects on the crop. The most common such mites are tarsonemid mites (family Tarsonemidae), mould mites (families Acaridae and Glycyphagidae), tydeid mites (family Tydeidae) and beetle or soil mites (suborder Oribatida).

Tarsonemid mites

Tarsonemid mites within the *Tarsonemus waitei* species-group are very common on citrus throughout the world, including in Australia. These mites feed on sooty mould and, although they can be found on leaves and branchlets infested with soft scale, are most common under the calyxes of fruit which are, or have been, infested with mealybugs.

Tarsonemids are very small, shiny, globular and transparent. They are fairly inactive, and often wedge themselves

*Plate 3.34 Tarsonemid mites (*Tarsonemus waitei*) feeding on sooty mould. Adults are orange, juvenile mites white, and eggs white.*

well under the calyx, where they may gather in large numbers, especially in winter. In the Riverland, the number of tarsonemid mites on navel oranges varies considerably from year to year. This variation is possibly linked to variations in mealybug populations.

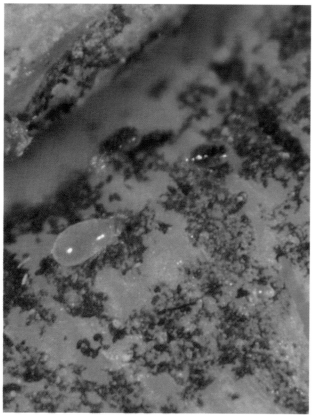

Plate 3.35 Mould mites (family Acaridae). These mites are feeding on sooty mould around and under the calyx of an orange.

Plate 3.36. Tydeid mites (Tydeus californicus) with eggs. These mites are scavengers.

Plate 3.37 Beetle mites (suborder Oribatida). These mites are scavengers.

Mould mites

Mould mites are common in citrus leaf litter, becoming more numerous on the leaves and fruit during winter. Like the tarsonemids, they are commonly associated with sooty mould around and under the calyx of navel oranges.

Adult mould mites are teardrop-shaped, off-white to transparent, and often have one or two dark spots on the body. They are quite large, and move about very slowly.

Tydeid mites

Tydeid mites are small to medium-sized, usually white to pale pink, and often have a conspicuous white stripe along the mid-line of the body. They typically make darting movements, and their forelegs vibrate as though the mites are nervous.

Tydeid mites are generally scavengers, feeding on debris on the surface of citrus leaves and fruit. *Tydeus californicus* is common on inland citrus, and can also be found on a variety of other tree crops throughout the world. This species is off-white, and coffin-shaped when seen from above. Large numbers occasionally gather in groups on the trunk and branches of citrus trees.

Beetle or soil mites

Beetle or soil mites are very common in citrus leaf litter. These large, slow-moving mites are mid-brown to dark brown and, as their name suggests, resemble small beetles. They are easily visible with the naked eye.

Beetle mites are scavengers, feeding on fungi, bacteria and decaying leaf material. They may become locally common on leaves and fruit in winter during periods of high humidity.

Insects

Insects have a separate head, three-segmented thorax, and abdomen. Each segment of the thorax typically has a pair of legs attached to it. Insects may have no wings, or one or two pairs of wings. When wings are present, they are attached to the thorax. The abdomen usually consists of 10 or 11 segments, with the terminal segments modified for mating or egg laying. Most insects lay eggs, but some types produce living young. Young insects moult from time to time as they develop into adults. Generally at each moult an increase in the insect's size or the development of a specific structure, e.g. the wings, takes place.

Most immature insects do not have wings. They can be like or unlike adults in their general appearance. There are two common types of life cycle in insects: holometabolous and hemimetabolous. In the holometabolous life cycle, the young (larvae or caterpillars) do not resemble the adults, and pass through different forms in their development, e.g. from larva to pupa to adult. Moths and beetles are examples of these kinds of insects. In the hemimetabolous life cycle, the young (nymphs) superficially resemble the adults, and with each moult they progressively develop more adult characteristics. Bugs and grasshoppers are example of these kinds of insects.

Ants (Hymenoptera)

Ants belong to the same group of insects (Hymenoptera) as wasps. They have a distinct waist or petiole between the thorax and the abdomen. The petiole bears one or two distinct, round or plate-like nodes. Ants are social insects, and nearly all species are organised into castes such as queens, workers and soldiers.

Ants and citrus pests

Ants do not cause direct damage to citrus in Australia. However, some species of ants at high densities can severely disrupt integrated pest management (IPM) programs, particularly those directed towards the control of red scale, honeydew-producing insects and sooty mould.

Soft scales, mealybugs, whiteflies, planthoppers and aphids excrete honeydew, a sugar-rich solution derived from the plant sap on which they feed. This honeydew is a favoured food source for several common ant species.

Ants entering citrus canopies in search of honeydew will interfere with the predators and parasites that are seeking out and destroying pest species, defending the honeydew-producing pests from attack.

Although ants remove some honeydew and dead scales from trees, and occasionally carry live scales back to the nest for food, these potential benefits are generally outweighed by the detrimental effects of high ant populations. High ant densities can greatly reduce the levels of parasitism and predation, while low ant densities do not appear to have much effect. Although red scale does not produce honeydew, ant foraging is known to reduce parasitism levels in this pest.

By protecting honeydew producers, ants also contribute to the development of sooty mould. Sooty mould results when fungi grow on the honeydew, forming a superficial coating of dark fungal growth on twigs, leaves and fruit. While not normally a serious problem, it can prevent fruit from colouring normally. Fruit may also need to be washed, which only partially removes the sooty mould, and also adds to costs.

Large numbers of ants, particularly of the meat ant (*Iridomyrmex purpureus*), can be a source of irritation to pickers and other workers. Ants can also block microsprinklers in the orchard.

Plate 4.1 *An ant searching for honeydew produced by green coffee scale.*

Plate 4.2 *An ant tapping hemispherical scale with its antennae. This induces the scale to produce honeydew.*

4 Ants (Hymenoptera)

Description

Ants can be easily differentiated from other citrus pests. They have a distinct waist, or petiole, between the thorax and abdomen. The petiole bears one or two distinct round or plate-like nodes (see plate 4.2).

Identification of ant genus or species can often only be accomplished by trained taxonomists.

Importance

Members of the genus *Iridomyrmex*, particularly species in the *rufoniger* species-group, are the most important ant species on citrus in eastern and southern Australia. They occur in over 80% of samples collected in citrus orchards in New South Wales, Victoria, South Australia, and southern Queensland.

The meat ant (*Iridomyrmex purpureus*) is also important in eastern and southern Australia, while the Argentine ant (*Linepithema humile*) is important in Western Australia.

Many other types of ants are found in citrus trees, but most are present only in very low numbers, and are of minor importance, causing minimal or no disruption to IPM systems. They include *Tetramorium bicarinatum*, *Pheidole* spp. and *Rhytidoponera* sp.

Natural enemies

Although ants may be parasitised by other insects such as strepsipterans and pteromalid wasps, there are currently no effective biological control agents available for controlling the ant species associated with citrus in Australia.

Management

▶ Monitoring

- Monitoring should be carried out from September to May in southern areas, and throughout the year in Queensland.
- Trees should generally be examined during the warmer part of the day, although there is evidence that certain ant species are active at night.

▶ Action level

When ant populations are very high, the numbers of soft scales, in particular, increase dramatically. Action should be taken to control ants if they are present on 50% or more of shoots examined for scales, mealybugs or other pests.

▶ Appropriate action

Trees must be skirted regularly and weeds kept under control. Ants can then access the tree only by climbing up the trunk.

Ground sprays are currently registered for ant control in most Australian states. They last only a short time, and subterranean colonies generally survive and rapidly return to pre-treatment levels. Spray drift and volatilisation from ground sprays may also harm beneficial insects within the tree canopies.

The only other form of ant control that is currently available and acceptable within an IPM system is the

Plate 4.3 Sooty mould on an orange. Soft scales, aphids, whiteflies, planthoppers and mealybugs produce honeydew on which the sooty mould fungus grows. Ants protect these pests, and therefore contribute to causing the sooty mould.

Plate 4.4 Ant nest at the base of a citrus tree.

application of either sticky bands or chemically impregnated barriers to the trunks of individual trees. Although barrier application is an extremely labour-intensive process, some barrier types will provide at least three years' protection from *Iridomyrmex* spp.

Controlling soft scales and other honeydew producers will indirectly help to control ants. Petroleum oil sprays (see chapter 27) or other appropriate means should be used to control these pests.

Scales

Scale insects are among the most significant pests of citrus in Australia. They are small, sap-sucking insects, usually with some sort of covering protecting their bodies. Young scale crawlers are mobile, while adult females are wingless, and sluggish or totally immobile.

Adult males are delicate and short-lived. They are mobile, with one pair of wings. The scales normally seen on scale-infested trees are females, and so most of the descriptions in the following section focus on the female scales. Immature males are less conspicuous, although in citrus snow scale, they are white and easily visible.

Soft scales (Hemiptera)

Soft scales are soft-bodied, and have no separate protective covering. However, in adults, the upper surface of the body is usually tough, or protected with a thick waxy or mealy secretion. Soft scales produce large amounts of sugary honeydew. The main economic damage caused by soft scales is from the downgrading of fruit quality due to sooty mould fungus growing on the honeydew.

Monitoring
The following points apply to the monitoring of all soft scales and their natural enemies:
- The number of trees from which samples are taken depends on block size (see chapter 25 for details).
- Additional care should be taken when monitoring blocks that have a history of economic damage.

Citricola scale

Coccus pseudomagnoliarum (Kuwana), Hemiptera: Coccidae

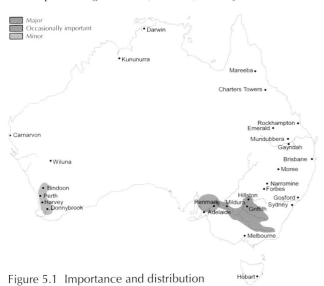

Figure 5.1 Importance and distribution of citricola scale.

Plate 5.1 Citricola scale on leaf. Note the dorsal ridge running the whole length of the scale.

Description

General appearance
The adult citricola scale is convex in shape, 3–4 mm long, and grey-brown. Young citricola scale are flat and transparent, but may appear yellow-green on leaves. When half-grown they are grey-brown, and darken as they get older.

Distinguishing features
Citricola scale may be confused with soft brown scale and black scale, particularly in the very young stages.

Second instar nymphs and adults of both citricola scale and soft brown scale have a dorsal longitudinal ridge, but while the ridge runs the whole length of the body on citricola scale, on soft brown scale it is shorter, stopping before the rear end of the body. This ridge is only visible under a microscope or ×10 hand lens.

The adult citricola female scale is grey-brown, in contrast to soft brown scale, which is dark brown and larger.

Black scales have a characteristic H-pattern on the back, which first becomes visible on young adults.

Life cycle
Female citricola scales produce up to 1500 eggs over 4–6 weeks in late spring to early summer. Eggs hatch after 2–3 days, and crawlers settle on leaves and twigs (rarely on fruit). Later in their development, the immature scales generally migrate to twigs.

Citricola scales pass through two nymphal (immature) stages. At any one time, citricola scales within a population are usually of similar age.

▶ Seasonal history
In all areas, there is one generation (or possibly two) each year.

▶ Habits
Citricola scale produces large amounts of honeydew, resulting in the growth of sooty mould fungus. The honeydew also attracts ants, which can interfere with the biological control of citricola and other scales.

▶ Hosts
In Australia, citricola scale is limited to citrus, but has been recorded on elm, walnut, nightshade, pomegranate and blackberry in other countries.

▶ Origin and distribution
Citricola scale also occurs in California, Japan, Mexico, Russia and Iran. It is very uncommon in Queensland.

Damage

Fruit
Superficial blemish from sooty mould.

Leaves
Leaf drop, if infestation is heavy.

Heavy infestations of citricola scale can reduce tree vigour, and the accompanying extensive sooty mould growth can reduce photosynthesis. However, the most significant effect is downgrading of fruit quality due to sooty mould.

▶ Varieties attacked
Most citrus varieties are susceptible to citricola scale.

Natural enemies

▶ Parasites
Parasitic wasps, e.g. *Coccophagus lycimnia* and *Coccophagus semicircularis*, are important parasites of citricola scale.

▶ Predators
The lacewings *Micromus tasmaniae*, *Mallada* spp., *Plesiochrysa ramburi*, and *Chrysopus* spp., and the ladybirds *Parapriasus australasiae*, *Rhyzobius lophanthae*, *Harmonia conformis*, *Cryptolaemus montrouzieri*, and *Diomus notescens* are important predators. The scale-eating caterpillar (*Catoblemma dubia*), also preys on citricola scale.

▶ Pathogens
Some pathogens may occur under humid conditions.

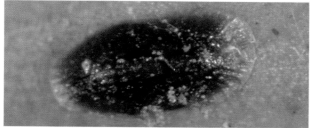
Plate 5.2 Citricola scale parasitised by a wasp Coccophagus sp.

Plate 5.3 Coccophagus semicircularis, a wasp parasite of citricola scale. Left, male; right, female.

Plate 5.4 Green lacewing (Plesiochrysa rambusi) adult and egg (inset).

Plate 5.5 Green lacewing (Plesiochrysa rambusi) larva. Note the remains of prey attached to the back of the larva.

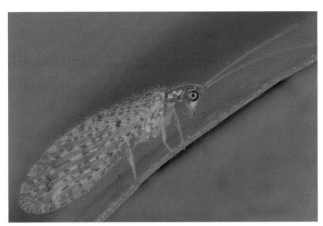

Plate 5.6 Brown lacewing (Micromus tasmaniae) *adult*.

Management

▶ Monitoring

- Citricola scale occurs on most citrus varieties.
- The scale and its natural enemies should be monitored monthly during October–November, to enable action to be taken, if necessary, when crawlers finish emerging. Some monitoring may also be beneficial in March–April.
- Using a hand lens, check for the presence or absence of scale on 5 randomly selected green twigs (bearing 5–10 leaves) per tree.

▶ Action level

The action level is 15% or more of green twigs infested with one or more scales. Noticeable honeydew and/or sooty mould are also important factors in making a decision.

▶ Appropriate action

Spray thoroughly with 1% petroleum spray oil (see chapter 27) when the majority of scales are young.

▶ Additional management notes

Numbers of citricola scale typically fluctuate from year to year. These fluctuations are probably related to climatic factors, especially heat waves. Hot, dry weather kills young scale.

Large populations of ants should be controlled to allow natural enemies to exert their full impact.

Sprays applied for mealybug and other scales will also control this pest depending on the timing of applications.

Green coffee scale

Coccus viridis (Green), Hemiptera: Coccidae

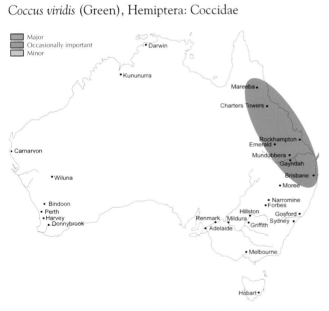

Figure 5.2 Importance and distribution of green coffee scale.

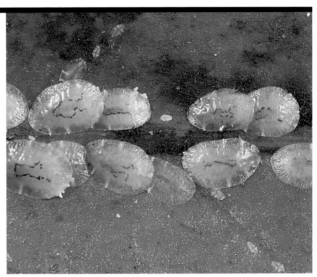

Plate 5.7 Green coffee scale along leaf midrib.

Description

▶ General appearance

The adult female green coffee scale is oval to elongate in shape, with a flattened profile. It is pale yellow-green in colour, and 3–4 mm long. The roughly U-shaped gut is visible through the partially transparent top of the scale as a line of black spots. At the anterior (head) end, there are two distinctive black eye spots. The scale have antennae and well-developed legs, and, unlike most other scales, can move around the host plant.

▶ Distinguishing features

Green coffee scale may be confused with other *Coccus* spp., e.g. soft brown scale. However, the pale yellow-green colour, oval shape, flattened profile, visible gut, and black eye spots are distinguishing features.

▶ Life cycle

The female green coffee scale reproduces without mating. Eggs hatch within the scale or immediately after laying, and the crawlers (first instar nymphs) emerge to spread over the host. Crawlers develop into second instar nymphs, and these develop into adult scales. The life cycle takes 6–9 weeks.

▶ Seasonal history

There are 3–4 generations each year.

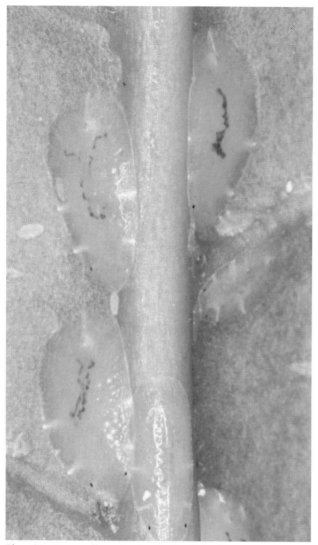

Plate 5.8 Green coffee scale, showing the distinctive eye spots.

▶ Habits
Green coffee scale infests mainly the young leaves and green twigs, but will move onto fruit in heavy infestations. The scales are commonly attended by ants that feed on honeydew the scales produce. Ants may reduce crawler deaths by preventing the accumulation of excess honeydew. Excess honeydew can cause disease in the crawlers, and they can also become trapped in it, because it is sticky.

▶ Hosts
Green coffee scale infests a wide range of hosts, including citrus, coffee, and ornamentals such as gardenia and ixora.

▶ Origin and distribution
Green coffee scale also occurs in East Africa, Central America, South-East Asia, Indonesia, Papua New Guinea and islands of the western Pacific, Hawaii and Florida.

Damage

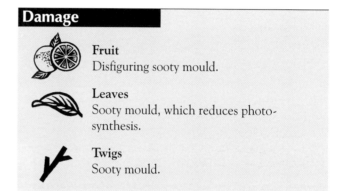

Fruit
Disfiguring sooty mould.

Leaves
Sooty mould, which reduces photosynthesis.

Twigs
Sooty mould.

▶ Varieties attacked
All varieties of citrus are attacked.

Natural enemies

▶ Parasites
There are one or two small parasitic wasps, e.g. *Coccophagus* near *rusti* and *Encarsia* sp., that periodically cause significant mortality. The Kenyan wasp *Diversinervus stramineus* is being investigated for release.

▶ Predators
The mealybug ladybird (*Cryptolaemus montrouzieri*) preys on green coffee scale.

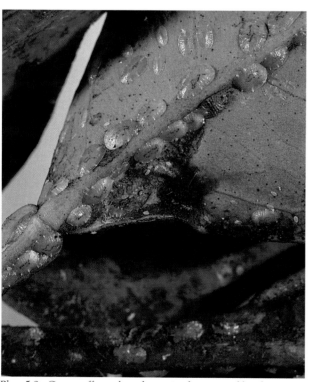

Plate 5.9 Green coffee scale and associated sooty mould on leaves.

Plate 5.10 Parasitised green coffee scale. Note the pinkish orange larva of the parasite Cocophagus *near* rusti, *which has been dissected out of the larger scale (centre). A parasite pupa is visible in the scale on the right. The smaller parasite* Encarsia *sp. is visible in the younger scales.*

Plate 5.11 The wasp Coccophagus *near* rusti, *a parasite of green coffee scale.*

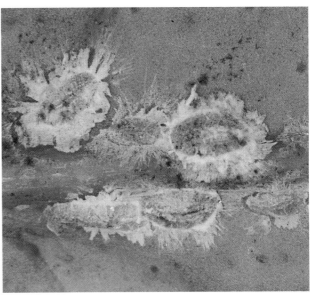

Plate 5.12 The fungus Verticillium lecanii *on green coffee scale.*

▶ Pathogens

The fungus *Verticillium lecanii* can cause up to 90% mortality of the scale during wet weather in late summer to autumn, particularly when populations of the scale are large.

Management

▶ Monitoring

- Green coffee scale occurs on all citrus varieties.
- The pest and its natural enemies should be monitored once in mid-October.
- Using a ×10 hand lens, check 5 randomly selected green twigs per tree.

▶ Action level

There will be a problem with sooty mould if 5% or more of green twigs are infested with one or more scales.

▶ Appropriate action

When the action level is reached, a 1% petroleum spray oil (see chapter 27) should be applied in a high-volume spray. Timing is critical. Spray in early to mid-November, immediately after most crawlers have emerged.

▶ Additional management notes

If scales are attended by large populations of ants, the ants should be controlled with a suitable treatment, e.g. basal trunk spray, or sticky bands around the trunk (see ants, page 38).

Soft brown scale

Coccus hesperidum L., Hemiptera: Coccidae

Figure 5.3 Importance and distribution of soft brown scale.

Plate 5.13 Soft brown scale on leaf.

Description

▶ **General appearance**

The adult female soft brown scale is flat, oval, 3–4 mm long, and yellow-green or yellow-brown, often mottled with brown spots. The colour of the female scale darkens with age. The scales have functional legs, and can move around until they become fully mature.

▶ **Distinguishing features**

Soft brown scale is similar to citricola scale in the juvenile stages. Second instar nymphs and adults of both species have a dorsal longitudinal ridge, but while the ridge runs the whole length of the body on citricola scale, on soft brown scale it is shorter, stopping before the rear end of the body. This ridge is only visible under a microscope or ×10 hand lens.

Mature soft brown scales are tan-brown and sometimes mottled (see plates 5.13 and 5.14), while citricola scales are grey-brown and mottled. The cover of soft brown scale becomes a uniform dark brown all over as the scale matures.

▶ **Life cycle**

Male soft brown scales are rare. Females can reproduce without mating, and give birth to a total of about 200 live young (crawlers). The life cycle takes approximately 2 months in summer. Development within generations is not synchronised, i.e. various stages are present at any given time, and the generations overlap.

▶ **Seasonal history**

In Queensland and the Northern Territory, there are 4–5 generations per year.

In New South Wales, Victoria and South Australia, there are 3–4 generations per year, and populations are highest in summer to autumn.

In Western Australia, there are 2–3 generations per year, and populations are highest in summer to autumn.

▶ **Habits**

Adult soft brown scales infest leaves and twigs, and occasionally green fruit. The scales produce large amounts of honeydew, resulting in the growth of sooty mould. The honeydew also attracts ants, which can interfere with biological control.

Newly hatched crawlers move onto younger growth and the stalks of fruit. Young scales move around until they are half-grown, and then migrate towards leaves and small green twigs where they settle and reproduce.

▶ **Hosts**

Soft brown scale occurs on a very wide range of hosts. Commercially grown crops infested include citrus, passionfruit, figs and numerous ornamentals such as oleander.

▶ **Origin and distribution**

Soft brown scale occurs worldwide.

Damage

Fruit
Superficial blemish from sooty mould.

Leaves
Leaf drop, if infestation is heavy.

Heavy infestations can reduce tree vigour, and extensive sooty mould growth can reduce photosynthesis. However, the most significant effect is downgrading of fruit quality due to sooty mould.

▶ **Varieties attacked**

All citrus varieties are attacked.

Plate 5.14 Diversinervus elegans, *a wasp parasite of soft brown scale.*

Plate 5.15 Metaphycus helvolus, *a wasp parasite of soft brown scale. Top, male; bottom, female.*

Natural enemies

Parasites

Soft brown scale is attacked by a number of parasitic wasps including: *Metaphycus helvolus*, *Coccophagus lycimnia*, *Coccophagus semicircularis*, *Microterys flavus* and *Diversinervus elegans*.

Predators

Predators of soft brown scale include the ladybirds *Parapriasus australasiae*, *Rhyzobius lophanthae*, *Rhyzobius* near *lophanthae*, *Harmonia conformis*, *Cryptolaemus montrouzieri*, *Diomus notescens* and the lacewings *Plesiochrysa ramburi* and *Micromus tasmaniae*. The scale-eating caterpillar (*Catoblemma dubia*) also feeds on soft brown scale.

Management

Monitoring

- Soft brown scale occurs on most citrus varieties.
- The pest and its natural enemies should be monitored once in mid-October — early December, and once or twice, if required, in February–March.
- Using a ×10 hand lens, check for the presence or absence of the pest on 5 randomly selected green twigs (with 5–10 leaves) per tree.

Action level

The action level is 15% or more of green twigs infested with one or more scales. Noticeable honeydew and/or sooty mould are also important factors in making a decision.

Appropriate action

Spray thoroughly with 1% petroleum spray oil in November–December and/or in February–March (see chapter 27). As mobile and newly settled crawlers are most susceptible to oil sprays, sprays should be applied when crawlers are active (soon after hatching).

Additional management notes

In the absence of large ant populations, natural enemies, particularly predators, usually provide sufficient control of soft brown scale. Large populations of ants must be controlled (see ants, page 38).

Young trees on boundary rows can often become infested.

Soft brown scale is usually not a problem in Queensland. Infestations may be an indication of inappropriate pesticide use, or pesticide drift from nearby stonefruit, avocados or macadamias, which has disrupted natural enemies.

Plate 5.16 *Microterys flavus, a wasp parasite of soft brown scale.*

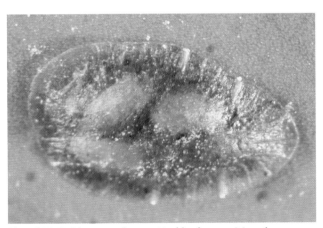

Plate 5.17 *Soft brown scale parasitised by the wasp Metaphycus sp. Note the three wasp pupae visible under the scale cover.*

Plate 5.18 *The ladybird* Harmonia conformis, *a predator of soft scales and aphids. These ladybirds have gathered together in a large group for the winter.*

Long soft scale

Coccus longulus (Douglas), Hemiptera: Coccidae

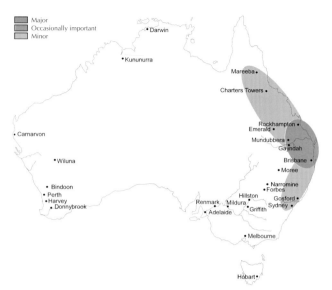

Figure 5.4 Importance and distribution of long soft scale.

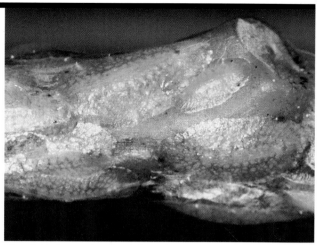

Plate 5.19 Long soft scale adults and nymphs on fruit stalk.

Damage

 Fruit
Disfigured by sooty mould.

 Leaves
Sooty mould, which reduces photosynthesis.

 Twigs
Sooty mould.

Description

▶ **General appearance**
The adult female long soft scale is an elongated oval in shape, 4–6 mm long, yellow to greyish-brown in colour, with visible eye spots.

▶ **Distinguishing features**
Long soft scale most closely resembles scale of other *Coccus* spp., e.g. soft brown scale, but can be distinguished by its greater length (up to 6 mm).

▶ **Life cycle**
Adult females produce living young (crawlers) over a period of about 4 weeks. These crawlers disperse and settle on twigs and leaves. They develop through 2 nymphal stages before reaching the adult stage. The complete life cycle takes about 2 months.

▶ **Seasonal history**
There are 4–6 generations each year.

▶ **Habits**
Long soft scale infests twigs and leaves, and commonly occurs on water shoots and fruit stalks. Crawlers move onto the fruit stalks in the spring where they develop and produce large amounts of honeydew. Sooty mould then grows on the honeydew. Heavy infestations of long soft scale can develop by early autumn.

▶ **Hosts**
Long soft scale occurs on a wide range of hosts, including custard apple, lychee, carambola, fig, leucaena, and many ornamentals.

▶ **Origin and distribution**
Long soft scale is distributed worldwide.

▶ **Varieties attacked**
All citrus varieties are attacked, but mandarins (particularly Imperial) are most susceptible.

Natural enemies

▶ **Parasites**
The small parasitic wasp *Coccophagus ceroplastae* attacks long soft scale.

▶ **Predators**
The mealybug ladybird (*Cryptolaemus montrouzieri*) preys on long soft scale.

▶ **Pathogens**
During prolonged wet weather, the fungus *Verticillium lecanii* can kill many long soft scale.

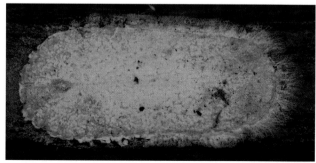

Plate 5.20 Long soft scale infected by *Verticillium lecanii*.

Management

▶ **Monitoring**

- Long soft scale occurs on all citrus varieties, but is more common on mandarins, especially Imperial mandarin.
- Monitor once in mid-October.
- Using a ×10 hand lens, check 5 randomly selected fruit stalks per tree. If blocks are infested, also check 5 randomly selected water shoots per tree in May to help predict the level of infestation in the following season.

▶ **Action level**

The action level is 5% or more of fruit stalks or water shoots infested with one or more scales.

▶ **Appropriate action**

Apply petroleum spray oil in November (see chapter 27).

▶ **Additional management notes**

Current levels of biological control of long soft scale are poor. Attempts are being made to collect more effective parasites from other countries.

Black scale

Saissetia oleae (Olivier), Hemiptera: Coccidae

Figure 5.5 Importance and distribution of black scale.

Description

▶ **General appearance**

The mature female black scale is dome-shaped, 3–5 mm long, and black or dark-brown. Ridges on the back form a raised H-pattern, particularly on young adult scale.

Young adults are dark mottled-grey, and are softer and less leathery than the mature adult. Male scales are more elongated than the females, and are not often seen.

▶ **Distinguishing features**

Black scale may be confused in the juvenile stages with citricola, hemispherical and soft brown scales. After the second moult, black scale can be distinguished by the characteristic H-pattern on the back. When reproducing, the scale is raised and rounded, and the H-pattern is less obvious. However, the characteristic black colour helps distinguish this scale from the grey-brown citricola scale and the tan-brown soft brown scale.

▶ **Life cycle**

The female black scale lays about 2000 pink eggs under her body. The newly hatched crawlers settle after a day or two on leaves, usually near the midrib or on young shoots.

Black scale develop through two nymphal stages before reaching young adult stage when they are 2–3 months old.

Plate 5.21 Black scale on leaf midrib. Note the light-brown crawlers and the wasp *Scutellista caerulea* (right), a parasite and an egg predator.

The scale cover is soft in the young adult, but later becomes more rigid and darkens.

The life cycle takes 4–8 months in southern districts, but is shorter in Queensland.

▶ **Seasonal history**

There are 2 generations per year in southern Australia, and 3–4 in northern Australia.

In southern areas, hatching occurs in December–January and again in autumn, but overlapping can occur. The second generation produces fewer eggs.

▶ **Habits**

Black scale can be found on green twigs, leaves and young fruit. After the second moult, young adult scales move to the place where they settle permanently—on leaves for the generation that hatches in autumn, and on twigs for the other generations.

Plate 5.22 Twig infested with black scale.

Hosts

Black scale infests a wide range of hosts, including commercial crops of citrus, olive, custard apple, duboisia, and ornamentals such as gardenia and oleander.

Origin and distribution

Black scale is believed to be native to Africa, but now occurs throughout the world.

Damage

Fruit
Superficial blemish from sooty mould.

Leaves
Leaf drop, if infestation is heavy.

Heavy infestations of black scale can reduce tree vigour, and extensive sooty mould growth can reduce photosynthesis. However, the most significant effect is downgrading of fruit quality due to sooty mould.

Varieties attacked

All citrus varieties are attacked.

Plate 5.23 The wasp Scutellista caerulea, *a parasite and egg predator of black scale.*

Plate 5.24 Metaphycus bartletti, *an imported wasp parasite of black scale.*

Plate 5.25 Metaphycus lounsburyi, *a wasp parasite of black scale.*

Plate 5.26 Larva (left) and adult (right) of Parapriasus australasiae, *a ladybird predator of black scale.*

Natural enemies

Parasites

A number of species of parasitic wasps attack black scale, including *Metaphycus lounsburyi*, *Metaphycus helvolus* and *Scutellista caerulea* (which is also an egg predator). The imported *Metaphycus bartletti* is being investigated for commercial breeding and release.

Predators

The ladybirds *Rhyzobius* near *lophanthae*, *Parapriasus australasiae*, the mealybug ladybird (*Cryptolaemus montrouzieri*), and *Diomus* spp., and the brown lacewing (*Micromus tasmaniae*), lacewings of *Mallada* spp., and the green lacewing (*Plesiochrysa ramburi*) are predators of black scale. A scale-eating caterpillar (*Catoblemma dubia*) also feeds on black scale. The larvae of the parasite wasp *Scutellista caerulea* prey on the eggs of black scale and some other soft scales.

Pathogens

In coastal regions, the fungus *Verticillium lecanii* can kill scales in high-density populations in humid conditions in late summer and autumn. White fungal growth can cover large areas of scale-infested branches.

Management

Monitoring

- Black scale occurs on most citrus varieties.
- The pest and its natural enemies should be monitored once in November–December, and again in February–March.
- Using a ×10 hand lens, check for the presence or absence of adult female scales on 5 randomly selected green twigs (with 5–10 leaves) per tree. The presence of honeydew and sooty mould also indicates the presence of scales. When monitoring tall trees, take 10% of samples from the tops of trees.

Action level

The action level for all citrus except mandarins is 10% or more of green twigs infested with one or more scales. On mandarins, the action level is 5%.

Appropriate action

Apply petroleum spray oil (see chapter 27). Timing is critical. Wait for the hatching of all eggs under adult scales. Egg hatch is complete when young scales are observed on leaves, and no liquid exudes when old scales are squashed.

Two sprays may be needed if warm humid conditions continue into late autumn and if generations overlap. Spray thoroughly (6000–10 000 L/ha for trees 4 m high) with petroleum spray oil (1 L in each 100 L of water).

If scales are present on lower branches and on tree trunks, make sure that the spray reaches these parts of the tree.

The recommended volumes are for oscillating booms with outriggers, air-blast sprayers with towers, and rotary atomisers. The 10 000 L rate is for moderate to heavy infestations. Higher volumes may be necessary if low-profile air-blast sprayers are used.

◗ Additional management notes
Large ant populations interfere with the biological control of black scale and other pests. Ant populations should not be allowed to build up.

Vigorous trees are more likely to suffer heavy infestations of black scale.

Hot, dry conditions kill eggs and young scale.

Hemispherical scale

Saissetia coffeae (Walker), Hemiptera: Coccidae

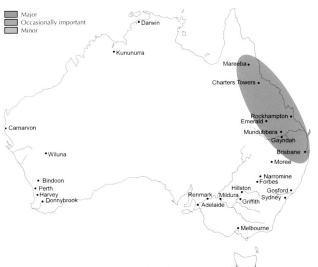

Figure 5.6 Importance and distribution of hemispherical scale.

Plate 5.27 Adult hemispherical scale on a fruit stalk.

Description

◗ General appearance
The adult female hemispherical scale is 2–4 mm long, glossy light to dark brown in colour, and roughly hemispherical in shape, with a smooth surface. Immature stages are lighter in colour and have a raised H-pattern on the back, similar to that on immature black scale.

◗ Distinguishing features
Younger stages may be confused with soft brown scale, black scale or citricola scale. However, the adult female can be distinguished by the high, rounded shape and smooth surface.

◗ Life cycle
Eggs are deposited under the scale and hatch into crawlers shortly afterwards. There are two nymphal stages before adulthood.

◗ Seasonal history
There are 4–6 generations per year.

◗ Habits
Crawlers commonly settle on the fruit stalk, leaves and green twigs, where they develop through the nymphal stages to become adults.

The spring to early summer generation of hemispherical scale, developing on the fruit stalks, causes the most trouble in citrus. Later generations are more heavily attacked by natural enemies.

Hemispherical scale does not do well in hot dry weather, and rarely attacks subcoastal or inland citrus.

◗ Hosts
Hemispherical scale also attacks avocado, coffee, ferns and palms.

◗ Origin and distribution
Hemispherical scale is found throughout the world.

Damage

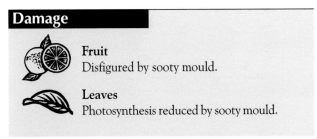

Fruit
Disfigured by sooty mould.

Leaves
Photosynthesis reduced by sooty mould.

◗ Varieties attacked
All citrus varieties are attacked, but particularly navel oranges.

Natural enemies

◗ Parasites
The most common wasp parasites of hemispherical scale are *Scutellista caerulea* (see pages 56–57 in the section on Florida wax scale) and *Encyrtus infelix*.

Plate 5.28 Hemispherical scale, showing adults and nymphal stages, including crawlers. Young scale are lighter in colour than adults.

Plate 5.29 Hemispherical scale parasitised by the wasp Scutellista caerulea *and also infected by the fungus* Verticillium lecanii. *Note the hole (top left) through which an adult wasp has emerged.*

▶ Predators

Larvae of the parasitic wasp *Scutellista caerulea* prey on eggs of hemispherical scale and some other soft scales. Other predators include ladybirds, e.g. the mealybug ladybird (*Cryptolaemus montrouzieri*), and the scale-eating caterpillar (*Catoblemma dubia*).

Plate 5.30 Encyrtus infelix, a wasp parasite of hemispherical scale.

▶ Pathogens

The fungus *Verticillium lecanii* can kill many scales in prolonged wet weather.

Management

▶ Monitoring

- Hemispherical scale occurs on all citrus varieties, but navel oranges are most commonly infested.
- The pest and its natural enemies should be monitored once in mid-October.
- Using a ×10 hand lens, check for the presence or absence of the pest on 5 randomly selected fruit stalks per tree.

▶ Action level

The action level is 5% or more of fruit stalks with one or more scales present.

▶ Appropriate action

If the action level is reached, apply petroleum spray oil during November (see chapter 27).

Timing is critical. Spray only when all eggs have hatched. The signs that hatching is complete are the presence of young scales on leaves and twigs, and older scales have finished laying eggs, and died. (Dead, empty scales are shrivelled and dry, and remain attached to the tree for several months.)

Nigra scale

Parasaissetia nigra (Nietner), Hemiptera: Coccidae

Figure 5.7 Importance and distribution of nigra scale.

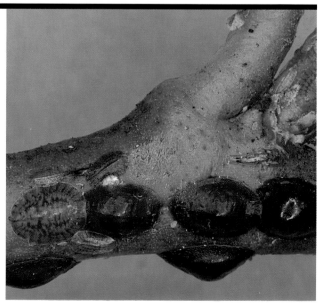

Plate 5.31 Adult and juvenile nigra scale on a twig.

Description

General appearance
The adult female nigra scale is 3–4 mm long, an elongated oval in shape, and a shiny dark brown to black in colour. Younger stages are light brown. The scale surface is smooth or very slightly wrinkled.

Distinguishing features
Younger stages of nigra scale may be confused with those of soft brown scale. However, the adult female is distinctively dark, shiny and smooth.

Life cycle
There seem to be no male nigra scale. Adult females each produce about 800 eggs which are protected under the scale body for about a fortnight before they hatch into crawlers. These young nymphs develop and pass through two moults before adulthood. The life cycle takes about 2 months in summer.

Seasonal history
There are 4–6 generations per year.

Habits
Young scales settle and feed on leaves, twigs and fruit stalks. However, they can move to new feeding sites until adulthood is reached. Then they settle permanently and produce eggs.

The spring to early summer generation of nigra scale, developing on the fruit stalks, causes the most trouble in citrus. Later generations are more heavily parasitised and affected by fungi.

Hosts
Nigra scale also attacks custard apple, avocado, guava and hibiscus.

Origin and distribution
Nigra scale is found throughout the world.

Damage

Fruit
Disfigured by sooty mould.

Leaves
Photosynthesis reduced by sooty mould.

Varieties attacked
All citrus varieties are attacked.

Natural enemies

Parasites
The most common parasite of nigra scale is the wasp *Scutellista caerulea* (see pages 56–57 in the section on Florida wax scale).

Predators
Ladybirds, such as the mealybug ladybird (*Cryptolaemus montrouzieri*), and larvae of the moth *Catoblemma dubia* prey on nigra scale. Larvae of the parasitic wasp *Scutellista caerulea* prey on eggs of nigra scale and some other soft scales.

Plate 5.32 Nigra scale parasitised by the wasp Scutellista caerulea. *Note the hole through which an adult wasp has emerged (bottom centre), the pupa under a scale cover (top centre) and the pink, legless larva of a wasp (top right) which has been dissected out of a scale.*

Plate 5.33 Nigra scale infected with the fungus Verticillium lecanii.

Pathogens
The fungus *Verticillium lecanii* commonly attacks nigra scale in wet weather during summer–autumn.

Management

Monitoring
- Nigra scale occurs on all citrus varieties.
- The pest and its natural enemies should be monitored once in mid-October.
- Using a ×10 hand lens, check for the presence or absence of the pest on 5 randomly selected fruit stalks per tree.

Action level
The action level is 5% or more of fruit stalks with one or more scales present.

Appropriate action
Apply petroleum spray oil during November (see chapter 27).

Additional management notes
When present in large numbers, ants interfere with the natural enemies of nigra scale, and should be controlled.

Nigra scale is more common in orchards where natural enemies have been disrupted by spray drift.

Cottony citrus scale (pulvinaria scale)

Pulvinaria polygonata Cockerell, Hemiptera: Coccidae

Figure 5.8 Importance and distribution of cottony citrus scale.

Plate 5.34 Cottony citrus scale adults and juveniles on twig. Note the female with cottony egg mass (bottom right).

Description

General appearance
The adult female cottony citrus scale is 3–5 mm long, yellow-brown, and studded with brown spots. The young scale are lighter in colour as well as smaller than mature scale.

Distinguishing features
The immature and early adult stages of cottony citrus scale may be confused with similar stages of *Coccus* species and *Saissetia* species. However, the white cottony egg mass at the rear end of the mature adult scale is a distinguishing feature. This egg mass also has a conspicuous longitudinal furrow.

Cottony cushion scale also produces a cottony egg mass, but it is much larger than that of cottony citrus scale, more fluted in shape, and with bright orange eggs.

Life cycle
The female cottony citrus scale deposits eggs in a protective cottony egg sac that remains attached to the rear end of her body. As the number of eggs in the sac increases, the female's body shrivels. Each female lays 200–300 eggs.

There are two nymphal stages before adulthood is reached. The complete life cycle takes 2–3 months.

Seasonal history
There are 2–3 generations per year.

Habits
Cottony citrus scale is found on leaves and twigs, and heavy infestations can appear in spring. Egg production occurs during November. Like many soft scales, cottony citrus scale is adversely affected by hot dry conditions and is hence a pest of citrus in moister coastal areas.

Hosts
Citrus species seem to be the only known hosts.

Origin and distribution
Cottony citrus scale is also found in South-East Asia.

Damage

Fruit
Disfigured by sooty mould.

Leaves
Sooty mould on leaves and leaf stalks; reduced photosynthesis; leaf drop.

Twigs
Sooty mould.

Varieties attacked
All citrus varieties are susceptible, but particularly Meyer lemon.

Natural enemies

Parasites
None have been yet been identified.

Predators
Ladybirds, particularly the mealybug ladybird (*Cryptolaemus montrouzieri*) and the steel-blue ladybird (*Halmus chalybeus*), normally control cottony citrus scale. However, infestations sometimes develop before predator numbers can build up.

The life cycle and habits of the mealybug ladybird are described on page 84.

The steel-blue ladybird is dark blue, and 4 mm across. Adults and larvae feed mainly on the egg sacs of cottony citrus scale and other hosts. Eggs are laid amongst egg sacs of the host. They hatch into larvae 1 mm long which grow and moult three times. They then change into pupae which are attached to leaves. The life cycle takes a month during the summer.

Pathogens
The fungus *Verticillium lecanii* kills many cottony citrus scale during prolonged wet weather.

Management

Monitoring
- Cottony citrus scale (pulvinaria scale) occurs on all citrus varieties, especially Meyer lemon.

5 Soft scales (Hemiptera)

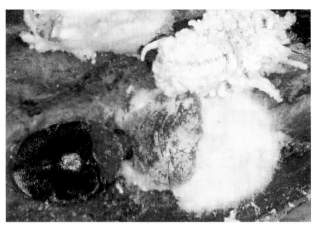

Plate 5.35 Predatory mealybug ladybird (Cryptolaemus montrouzieri), feeding on cottony citrus scale. Lower left, adult; upper right, larva.

- The pest and its natural enemies should be monitored once or twice from mid-October to late November.
- Using a ×10 hand lens, check for the presence or absence of the pest on 25 randomly selected leaves per tree.

▶ Action level
Spray if 5% or more of leaves are infested with at least one scale, and the main predator, mealybug ladybird (*Cryptolaemus montrouzieri*) is not present on 50% or more of the sample trees. However, spraying is rarely required.

▶ Appropriate action
Apply petroleum spray oil during November (see chapter 27), when nearly all of the adult scales have finished producing eggs, and the eggs have hatched. The petroleum spray oil must be applied thoroughly to achieve good control.

▶ Additional management notes
The two main predators should be conserved by avoiding the use of disruptive broad-spectrum pesticides.

Mealybug ladybirds (*Cryptolaemus montrouzieri*) can be purchased and released into orchards.

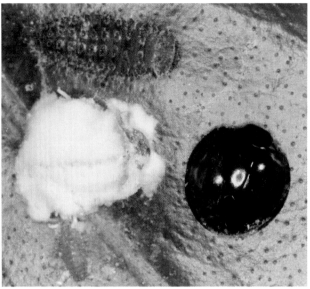

Plate 5.36 The predatory steel-blue ladybird (Halmus chalybeus). An adult and two larvae surround a cottony citrus scale.

Plate 5.37 Cottony citrus scales infected with the fungus *Verticillium lecanii*. Left, two twigs with healthy scales; right, two twigs with infected scales.

Pink wax scale

Ceroplastes rubens Maskell, Hemiptera: Coccidae

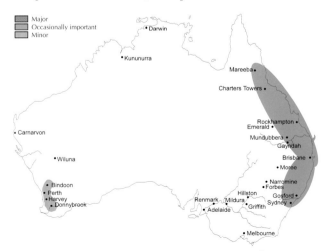

Figure 5.9 Importance and distribution of pink wax scale.

Plate 5.38 Pink wax scale and wasp parasite *Anicetus beneficus*.

Description

General appearance
The adult female pink wax scale is protected by a hard wax covering, pink to red in colour. It is about 3–4 mm long, and globular and smooth in shape, with two lobes on either side and a depression at the top.

Distinguishing features
Pink wax scale most resembles Florida wax scale and hard wax scale. However, Florida wax scale is much paler in colour, and hard wax scale is mostly white and larger than pink wax scale.

Life cycle
Reproduction occurs without fertilisation (male scales are very uncommon).

The number of eggs laid is related to the size of females, which in turn is related to the levels of nitrogen in the leaves on which the scales feed. While one female can produce up to 900 eggs, the average number is probably below 200 per female. The eggs are brick-red, and the female holds them in a cavity beneath her body.

Eggs hatch into crawlers, which are small and mobile, with three pairs of legs, two eye spots and paired antennae. They spread out over the tree, settling mostly along the midribs of leaves and on green twigs. Caught by the wind, they are dispersed throughout the orchard.

Once the crawlers settle, they lose their appendages and eye spots, and begin secreting the waxy cap (first white, then pink) over their bodies.

Seasonal history
There are 2 generations per year in Queensland and the Northern Territory, and one generation per year elsewhere.

In Queensland, crawlers of the first generation emerge from mid-September until early December, but mostly from mid-October to mid-November. Crawlers of the second generation emerge from February until late April.

On the central and south coast and in inland areas of New South Wales, crawlers emerge from late October until late December.

Habits
Pink wax scale occurs on leaves (mostly along the midrib on both surfaces) and young twigs, but all stages favour leaves.

This scale produces large amounts of honeydew, and as a result, sooty mould levels can be high on both fruit and leaves.

The hot dry conditions in southern inland regions are unfavourable for the scale and so it occurs mainly in coastal areas. If heat waves occur in coastal areas when first and second instar scales are present, scale populations can be drastically reduced.

Hosts
Pink wax scale infests a wide range of other hosts including mango, avocado, custard apple and many native shrubs (e.g. lillypilly, pittosporum and umbrella tree).

Origin and distribution
Pink wax scale also occurs in South-East Asia, China, Japan, islands of the Western Pacific, Spain and Florida. It possibly originated in Africa.

Damage

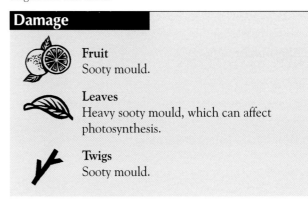

Fruit
Sooty mould.

Leaves
Heavy sooty mould, which can affect photosynthesis.

Twigs
Sooty mould.

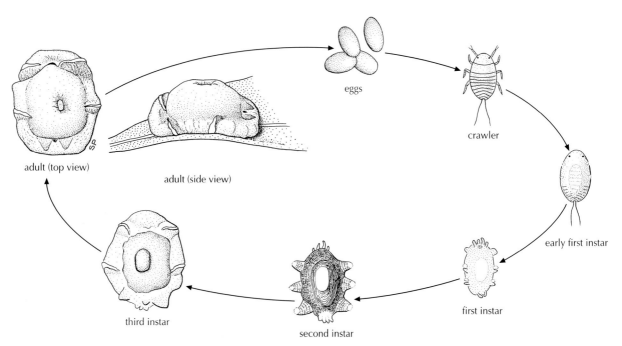

Figure 5.10 Life cycle of pink wax scale.

Plate 5.39 *Sooty mould, a dark fungal growth, on leaves and fruit. The fungus grows on honeydew excreted by pink wax scales and other scales.*

Plate 5.40 *The wasp* Anicetus beneficus *parasitising a pink wax scale.*

▶ Varieties attacked

All citrus varieties are attacked, with Emperor mandarin the most susceptible, and Imperial mandarin also favoured.

Natural enemies

▶ Parasites

The most important parasite of pink wax scale is a small wasp *Anicetus beneficus*, introduced from southern Japan to Queensland in 1977. *A. beneficus* now occurs in most areas of Queensland and is effective in controlling the scale where no disruptive pesticides are used.

The wasp is 3 mm long. The female is honey-coloured with large lamellate antennae, and the male is black.

The main emergence periods for *A. beneficus* are during November and February–March. The parasite completes its life cycle within the body of the scale, and the adult wasp, after emerging from its pupa, cuts a circular exit hole through the waxy scale cover.

Just before emergence of an adult wasp, parasitised scales are dark in colour and the skin across the bottom of the scale is black instead of the usual pink.

Other common parasites of pink wax scale are the wasps *Metaphycus varius* (3 mm long, tan coloured), *Scutellista caerulea* (2 mm long, blue-black) and *Coccophagus ceroplastae* (2 mm long, black).

Most larvae of *S. caerulea* feed on pink wax scale eggs, although some feed ectoparasitically on immature adult female scales. The effectiveness of this parasite is reduced as the size of the host decreases.

The wasp parasites *Coccobius atrithorax*, *Encarsia australiensis* and *Moranila californica* have also been recorded.

▶ Predators

The ladybird *Rhyzobius ventralis*, the mealybug ladybird (*Cryptolaemus montrouzieri*), the ladybirds *Diomus notescens* and *Diomus* sp., and the common spotted ladybird (*Harmonia conformis*), together with the scale-eating caterpillar (*Catoblemma dubia*) and the lacewings *Mallada* spp. prey on pink wax scale. Larvae of the wasp parasite *Scutellista caerulea* prey on the eggs of pink wax scale and some other soft scales.

▶ Pathogens

The fungus *Verticillium lecanii* infects pink wax scale in high-density populations under humid conditions.

Management

▶ Monitoring

- Pink wax scale occurs on most citrus varieties, especially mandarins.
- The pest and its natural enemies should be monitored once or twice in mid-October to early December, and once or twice, if required, in February-March.
- Using a ×10 hand lens, check for the presence or absence of the pest on 25 mature leaves per tree (5 randomly selected twigs each with 5 leaves).

▶ Action level

Serious infestations of pink wax scale are much less common since the introduction of *Anicetus benificus*, but can still occur. Action is required when more than 5% of leaves are infested with one or more scales.

▶ Appropriate action

Apply petroleum spray oil (see chapter 27). Correct timing of spraying and thorough application is essential for good results.

Spray only when all eggs under adult scales have hatched. Egg hatch is complete when young scales are noted on leaves, and no liquid exudes when old adult female scale are squashed. At this point the highly susceptible young scales line the midribs of outer foliage.

Florida wax scale

Ceroplastes floridensis Comstock, Hemiptera: Coccidae

Figure 5.11 Importance and distribution of Florida wax scale.

Plate 5.41 Florida wax scale on twig.

Description

▶ General appearance
The adult female Florida wax scale is 2–4 mm long, and protected by a hard wax covering, pale pink to white in colour.

▶ Distinguishing features
Florida wax scale is similar in appearance to pink wax scale, and hard wax scale. However, it is paler in colour than pink wax scale, and considerably smaller than hard wax scale.

▶ Life cycle
The female lays up to 1400 brick-red eggs and holds them in a cavity beneath her body. Crawlers of the first generation emerge from eggs in late October to early December, and crawlers of the second generation emerge from February until April.

▶ Seasonal history
There are two generations of Florida wax scale per year in southern Queensland. The seasonal history is similar to that of pink wax scale, but the first generation of Florida wax scale tends to start hatching about a fortnight later in the spring, i.e. at the end of September, continuing into December. The second generation hatches from February until late April.

▶ Habits
The habits of Florida wax scale are similar to those of pink wax scale. However, about two-thirds of the scales settle on the twigs, and one-third on the leaves, while pink wax scales prefer the leaves. Florida wax scales produce large amounts of honeydew, resulting in heavy growths of sooty mould.

▶ Hosts
Florida wax scale has a wide host range, but citrus is the only commercial crop infested in Australia.

▶ Origin and distribution
Florida wax scale also occurs in Israel, South-East Asia, Hawaii, Florida, and Central and South America.

Damage

 Fruit
Sooty mould.

 Leaves
Sooty mould, which reduces photosynthesis.

▶ Varieties attacked
All citrus varieties are susceptible, particularly Meyer lemon and Valencia orange.

Natural enemies

▶ Parasites
The most important biological control agents are the parasitic wasps *Scutellista caerulea*, *Moranila* sp., *Diversinervus elegans* and *Microterys flavus*. Other species are of lesser importance.

Adult *S. caerulea* are blue-black, stubby and about 2 mm long. The female lays its egg through the wax covering of mature scales. The egg hatches into a larva which develops and then pupates. The adult emerges through a round hole that it cuts in the scale's waxy covering.

Plate 5.42 *Moranila* sp. a wasp parasite of Florida wax scale.

Most larvae of *S. caerulea* feed on eggs laid by the Florida wax scale, although some feed ectoparasitically on immature adult female scales. The effectiveness of this parasite is reduced as the size of the host decreases.

Another important parasite is *Tetrastichus ceroplastae* (originally from South Africa and known to attack white wax scale in New South Wales). Attempts are being made to establish it in Queensland.

▶ Predators
Ladybirds, such as the mealybug ladybird (*Cryptolaemus montrouzieri*), and the scale-eating caterpillar (*Catoblemma dubia*) prey on Florida wax scale.

▶ Pathogens
The fungus *Verticillium lecanii* infects Florida wax scale in high-density populations under humid conditions.

Management

▶ Monitoring
- Florida wax scale occurs on most citrus varieties, especially Meyer lemon and Valencia orange.
- The pest and its natural enemies should be monitored once or twice in mid-October – early December, and once or twice, if required, in February–March.
- Using a ×10 hand lens, check for presence or absence of the pest on 25 mature leaves (5 randomly selected twigs each with 5 leaves).

▶ Action level
The action level is 5% or more of mature leaves infested with one or more scales.

▶ Appropriate action
Apply 1% petroleum spray oil (see chapter 27).

Correct timing of spraying is essential for good results. Wait until eggs have hatched, and adults have died. At this point, the highly susceptible young scales line the midribs of outer foliage.

Usually two applications are necessary for a heavy infestation, one in early November and another 3–4 weeks later.

A light infestation can be treated with a single spray in late November.

White wax scale

Ceroplastes destructor Newstead, Hemiptera: Coccidae

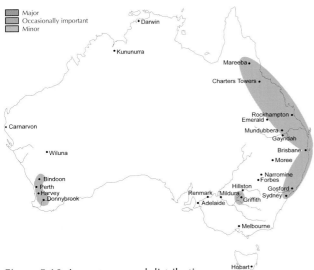

Figure 5.12 Importance and distribution of white wax scale.

Plate 5.43 Adult white wax scale on twig.

Description

▶ General appearance
The adult female white wax scale has a soft, white, waxy covering about 6 mm in diameter. The body of the insect beneath the wax is light red to dark brown, plump and soft. Its hind end narrows into a triangular tail (pygidium) that lies flat on the leaf or twig on which the scale has settled.

▶ Distinguishing features
White wax scale is larger and paler than pink wax scale. Its wax covering is soft and moist, unlike the harder, drier waxes of hard wax scale, pink wax scale and Florida wax scale.

▶ Life cycle
Males are unknown and reproduction is achieved without mating.

Each adult female produces up to 3000 orange eggs which accumulate in a thick mass under her body. The eggs hatch into crawlers which settle mainly along the midribs of leaves, on leaf stalks and on twigs. Within a few hours of settling, these nymphs produce white marginal rays of felted, waxy material and a pad on the dorsal surface.

Crawlers remain on the leaves for 4–5 weeks, then move to twigs and settle in permanent positions, usually on new or one-year-old wood. A conical cap of wax is then formed, and the insect is at the 'peak' stage, which is followed a few weeks later by the 'dome' stage. The wax increases until the scale is fully developed.

The adult stage is reached after two moults.

▶ Seasonal history
There are two generations of white wax scale per year in Queensland and northern New South Wales. Crawlers of the

first generation first appear in mid-October and continue emerging until early February. Crawlers of the second generation first appear in early April, and continue emerging until September.

There is one generation per year in other parts of New South Wales. Climate and aspect determine the date of crawler hatching. In any one situation the time of hatching is fairly constant from year to year. Hatching begins during late October in warm areas, and during December in cool areas, and is complete by early January.

▸ Habits

Adult white wax scales are nearly always found on twigs less than two years old. The scales produce honeydew and deposits may be abundant on foliage and fruit by late summer.

Within orchards the crawlers are transported by wind, insects, birds and on the clothing of orchard workers.

The hot, dry conditions which prevail in southern inland regions are unfavourable for white wax scale, and so it occurs mostly in moister coastal areas. Heat waves in coastal areas can drastically reduce populations of white wax scale if first and second instar scales are present.

In the Murrumbidgee Irrigation Area, white wax scale is known to occur in some orchards with overhead irrigation.

▸ Hosts

White wax scale can occur on numerous other hosts besides citrus, including groundsel, white cedar, guava, persimmon, mango, pear, quince, gardenia, lillypilly, pittosporum, pepperina and turkey bush.

▸ Origin and distribution

White wax scale originated in southern Africa, from where it spread to Australia.

Damage

Fruit
Heavy growth of sooty mould.

Leaves
Heavy growth of sooty mould, reducing photosynthesis.

Direct effects of scale feeding are negligible, but in humid conditions sooty mould fungi develop on the honeydew. Sooty mould may interfere with photosynthesis in leaves, retard colour development of new fruit and is an unsightly blemish on fruit at harvest. Mould can be difficult to remove completely, even if fruit are washed and brushed before packing.

▸ Varieties attacked

All citrus varieties are attacked, with grapefruit being most susceptible.

Natural enemies

▸ Parasites

The wasp parasite *Paraceraptrocerus nyasicus* was introduced from South Africa in 1968. It was established in Queensland and New South Wales by the early 1970s and since then, white wax scale has been a very minor pest in Queensland. Where the scale occurs, parasitism levels often reach 90%.

P. nyasicus is a dark-blue wasp, 5 mm long. It lays eggs in young scales 1–2 months old, and completes its life cycle within the host. The adult wasp cuts a large neat exit hole through the wax covering of the scale.

In central coastal New South Wales, three other parasites from South Africa have also been established: *Anicetus communis*, *Tetrastichus ceroplastae* and *Scutellista caerulea*. As a result, white wax scale has been much less abundant since the early 1970s. The dominant parasite in central coastal New South Wales is *A. communis* (responsible for up to 85% of endoparasitism).

Parasitism by *S. caerulea* can exceed 60% in young adult scales, but a large proportion of parasite larvae die. This is the result of high scale mortality which may be associated with diseases. *S. caerulea* also attacks hard wax scale, and the differences in the seasonal histories of the two hosts allows the parasite to move from one species of scale to the other.

Adult *S. caerulea* are blue-black, stubby and about 4 mm long. The female lays its egg through the wax covering of mature scales. The egg hatches into a larva which develops and then pupates. The adult emerges through a round hole it cuts in the scale cover.

Most larvae of *S. caerulea* feed on eggs of white wax scale, although some feed ectoparasitically on immature adult female scales. The effectiveness of this parasite is reduced as the size of the host decreases.

Plate 5.44 Heavy infestation of white wax scale.

Plate 5.45 The small wasp Paraceraptrocerus nyasicus *laying eggs in white wax scale.*

Plate 5.46 *The small wasp* Anicetus communis *laying eggs in white wax scale.*

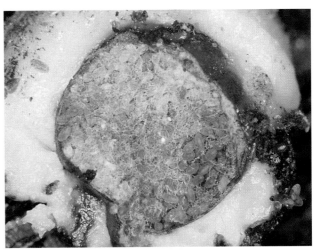

Plate 5.47 *White wax scale egg mass infected with the fungus* Fusarium stilboides.

▶ Predators
The most common predators of white wax scale are the steel-blue ladybird (*Halmus chalybeus*), and the ladybirds *Scymnodes lividigaster*, *Serangium bicolor* and *Micraspis frenata*.

Birds also prey on mature scales and bees often remove the soft moist wax. Lacewings, e.g. *Mallada* spp., also attack white wax scale.

▶ Pathogens
The fungus *Fusarium stilboides* can destroy entire egg masses in up to 10% of mature scales. Species of beetles and mites normally known for infesting stored products (such as grain) may play a role in spreading the disease within scale populations.

Although there are probably other diseases of immature scales, none have yet been identified.

Management

▶ Monitoring
- White wax scale occurs on most citrus varieties.
- The pest and its natural enemies should be monitored once or twice in mid-October – early December, and once or twice, if required, in February–March.
- Using a ×10 hand lens, check for the presence or absence of the pest on 5 randomly selected green twigs per tree.

In Queensland and New South Wales, numbers of white wax scale now rarely reach a significant level. Infestations are rare and tend to be very localised—on one or two branches only—where they are rapidly located and controlled by wasp parasites.

▶ Action level
Action is required when 5% or more of green twigs are infested with one or more scales.

▶ Appropriate action
Apply petroleum spray oil (see chapter 27). Correct timing of spraying is essential.

Wait until all eggs under adult scales have hatched. Egg hatch is complete when young scales are present on leaves, and no liquid exudes when old adult female scales are squashed. At this point, the highly susceptible young scales line the midribs of outer foliage.

Hard wax scale (Chinese wax scale)

Ceroplastes sinensis Del Guercio, Hemiptera: Coccidae

Figure 5.13 Importance and distribution of hard wax scale.

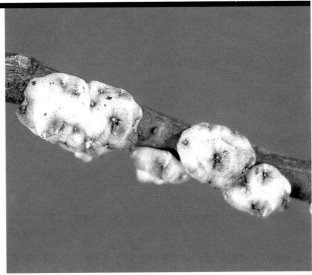

Plate 5.48 *Mature hard wax scale on twig. Note the distinctive dirty-white colour.*

Description

General appearance
The adult female hard wax scale has a hard, dry, dirty-white waxy covering, which is up to 7 mm long. The body of the insect beneath the wax ranges in colour from a medium red to dark red.

Distinguishing features
This scale is the largest wax scale found on citrus in Australia, and its waxy covering is quite characteristic. The immature instars and immature adults have a prominent cover of hard, red-pink wax.

Life cycle
Males are rare. Reproduction is generally achieved without mating.

Each adult female produces as many as 5000 orange-coloured eggs. The eggs hatch into crawlers, which disperse and settle into their permanent positions several weeks later. The adult stage is reached after two moults.

Seasonal history
There is one generation per year in the central and south coast regions of New South Wales. Hatching begins in early January, but peaks during February and early March. It is generally completed by mid-March.

Little is known of the scale's seasonal history in Queensland and northern New South Wales, but 2 generations may occur, with the timing being similar to that of life cycles of other wax scales in those areas.

Habits
Newly hatched crawlers settle mainly along the midribs of leaves. Several weeks later, most migrate from the leaves to permanent positions on the twigs and minor branches, where they remain as adults, producing honeydew in large amounts. Serious hard wax scale infestations can develop within two years.

Hosts
Hard wax scale occurs on numerous hosts other than citrus, including native melaleucas.

Plate 5.49 *Immature hard wax scales on twig. Immature scales are pink in autumn–winter.*

Origin and distribution
Despite the former common name 'Chinese wax scale', hard wax scale is not native to China and does not occur in China. Its species name *sinensis* was derived from that of the orange (*Citrus sinensis*).

Hard wax scale is native to South America, and is now also found in the southern USA, Mexico, New Zealand, some Pacific islands, and in some citrus-producing regions around the Mediterranean.

Damage

 Fruit
Heavy growth of sooty mould.

 Leaves
Very heavy growth of sooty mould, which can reduce photosynthesis.

 Twigs
Heavy growth of sooty mould.

Hard wax scale produces large amounts of honeydew. The resulting heavy growth of sooty mould can form a black cellophane-like layer 1–2 mm thick, covering entire leaves.

Varieties attacked
All varieties of citrus are attacked. The heaviest infestations occur on lemon, orange, and Seminole tangelo.

Natural enemies

Parasites
The most common endoparasite of hard wax scale in New South Wales is the wasp *Tetrastichus ceroplastae*, but the level of parasitism is very low (less than 10% of the total number of hard wax scales).

Parasitism by the wasp *Scutellista caerulea* ranges from 10% to 70%, and appears to be important in reducing scale numbers. This wasp also attacks white wax scale, and the differences in the seasonal histories of the two hosts allows the parasite to move from one to the other.

Predators
The most common predators of hard wax scale are the steel-blue ladybird (*Halmus chalybeus*) and the ladybirds *Scymnodes lividigaster*, *Serangium bicolor* and *Micraspis frenata*.

Pathogens
The fungus *Fusarium moniliformae* var. *subglutinans* can destroy entire egg masses of hard wax scale. *Verticillium lecanii* can drastically reduce high-density populations of the scale in warm moist autumns. Species of beetles and mites normally known for infesting stored products (such as grain) may play a role in spreading fungal diseases of this scale.

Management

Monitoring
- Hard wax scale occurs on most citrus varieties, especially lemon, orange and tangelo.
- Monitor once or twice in February in central coastal New South Wales.

- Using ×10 hand lens, check for the presence or absence of the pest on 5 randomly selected green twigs (with 5–10 leaves) per tree.

▶ Action level
The action level is 5% or more of green twigs infested with one or more scales.

▶ Appropriate action
Apply petroleum spray oil (see chapter 27). Correct timing of spraying is essential. Wait until all eggs have hatched. Egg hatch is complete when young scales are noted on leaves, and no liquid exudes when old adult female scales are squashed. At this point, the highly susceptible young scales line the midribs of outer foliage.

Plate 5.50 *The wasp* Tetrastichus ceroplastae, *a parasite of hard wax scale.*

Cottony cushion scale

Icerya purchasi Maskell, Hemiptera: Margarodidae

Figure 5.14 Importance and distribution of cottony cushion scale.

Description

▶ General appearance
The adult female cottony cushion scale is red-brown, about 5 mm long, and may be covered with a white mealy secretion. An egg sac, which has a white fluted surface, adds about 10 mm to the length of the insect.

Adult males are delicate and have wings.

▶ Distinguishing features
The large, white, fluted cottony egg sac, containing bright orange eggs, is very distinctive.

Cottony citrus scale (*Pulvinaria polygonata*) also produces a cottony egg sac, but is much smaller.

▶ Life cycle
Each female produces up to 1000 eggs. Crawlers are bright red, and have a series of yellow tufts on the back and edges of the body. The life cycle takes about 2 months in summer.

▶ Seasonal history
There are at least 2 generations per year.

Plate 5.51 *Cottony cushion scale, with egg mass and crawlers.*

Plate 5.52 *Lemon twig infested with cottony cushion scale.*

Habits

All younger stages are mobile. After hatching, crawlers settle along the midribs on the undersurfaces of leaves. Older stages migrate to twigs or branches, or even to the trunk. The adult female cottony cushion scale is mobile until the egg sac develops.

Cottony cushion scales produce honeydew, on which sooty mould grows.

Hosts

Citrus is a common host, but other plants, e.g. acacias, pittosporums and grevilleas, are also attacked.

Origin and distribution

Cottony cushion scale originated in Australia, but now occurs worldwide on citrus.

Damage

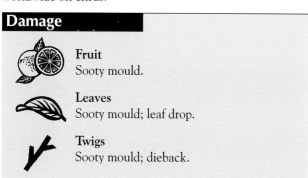

Fruit
Sooty mould.

Leaves
Sooty mould; leaf drop.

Twigs
Sooty mould; dieback.

Heavy infestations of cottony cushion scale are rare, but if they occur, can result in heavy growth of sooty mould, with associated leaf drop and twig dieback.

Varieties attacked

All citrus varieties are attacked.

Parasites

Small cryptochaetid flies, e.g. *Cryptochaetum iceryae*, are important parasites of cottony cushion scale. *Ophilosia* spp. wasps are also important parasites. Adults lay eggs in the egg masses of the scale, and larvae feed on the scale eggs.

Plate 5.53 *Cryptochaetum iceryae*, *a fly parasite of cottony cushion scale.*

Predators

Native ladybirds, *Rodolia* spp., including *Rodolia cardinalis* and *Rodolia koebeli*, and the mealybug ladybird (*Cryptolaemus montrouzieri*), prey on this scale.

Rodolia cardinalis is a small, red and black ladybird about 4 mm long. It lays its eggs on or near all stages of cottony cushion scale, and adults and larvae feed voraciously on all stages of the scale. After developing through four juvenile stages, the ladybird larva forms a pupa attached to a leaf or twig. The life cycle takes about 4 weeks.

Lacewings are also very important predators of cottony cushion scale.

Management

Monitoring

- Cottony cushion scale occurs on all citrus varieties.
- The pest and its natural enemies should be monitored once in February–March.
- Check for the presence or absence of the pest on 5 randomly selected green twigs per tree. Examine each twig for the presence of adult scales with egg sacs.
- Extra sampling may be required in parts of blocks with a history of scale problems, e.g. areas near dirt roads.

Plate 5.54 Cottony cushion scale and the predatory ladybird Rodolia cardinalis.

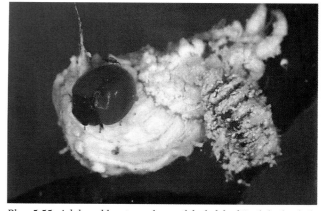

Plate 5.55 Adult and late instar larva of the ladybird Rodolia koebeli *attacking a cottony cushion scale.*

▶ Action level
The action level is 5% or more of green twigs infested with one or more scales. Control is rarely required.

▶ Appropriate action
If necessary, spray thoroughly with 1% petroleum spray oil in mid- to late March (see chapter 27).

▶ Additional management notes
Excellent biological control of cottony cushion scale can be achieved in all regions. Only very small infestations develop (on one branch or a few twigs), and these are quickly controlled by predators.

Problems will arise only if biological control is disrupted by the use of pesticides or by ant infestations.

Plate 5.56 *Larvae of the mealybug ladybird* (Cryptolaemus montrouzieri) *(left) feeding on cottony cushion scale.*

6 Hard or armoured scales (Hemiptera)

Most armoured scales are very small, and circular, oval or mussel-shaped, with a thin, hard, waxy cover (the 'armour'). Depending on the stage of development of the insect, the cover may be separated from, or attached to, its body.

Armoured scales do not produce honeydew, but their feeding can blemish fruit or cause leaf drop. More importantly, they inject toxins into plant tissues, and high populations can cause the death of trees.

Monitoring
The following points apply to the monitoring of all hard scales and their natural enemies:
- The number of trees from which samples are taken depends on block size (see chapter 25 for details).
- Additional care should be taken when monitoring blocks that have a history of economic damage.

Red scale

Aonidiella aurantii (Maskell), Hemiptera: Diaspididae

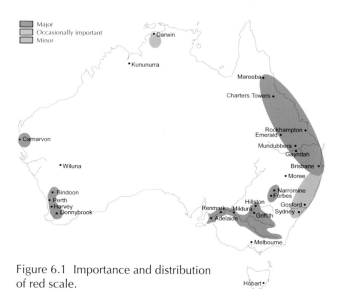

Figure 6.1 Importance and distribution of red scale.

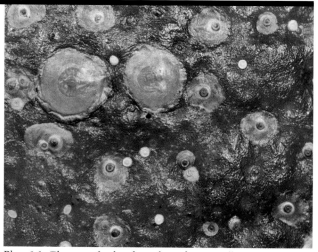

Plate 6.1 Close-up of red scale on fruit, showing yellow crawlers (bottom left), white-capped first instars, older first instars, second instar males and females, and mature females.

Description

▶ General appearance

Red scale has a thin, circular, leathery covering over the soft, flattened, shield-shaped insect. The reddish-brown scale cover of the adult female is about 2 mm across. The cover of the male scale is elongated, and smaller and paler than that of the female.

▶ Distinguishing features

Red scale can be confused with yellow scale. However, red scale generally occurs on the outer canopy and on fruit, while yellow scale occurs on the inside of the tree canopy and mainly on leaves. Yellow scale is paler in colour and flatter in shape than red scale. In addition, yellow scale is very rare in Australia.

▶ Life cycle

The adult female gives birth to 100–150 mobile young, called crawlers, at a rate of 2–3 per day over a 6–8 week period. The crawlers emerge from under their mother's scale cover, and search for a suitable feeding site on leaves, shoots or fruit. Crawlers wandering on the tree canopy can be blown by the wind into neighbouring trees or orchards.

Once a crawler settles, it inserts its mouthparts into the plant and starts feeding. It secretes a white waxy covering, and at this stage is called a 'whitecap'. After a period of feeding and growth, the insect moults. The cast skin is attached to the scale cover, giving the cover its typical red colour.

The development stage and sex of red scale can be determined by the shape and size of the scale cover. After the second stage, scales can be identified as male or female. The scale cover of males is elongated, while the scale cover of females is circular.

The male develops through a pre-pupal and pupal stage under a scale cover, before emerging as a delicate, winged insect. It is attracted to the female by a pheromone, and dies after mating.

In first-stage and second-stage females, and in third-stage (i.e. adult) unmated females, the scale cover is not attached to the body of the scale. When first-stage and second-stage females are moulting, and when third-stage females have mated, the scale cover is attached to the body of the scale.

▸ Seasonal history
In South Australia, Victoria, New South Wales and Western Australia, there are 2–5 generations per year.

In Queensland and the Northern Territory, there are 5–6 generations per year.

▸ Habits
Red scale does particularly well at temperatures between 30°C and 38°C, even with low humidity. During winter, especially in the cooler districts, development of all scales progresses very slowly up to the adult stage for females, and to the pupal stage for males, at which time development stops until the onset of warmer weather in September.

With the onset of warmer weather, adult males emerge and fertilise the adult females. Females start reproducing in September or October, and crawlers move onto young, new-season fruit after petal fall in mid-October and continue to do so for several weeks.

From this time until midsummer, the population tends to be at the same stage of development. As the season progresses, generations start to overlap.

Larger numbers of red scale are found on the north-eastern aspect of trees and on the tops of trees.

▸ Hosts
Red scale occurs on a wide range of hosts other than citrus, including passionfruit and roses.

▸ Origin and distribution
Red scale is probably Chinese in origin, but is now distributed worldwide.

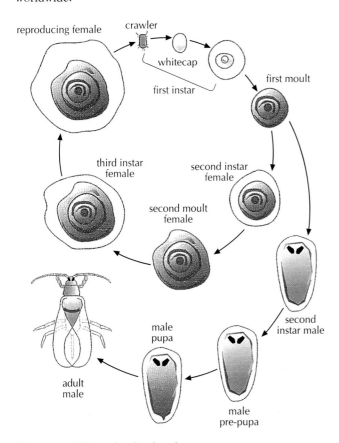

Figure 6.2 Life cycle of red scale.

Plate 6.2 Red scale on Murcott mandarins.

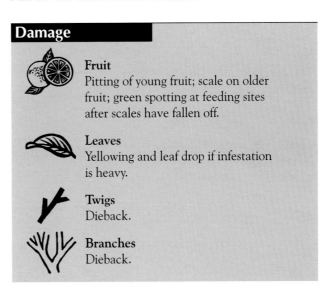

Young and thinly foliaged trees are more prone to heavy infestations than mature, densely foliaged trees.

Scales on fruit intended for the fresh market cause downgrading.

Serious infestations reduce tree vigour, cause yellowing of foliage, leaf drop and twig dieback. Very serious infestations can kill trees.

▸ Varieties attacked
All varieties of citrus are attacked by red scale. Ellendale tangor is less affected than other varieties.

Natural enemies

Red scale has been a very serious pest of citrus in all areas of Australia. Since the 1940s, a range of introduced natural enemies has substantially reduced its impact. However, it remains a major pest.

▸ Parasites
The most important natural enemies of red scale are the parasitic aphytis wasps (*Aphytis melinus* and *Aphytis lingnanensis*) and the wasp *Comperiella bifasciata*. *A. melinus* predominates in the drier inland citrus areas, while *A. lingnanensis* predominates in the subtropical regions. These two aphytis wasps are the only red scale parasites which are commercially sold in Australia. *Aphytis chrysomphali* used to occur in most citrus areas of Australia, but has mostly been displaced by other aphytis.

Plate 6.3 The aphytis wasp Aphytis lingnanensis. *Left, wasp feeding on red scale; centre, tapping scale with antennae before laying an egg; right, laying an egg on the body of an unmated adult female scale.*

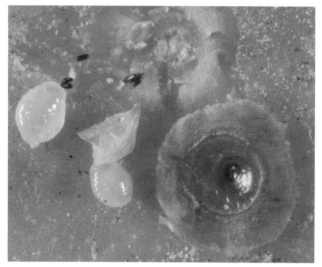

Plate 6.4 Aphytis larvae (left) and adult female red scale cover. Note the brown faecal pellets excreted by the aphytis larvae, and the remains of the female scale (top centre). (These larvae have been dissected out so they could be photographed. They would normally be hidden from view under the scale cover.)

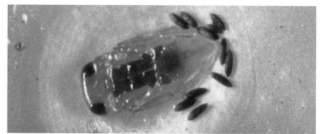
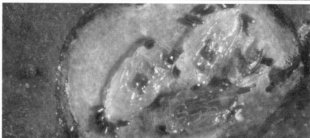
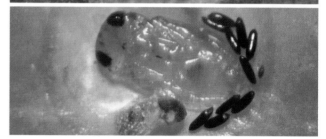

Plate 6.5 Pupae of three different aphytis wasps, showing their different pigmentation. Top, the pupa of A. lingnanensis *has a large area of dark colouring; centre, the pupa of* A. melinus *has a small dark patch on the thorax; bottom, the pupa of* A. chrysomphali *has no dark pigmentation at all. The seed-like objects are faecal pellets.*

Aphytis wasps

Aphytis are ectoparasites, laying microscopic eggs on the body of the red scale beneath the scale cover, at stages when the scale cover is not attached to the body, i.e. second instar males and females, and unmated adult females. Other stages are not attacked by aphytis.

The larva feeds externally on the body of the scale, and then pupates beneath the scale cover. The characteristic presence of brown faecal pellets produced by the larva indicates that a scale has been parasitised by aphytis, even after the parasite has emerged.

Two or three *A. melinus* wasps may develop on a single scale, whereas only one *A. lingnanensis* usually develops on a single scale.

The pupae of different aphytis species can be distinguished from each other by their colouring. This is a useful way of identifying which species of wasps are present, as the adults are difficult to distinguish from each other.

Adult female aphytis need protein to produce eggs, and commonly obtain this by feeding on red scale. Up to 50% of all red scale can be destroyed by aphytis feeding. Up to 80% of susceptible stages of red scale are commonly parasitised.

The whole life cycle is completed in about 17 days at 25°C. The adult can live for 2–6 weeks. There are usually two generations of aphytis for each scale generation.

Comperiella bifasciata

The most important endoparasite of red scale in Australia is *Comperiella bifasciata*, a shiny-black wasp about 1.5 mm long. It was deliberately introduced in the 1940s from China, but may also have been accidentally introduced before that. This wasp can parasitise up to 80% of adult mated female red scales, preventing them from reproducing, and eventually killing them.

Adult female wasps lay eggs inside the body of the scale (one larva develops per scale). About 50 eggs are laid by each wasp. Parasite development is synchronised with that of the scale. For example, if the egg has been laid in the first instar scale, the parasite larva develops slowly at first, allowing the scale to grow. If the egg is laid in a mature scale, the parasite develops faster. At 26°C the life cycle can vary from 3 to 6 weeks, depending on the stage of development of the host.

The larva changes to a black pupa about 1.5 mm long. After pupation, the adult emerges by cutting a characteristic hole in the scale. Adults live for 3–4 weeks.

Plate 6.6 The small parasitic wasp Comperiella bifasciata. *Note the emergence hole where the wasp has chewed through the scale cover.*

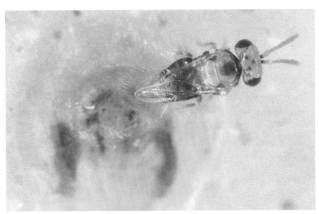

Plate 6.7 The small parasitic wasp Encarsia citrina *laying eggs in a young red scale.*

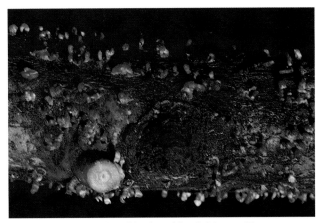

*Plate 6.8 Red scale infected by the red-headed fungus (*Fusarium coccophilum*). The small, raised, red-orange objects are fruiting bodies of the fungus.*

Other endoparasites

There are a number of other endoparasites of red scale, the most common being *Encarsia citrina* and *Encarsia perniciosi*. *E. citrina* is a small yellow and brown internal parasite of second-stage male and female scales. Parasitism levels in second instar scales often reach 25–50%. *E. perniciosi* parasitises second moult and third-stage scales. Parasitism levels can exceed 20%. A characteristic horseshoe-shaped darkening indicates the presence of encarsia larvae or pupae inside the scale.

▶ Predators

Predators help to control red scale, and a number of native Australian species have been used in IPM programs in other countries. Predators include ladybirds (most commonly the scale-eating ladybird (*Rhyzobius lophanthae*), the steel-blue ladybird (*Halmus chalybeus*) and chilocorus (*Chilocorus circumdatus*)), and also lacewing larvae (*Mallada* sp.).

The predatory mite *Eupalopsis jamesi* has been recorded feeding on red scale in inland New South Wales. Phytoseiid mites, e.g. *Euseius elinae* and *Euseius victoriensis*, also prey on red scale.

▶ Pathogens

Red-headed fungus (*Fusarium coccophilum*) can infect and kill large numbers of red scale during prolonged wet weather.

Management

The synchrony in population development of red scale is an important factor in the control of this pest. When spraying with petroleum oil, it is best to choose a time when the population consists mainly of whitecaps and immature scales, because these young stages are more susceptible to the spray. In late November, when young fruit are developing, young stages of the scale predominate. In addition, effective spray coverage is easier to achieve in all citrus varieties at this time of year rather than later in the season.

▶ Monitoring

- Red scale occurs on all citrus varieties, although Ellendale tangor is less affected.
- Red scale and its parasites should be monitored from fruit set until harvest.
- Using a ×10 hand lens, check for the presence or absence of scales on 5 randomly selected fruit per tree. On tall trees, take 20% of samples from the tops of trees.

*Plate 6.9 Closeup of the red-headed fungus (*Fusarium coccophilum*) on red scale. Note the red fruiting bodies of the fungus raised above the scale.*

- Assessments should also be made of wasp parasitism levels (particularly if scale infestation is becoming heavier, or to gauge the effects of parasite releases). Collect 20 infested fruit at random from a block. Using a binocular microscope, check 100 third instar unmated and mated female scales for parasitism.

▶ Action levels

Action is required if the following levels of infestation are reached:

- on early varieties in early to mid-season, one or more scales of any age on 10% or more fruit
- on early varieties in late season, 3 or more adult female scales on 10% or more fruit
- on later varieties in early to mid-season, one or more scales of any age on 15–20% or more fruit
- on later varieties in late season, 3 or more adult female scales on 15–20% or more fruit.

The levels of parasitism of the scale are also important in making a decision. In Queensland, 20% or more of unmated adult females should be parasitised by December, and 50% or more by late January. In inland areas, levels should be 20% by January, and 50% by early March.

▸ Appropriate action
If the action level is reached, and parasitism is low between October and March, release aphytis wasps at 25 000–50 000 per hectare. More than one release may be required, especially in blocks with a history of red scale problems.

The other option is to apply petroleum spray oil, targeting young scales (see page 67 and chapter 27).

Combinations of spraying and parasite release are often advisable, especially in orchards growing early varieties for the fresh fruit market.

▸ Additional management notes
High levels of activity by the parasitic wasps aphytis and *Comperiella bifasciata* are essential for effective control of red scale. Broad-spectrum pesticides and dust are common causes of parasite disruption.

Heatwaves can also severely reduce parasite activity and red scale populations can flare. This is especially so when eight or more days over 40°C are recorded between November and March. When this happens, extra care should be taken with monitoring and control.

Yellow scale

Aonidiella citrina (Coquillett), Hemiptera: Diaspididae

Figure 6.3 Importance and distribution of yellow scale.

Plate 6.10 Yellow scale. Lower left, second instars; centre left, male; right, adult females.

Description

▸ General appearance
The adult female yellow scale is about 2.5 mm in diameter, circular with a greyish semi-transparent scale covering. The yellow, shield-shaped body gives the scale its pale-yellow colour. The cover of the male scale is smaller and more elongated than that of the female.

▸ Distinguishing features
Yellow scale can be hard to distinguish from red scale in the field. However, yellow scale is paler in colour and flatter in shape than red scale. It prefers the inside of the tree canopy, and occurs mainly on leaves, only occasionally on fruit, and rarely on twigs and older wood. Red scale, on the other hand, generally occurs on the outer canopy and on fruit. Red scale is much more common than yellow scale in Australia, as in the rest of the world.

▸ Life cycle
The life cycle of yellow scale is very similar to that of red scale (see pages 64–65), with females producing live crawlers.

▸ Seasonal history
There are 2–4 generations each year.

▸ Habits
Yellow scale prefers temperate, shady conditions. It occurs in the lower inside part of the tree, mainly on the underside of leaves, and occasionally on fruit.

This scale is an uncommon pest (rare in Queensland because of its climate preferences), and does less damage to the tree than red scale does.

▸ Hosts
Yellow scale also occurs on ornamentals, e.g. palms, ivy and privet.

▸ Origin and distribution
Yellow scale also occurs in North and South America, China and Japan.

Damage

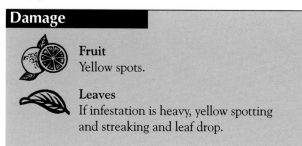

Fruit
Yellow spots.

Leaves
If infestation is heavy, yellow spotting and streaking and leaf drop.

▸ Varieties attacked
All citrus varieties are susceptible.

Natural enemies

▶ **Parasites**

The wasp *Comperiella bifasciata* (yellow scale strain) is a very effective parasite of yellow scale, and since its introduction, the scale has been only a minor pest. (Before the introduction of this parasite, yellow scale was an occasionally important pest.)

The wasps *Aphytis chrysomphali* and *Encarsia citrina* also attack yellow scale.

▶ **Predators**

Ladybirds, including the scale-eating ladybird (*Rhyzobius lophanthae*), and lacewings prey on yellow scale.

Management

▶ **Monitoring**
- Yellow scale can occur on most citrus varieties, but is uncommon and rarely causes problems.
- Monitor the pest and parasites from fruit set until harvest.
- Using a ×10 hand lens, check for the presence or absence of scales on 5 randomly selected fruit per tree.

▶ **Action level**

Take action if 20% or more of fruit are infested with one or more scales.

▶ **Appropriate action**

Spray trees very thoroughly with petroleum spray oil (see chapter 27).

Circular black scale (Florida red scale)

Chrysomphalus aonidum (L.), Hemiptera: Diaspididae

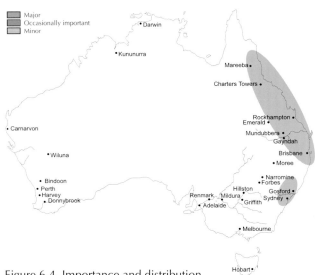

Figure 6.4 Importance and distribution of circular black scale.

Plate 6.11 *Circular black scale and wasp parasite* Aphytis holoxanthus *(centre).*

Description

▶ **General appearance**

The adult female circular black scale is circular, dark reddish-brown to black and 2 mm across. The male scale is oval, and 1 mm long.

▶ **Distinguishing features**

Circular black scale is a similar size to red scale, but the scale cover is distinctly darker. The bodies of mature, mated female circular black scales are bright yellow, and never become attached to their scale covers. Circular black scale females lay eggs, whereas red scale females produce live young, known as crawlers.

▶ **Life cycle**

Females lay up to 300 eggs and the emerging crawlers tend to settle close to the mother scale on leaves and fruit. Crawlers pass through two nymphal (juvenile) stages before becoming adults. The total life cycle takes 6–8 weeks.

▶ **Seasonal history**

In Queensland and the Northern Territory, there are 5–6 generations per year.

In New South Wales, there are 2–4 generations per year.

▶ **Habits**

Circular black scale infests leaves (both surfaces) and fruit. Crawlers move onto the young fruit after petal fall.

▶ **Hosts**

Circular black scale has a wide host range, including acacia, custard apple, avocado, banana, oleander and palms.

▶ **Origin and distribution**

Circular black scale is found worldwide.

Damage

 Fruit
Disfigured by the presence of the scale.

 Leaves
Leaf drop in heavy infestations.

 Twigs
Dieback associated with leaf drop in heavy infestations.

▶ Varieties attacked
All citrus varieties are susceptible.

Natural enemies

▶ Parasites
The main parasite of circular black scale is the small Chinese wasp *Aphytis holoxanthus*, introduced in 1975. It is extremely effective and the scale is now scarce. Other wasp parasites, e.g. *Encarsia* spp., and *Aphytis columbi*, also attack circular black scale.

▶ Predators
Ladybirds, such as *Rhyzobius ventralis*, and the steel-blue ladybird (*Halmus chalybeus*), feed on the scale.

Management

▶ Monitoring
- Circular black scale occurs on most citrus varieties.
- The pest should be monitored from fruit set until harvest.
- Using a ×10 hand lens, check for the presence or absence of scales on 5 randomly selected fruit per tree.
- Serious infestations of circular black scale are rare. When they do occur, they tend to be localised on one or two branches, and are rapidly controlled by the small wasp *Aphytis holoxanthus*.

▶ Action level
The action level is 10% or more fruit infested with one or more scales, but since the introduction of *A. holoxanthus*, such infestation has never occurred.

▶ Appropriate action
Make sure that disruptive pesticides are not used in the orchard.

Plate 6.12 Circular black scale on an orange.

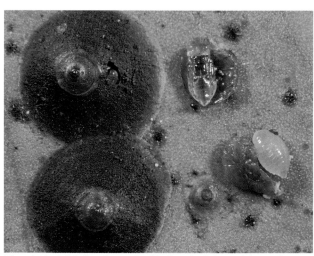

Plate 6.13 Adult female circular black scale (left), with larva (bottom right) and pupa (top right) of the wasp parasite *Aphytis holoxanthus*. Note the adult wasp chewing an emergence hole in the top scale. (Scale covers have been removed to show the larva and pupa.)

Citrus snow scale (white louse scale)

Unaspis citri (Comstock), Hemiptera: Diaspididae

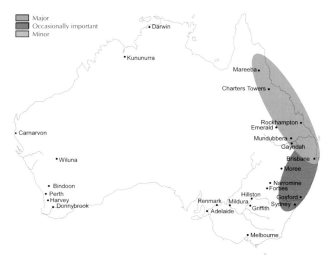

Figure 6.5 Importance and distribution of citrus snow scale.

Plate 6.14 Citrus snow scale on twig. Note brown female and white male scales, crawlers, and wasp emergence holes in both male and female scales.

Description

General appearance
The white cover of the male citrus snow scale gives the insect its common name. The cover is about 1 mm long, narrow, and rectangular, with three longitudinal ridges. Male scales form a pre-pupa and a pupa before the short-lived winged adult emerges. The adult male has an orange body and transparent wings.

The adult female scale cover is mussel-shaped, 2 mm long and dark-brown in colour. The scale body underneath is orange.

Distinguishing features
The presence of white male scales distinguishes this scale from mussel scale and Glover's scale. In addition, the adult female citrus snow scale has a longitudinal ridge along the full length of the scale cover, while female mussel scale (or purple scale) and Glover's scale do not have such a ridge. The latter two scales tend to be light rather than dark brown, and are larger than citrus snow scale. Mussel scale is 3–4 mm long, and Glover's scale 5–6 mm long.

Life cycle
Each female produces up to 150 eggs over 2–3 months. The eggs hatch almost immediately, and crawlers move over the tree. The life cycle takes about 8 weeks in summer.

Crawlers may be produced all year round, with the largest numbers appearing in autumn. They are dispersed mainly by wind, but also on farm machinery, clothing and plants.

Seasonal history
In Queensland and the Northern Territory, there are 5–6 generations per year.

In New South Wales, there are 3–4 generations per year.

Habits
While citrus snow scale can infest any part of the tree, it is more common to find it on the trunk and main limbs of mature trees. The white male scales give the trunk a white-washed, speckled appearance.

Hosts
Citrus snow scale attacks citrus only.

Origin and distribution
Citrus snow scale originated in South-East Asia, and now also occurs in eastern Australia and Florida.

Damage

Fruit
Yellowing and disfigurement.

Leaves
Yellow spotting; leaf drop if infestation is heavy.

Twigs
Dieback.

Branches
Splitting of the bark if infestation is heavy.

Main trunk
Longitudinal bark splitting.

Heavy infestations of citrus snow scale can cause extensive drying and splitting of the bark on the trunk and main limbs of the tree. Heavy infestations are more common on older trees.

Varieties attacked
All citrus varieties are attacked, but Imperial mandarin is the least affected.

Natural enemies

Parasites
There are several wasp parasites which attack citrus snow scale: *Encarsia citrina* which parasitises second instar scales; an *Aphytis* sp. wasp introduced from Thailand in 1988; and strains of *Aphytis lingnanensis*.

Predators
The most important natural enemy of citrus snow scale is *Chilocorus circumdatus*, an introduced ladybird. This ladybird now occurs in all citrus areas of Queensland and northern New South Wales, and has reduced scale populations to a low level in many orchards.

The adult ladybird is bright orange with a thin black band at the edge of its wing covers, and is about 6 mm long. Adults are strong dispersers and can fly several kilometres in search of suitable hosts.

Plate 6.15 Extensive splitting of the bark caused by a heavy infestation of citrus snow scale. Note the white speckled appearance of areas covered by male scales.

Plate 6.16 Chilocorus ladybirds (Chilocorus circumdatus) mating amongst citrus snow scale.

Plate 6.17 Chilocorus circumdatus *larva and adult feeding on citrus snow scale.*

Plate 6.19 Larva of the predatory moth Batrachedra arenosella *feeding on citrus snow scale. Note the brown pupa (to the left of the larva).*

Plate 6.18 Clusters of Chilocorus circumdatus *pupae on a scale-infested branch. Note the newly emerged adults.*

The eggs are white and 1 mm long. They are deposited under the covering of eaten scales. Four juvenile stages (larval instars) occur, and fourth-stage larvae aggregate to pupate in groups on major limbs or dead twigs. The complete life cycle takes 4–8 weeks at temperatures ranging from 19° to 31°C.

Other predators of citrus snow scale include the scale-eating caterpillar (*Batrachedra arenosella*) which spins a fine mat or webbing. The adult is a small, nondescript, grey-brown moth about 10 mm long. The beetles *Telsimia* sp., *Rhyzobius* sp. and *Cybocephalus* sp. also prey on citrus snow scale. In coastal areas, the mite *Hemisarcoptes* sp. (often associated with chilocorus) feeds on adult scales and crawlers.

Pathogens

In coastal areas during prolonged wet spells, fungi, e.g. the red-headed fungus (*Fusarium coccophilum*), also help to control citrus snow scale.

Lichen and moss on the trunks and limbs of coastal trees inhibit the growth of scale populations.

Management

Monitoring

- Citrus snow scale occurs on most citrus varieties.
- Sample once in November–December.
- Rate infestations using the following system:

  0 no scale

  1 patch of scale (up to 100 cm^2) on the trunk or on one limb

  2 scale more evident on the trunk and main limbs, with little on the leaves or fruit

  3 scale on most parts of the tree, but still with very little on fruit, and no bark splitting or leaf drop

  4 heavy scale infestation, making the trunk and limbs look white, with some bark splitting and leaf drop; scales on fruit and leaves

  5 very heavy scale infestation with noticeable bark splitting, leaf drop and dieback; fruit and leaves heavily infested

Action level

Action should be taken if the average tree rating on the above scale is 2.5 or more.

Appropriate action

Biological control is the preferred option. If chilocorus is not present, it should be released into the orchard.

If the average tree rating is 2.5 or more, and there are less than 5 chilocorus per tree, high-volume sprays of a suitable pesticide must be very thoroughly applied. Two sprays (separated by about 4–6 weeks) are usually necessary. Apply these in November–December. Other scales may simultaneously be controlled by these sprays.

Additional management notes

Natural enemies of citrus snow scale must be preserved by avoiding the use of disruptive sprays. Spray drift from neighbouring blocks can be a serious problem.

Fruit fly baits must also be applied low on the tree skirts to minimise their impact on chilocorus and other natural enemies of pests. Other factors that can reduce numbers of chilocorus are extremely hot, dry weather, and the fungus *Beauvaria bassiana* which attacks chilocorus larvae and pupae in warm, humid weather. This fungus may be more prevalent in blocks with overhead irrigation.

Mussel scale (purple scale)

Lepidosaphes beckii (Newman), Hemiptera: Diaspididae

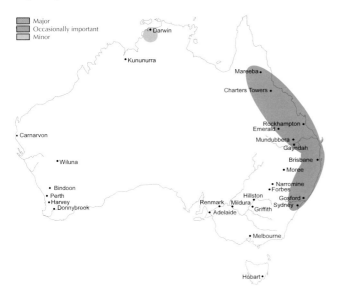

Figure 6.6 Importance and distribution of mussel scale.

Description

▶ General appearance
The scale cover of the adult female mussel scale is shaped like a mussel shell, 3–4 mm long, and light brown. The scale cover of the male is half as long, narrower, and lighter in colour. The body of the female scale is white, and held in place under the scale by a thin white membrane.

▶ Distinguishing features
Citrus snow scale and Glover's scale are similar in appearance to mussel scale. However, the adult female citrus snow scale is dark brown and smaller (2 mm long), with a longitudinal ridge, and Glover's scale is longer (5 mm long) and thinner (see plate 6.24, page 75).

▶ Life cycle
A mass of about 50–100 pearly white eggs is deposited in two rows under the female scale cover. They hatch after 2 weeks.

The crawlers settle in sheltered sites, on older leaves and beneath fruit calyx lobes. The life cycle takes 6–8 weeks.

▶ Seasonal history
In Queensland and the Northern Territory, there are 5–6 generations per year.

In New South Wales there are 2–5 generations per year.

▶ Habits
Single scales may be scattered over infested fruit, or clumps of scales may gather where two or more fruit touch. Scales will often be found under the fruit calyx lobes, even in lightly infested trees.

Typically, mussel scale enters an orchard on young trees at planting. Infested buds or grafts can carry scale to new trees in nurseries.

Mussel scale is adversely affected by extremely hot, dry weather, and is more common in coastal areas.

Plate 6.20 Mussel scale. Males are thin and females broad. Note the emergence holes made by the small parasitic wasp *Aphytis lepidosaphes*.

▶ Hosts
Hosts include avocado, eucalyptus, fig, pecan and many ornamental plants.

▶ Origin and distribution
Mussel scale is found throughout the world.

Damage

Fruit
Green marks where scales were attached; dry, brown calyx lobes; unsightly appearance of infested rind.

Leaves
Bright yellow patches; leaf drop if infestation is heavy.

Twigs
Dieback.

Branches
Bark cracking if infestation is heavy.

Plate 6.21 Mussel scale on a mandarin. This cluster of scales has gathered in the place where the fruit has been touching another fruit.

Mussel scale occurs on twigs, leaves and fruit causing disfigurement and dieback. Wood up to 25 mm in diameter is most severely damaged. Large limbs weakened by mussel scale are often invaded by citrus snow scale, and part or all of the limb may die.

Infested fruit are difficult to clean. The calyx lobes become dry and brown, and detract from the appearance of harvested fruit. Infested fruit are prone to develop stem-end rot, a disease caused by the fungus *Diaporthe citri*.

▸ Varieties attacked
All citrus varieties are attacked, but particularly oranges and grapefruit.

Natural enemies

▸ Parasites
The main parasite of mussel scale is the small yellow wasp *Aphytis lepidosaphes*, which originated in China. It attacks mainly third instar scales, and usually achieves adequate control of the pest.

▸ Predators
Predatory ladybirds, such as the steel-blue ladybird (*Halmus chalybeus*), feed on mussel scale.

▸ Pathogens
The red-headed fungus (*Fusarium coccophilum*) and other unidentified pathogens attack mussel scale.

Management

▸ Monitoring
- Mussel scale can occur on all citrus varieties.
- The pest and its parasites should be monitored from fruit set until harvest.

Plate 6.22 *Mussel scale parasitised by the small parasitic wasp* Aphytis lepidosaphes. *Note the emergence holes and, in overturned scale covers, a pupa (left), and adult about to emerge (right).*

- Using a ×10 hand lens, check for the presence or absence of scales on 5 randomly selected fruit per tree. On tall trees, take 20% of samples from the tops of trees.

▸ Action level
Action is required if more than 20% of the fruit are infested with one or more scales.

▸ Appropriate action
Spray very thoroughly with 1% petroleum spray oil (see chapter 27).

▸ Additional management notes
Minimise the use of disruptive pesticides, particularly carbamates and synthetic pyrethroids.

Glover's scale

Lepidosaphes gloverii (Packard), Hemiptera: Diaspididae

Figure 6.7 Importance and distribution of Glover's scale.

Plate 6.23 *Closeup of Glover's scale on fruit.*

6 Hard or armoured scales (Hemiptera)

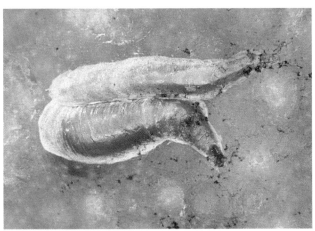

Plate 6.24 Glover's scale (top), mussel scale (bottom). Note that Glover's scale is longer and thinner than mussel scale.

Plate 6.25 Glover's scale on an orange.

Description

▶ **General appearance**
Glover's scale is long (5–6 mm) and slender, with parallel sides. The scale cover is light to dark brown. The body of the female scale is white.

▶ **Distinguishing features**
Glover's scale is similar in colour to mussel scale, but the scale cover is noticeably longer and thinner.

▶ **Life cycle**
The life cycle of Glover's scale is similar to that of mussel scale. A mass of about 50–100 pearly white eggs is deposited in two rows under the female scale cover. They hatch after 2 weeks.
  The crawlers settle in sheltered sites, on older leaves and beneath fruit calyx lobes. The life cycle takes 6–8 weeks.

▶ **Seasonal history**
In Queensland, there are 5–6 generations per year. In New South Wales, there are 2–4 generations per year.

▶ **Habits**
Glover's scale can occur on most parts of the tree—on fruit, leaves, twigs and sometimes larger limbs.

▶ **Hosts**
This scale occurs on a range of hosts other than citrus, including mango, coconut and palms.

▶ **Origin and distribution**
Glover's scale is found worldwide.

Damage

 Fruit
Disfigured by the appearance of the scales and damage to the skin.

 Leaves
Leaf drop if infestation is heavy.

 Twigs
Twig dieback if infestation is heavy.

▶ **Varieties attacked**
All varieties of citrus are attacked.

Natural enemies

▶ **Parasites**
The tiny wasp parasites *Aphytis* spp. are important in controlling Glover's scale.

▶ **Predators**
The chilocorus ladybird (*Chilocorus circumdatus*) is the most common predator in Queensland.

▶ **Pathogens**
The red-headed fungus (*Fusarium coccophilum*) is known to attack Glover's scale.

Management

▶ **Monitoring**
- Glover's scale can occur on all citrus varieties, but is very uncommon in Australia.
- The pest and its parasites should be monitored from fruit set until harvest.
- Using a ×10 hand lens, check for the presence or absence of scales on 5 randomly selected fruit per tree. When monitoring tall trees, take 20% of samples from the tops of the trees.

▶ **Action level**
Action is required when 20% or more of fruit are infested with one or more Glover's scales.

▶ **Appropriate action**
Spray trees thoroughly with 1% petroleum spray oil (see chapter 27).

Chaff scale

Parlatoria pergandii Comstock, Hemiptera: Diaspididae

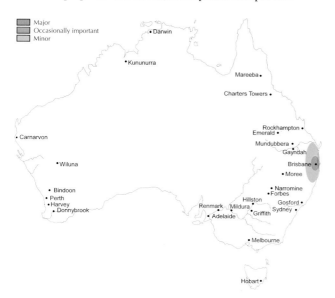

Figure 6.8 Importance and distribution of chaff scale.

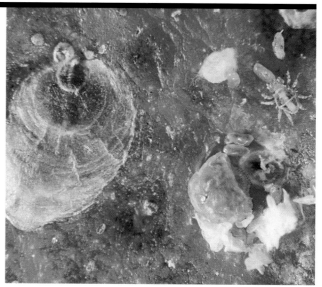

Plate 6.26 Chaff scale. Note the cheyletid mite (upper right) feeding on the egg from the purple adult female chaff scale (lower right). (The scale cover has been removed from this adult for the photograph.)

Description

▶ **General appearance**
The cover of the adult female chaff scale is an irregular oval shape, about 1.5 mm in diameter, and grey-brown. The male scale is smaller than the female.

▶ **Distinguishing features**
Chaff scale often occurs with citrus snow scale, red scale and mussel scale, mainly on the limbs and trunk but also on twigs, leaves and fruit. Chaff scale can be distinguished from other scales by the irregular oval shape of its cover, and the purple colour of the body.

▶ **Life cycle**
Adult females produce purple eggs. These hatch into crawlers, which develop through two juvenile stages. The life cycle takes about 2 months. Populations of this scale build up slowly, because it has a longer development period and produces fewer eggs than other armoured scales.

▶ **Seasonal history**
In Queensland, there are 5–6 generations per year.

▶ **Habits**
Chaff scale most commonly occurs on the tree trunk and main limbs, although twigs, leaves (along the midrib) and fruit (near or underneath the calyx) can also be infested. Fruit in clusters, or in protected positions, are more likely to be infested by chaff scale.

▶ **Hosts**
Chaff scale also occurs on a wide range of ornamentals, such as viburnum and jasmine.

▶ **Origin and distribution**
Chaff scale is found throughout the world.

Damage

 Fruit
Disfigured by the appearance of the scales and damage to the skin.

 Leaves
Leaf drop, if infestation is heavy.

 Twigs
Dieback, if infestation is heavy.

 Branches
Decline in tree health, if infestation is moderate or heavy.

 Trunk
Decline in tree health, if infestation is moderate or heavy.

▶ **Varieties attacked**
All citrus varieties are susceptible.

Natural enemies

▶ **Parasites**
The small yellow aphytis wasps are important parasites of chaff scale.

▶ **Predators**
Ladybirds, e.g. *Chilocorus circumdatus*, prey on the scale. Cheyletid mites feed on the eggs.

▶ **Pathogens**
The red-headed fungus (*Fusarium coccophilum*) attacks chaff scale.

Plate 6.27 Chaff scale parasitised by a small aphytis wasp. At top right is an unparasitised female producing purple eggs (the scale cover has been removed for the photograph). Note also the intact unparasitised female scale (bottom centre), the wasp emergence hole in a juvenile scale (top centre), the wasp pupa exposed by removing an adult female scale cover (left) and two parasitised male scales (bottom left).

Plate 6.28 A cheyletid mite with chaff scale eggs. This predator feeds on the eggs.

Management

▶ Monitoring
- Chaff scale can occur on all citrus varieties, but serious infestations are rare.
- The pest should be monitored from fruit set until harvest.
- Using a ×10 hand lens, check for the presence or absence of scales on 5 randomly selected fruit per tree. On tall trees, take 20% of samples from the tops of trees.

▶ Action level
Action is required if 20% or more of fruit are infested with one or more scales.

▶ Appropriate action
Spray thoroughly with 1% petroleum spray oil (see chapter 27).

7 Mealybugs (Hemiptera)

Mealybugs are slow-moving, soft-bodied, oval-shaped insects. They are covered with a thin coating of white, mealy wax, which extends into filaments around the edge of the body. Adults are generally 3–4 mm long.

Mealybugs commonly crowd together in sheltered sites. They feed on plant sap, using their sucking mouthparts, and excrete a sugary liquid called honeydew. The main economic damage caused by mealybugs is from the downgrading of fruit quality due to sooty mould fungus growing on the honeydew.

Monitoring

The following points apply to the monitoring of all mealybugs and their natural enemies:

- The number of trees from which samples are taken depends on block size (see chapter 25 for details).
- Additional care should be taken when monitoring blocks that have a history of economic damage.

Citrophilous mealybug

Pseudococcus calceolariae (Maskell), Hemiptera: Pseudococcidae

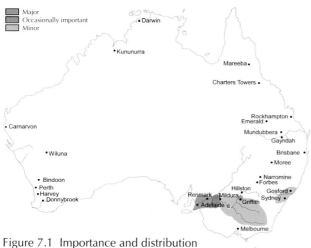

Figure 7.1 Importance and distribution of citrophilous mealybug.

Plate 7.1 Citrophilous mealybugs and their cottony egg sacs on fruit.

Description

▶ General appearance

The adult female citrophilous mealybug is a slow-moving, oval insect 3–4 mm long. It is covered by a thin coating of white, mealy wax which extends into filaments around the edge of the body. There are four tail filaments, the longest pair being about one-third the length of the mealybug's body. On top of the body is a pattern of four, dark-claret, longitudinal lines where the wax covering is thinner.

The crawlers are minute, pink and mobile. Later stages are similar in appearance to adult females.

Adult males are tiny, delicate winged insects with long tail filaments.

▶ Distinguishing features

Body fluids of the citrophilous mealybug are a dark claret-red, whereas those of longtailed mealybug and citrus mealybug are pale yellow. (Body fluid colour can be detected by squashing the mealybug.)

The waxy tail filaments of the citrophilous mealybug are much shorter and thicker than those of the longtailed mealybug, but longer than those of the citrus mealybug.

Plate 7.2 Citrophilous mealybugs on fruit stalk.

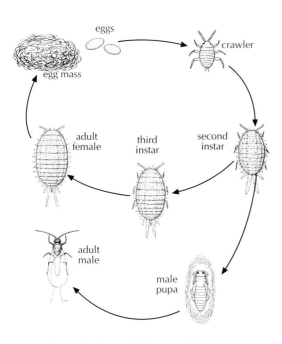

Figure 7.2 Life cycle of citrophilous mealybug.

▶ Life cycle
Up to 500 eggs are laid in a cottony egg sac, and hatch within a few days. There are three moults in females and four in males. The complete life cycle takes about 2 months in midsummer and 3–4 months in winter.

▶ Seasonal history
In New South Wales, Victoria and South Australia, there are 3–4 generations per year.

▶ Habits
The spring generation of crawlers migrates to young fruit in late November and early December. By the third generation, there is considerable overlap of all stages of development.

Juveniles and adults seek out sheltered sites, such as under the fruit calyx, inside the navel of navel oranges, between touching fruit and leaves, in curled leaves attacked by citrus leafminer, and inside cracks and crevices in bark.

Many adult females reproduce in these sheltered sites, although others migrate down the trunk to reproduce on tree skirts, or on some broadleaf weeds.

Most over-wintering citrophilous mealybugs are juveniles at various stages, which develop slowly and reach adulthood by August–September.

Mealybugs excrete sticky honeydew that is a good medium for the growth of sooty mould and other fungi.

▶ Hosts
Citrophilous mealybug is found on a wide range of hosts, including the following: pome, stone and berry fruits; ornamentals such as roses, grevillea and acacia; and weeds, most importantly nightshade, three-corner jack, bridal creeper, and caltrop.

▶ Origin and distribution
Citrophilous mealybug is probably native to eastern Australia. It now also occurs in California, South America, New Zealand, South Africa and Europe.

Damage

 Fruit
Sooty mould.

 Leaves
Sooty mould inside curled and distorted leaves, and where leaves and fruit touch.

Honeydew and sooty mould accumulate under and around the fruit calyx and inside the navel, and on the surface where fruit and foliage touch. Colonies of insects continue to infest the fruit after harvest and can survive to the marketplace. The sticky sooty mould gives the plant, especially the fruit, an unsightly appearance.

Other fungi may also grow on the honeydew and cause rotting and breakdown of fruit.

▶ Varieties attacked
All citrus varieties are attacked. Grapefruit and navel oranges are often more severely affected. Mandarins are less affected, possibly because of their smaller calyx.

Natural enemies

▶ Parasites
The encyrtid wasps Tetracnemoidea brevicornis and Anagyrus fusciventris are major parasites of citrophilous mealybug in most citrus-growing districts of south-eastern Australia.

The aphelinid wasp Coccophagus gurneyi is an important parasite on the east coast of Australia, and attempts are being made to establish it in inland citrus districts.

Other common wasp parasites of citrophilous mealybug in south-eastern Australia are the platygasterid Allotropa sp., and the pteromalid Ophelosia sp.

Plate 7.3 The wasp Tetracnemoidea brevicornis, a parasite of citrophilous mealybug. Male (left), female (right). Note the branched antennae of the male.

Plate 7.4 Female of Coccophagus gurneyi, a wasp parasite of citrophilous mealybug.

Parasitised mealybugs form 'mummies' which are cigar-shaped and brittle, whereas live mealybugs are oval, flattened and soft. They are thus easy to distinguish and can be gathered, and retained to collect the emerging adult wasps. If the parasite has already emerged, an exit hole will be apparent.

▶ Predators
Predators include lacewing larvae (e.g. *Mallada* spp.), ladybirds, such as the mealybug ladybird (*Cryptolaemus montrouzieri*), *Rhyzobius ruficollis* and *Scymnus* spp. and the predatory cecid fly *Diadiplosis koebelei*. (See citrus mealybug, pages 83–84, for more information on predators.)

Management

▶ Monitoring
- Citrophilous mealybug occurs on all citrus varieties, but is more common on grapefruit and navel oranges.
- Monitor fortnightly in November–December, and monthly thereafter until early June.
- Check 5 fruit per tree from randomly selected trees. With a ×10 hand lens, check the calyx for the presence of mealybugs. In autumn, also check the navels of navel oranges.
- In January–March, collect 50–100 mealybugs to assess parasitism levels.

▶ Action level
In navel oranges and grapefruit, action should be taken when the percentage of fruit with mealybugs reaches 10% or more. In other varieties, action should be taken when the percentage of fruit with mealybugs reaches 20%. However, some pest scouts prefer to use a lower action level earlier in the season.

Control is rarely required in mandarins.

If blocks have a history of good biological control, e.g. they are not being sprayed and the parasitism level is 40% or more in January–March, action levels may be raised.

▶ Appropriate action
In late November and early December, control is required if the action level is reached before calyx closure. Trees should be very thoroughly sprayed with an appropriate pesticide, or petroleum spray oil (see chapter 27), or a combination of oil and pesticide, as soon as possible. (Note that petroleum spray oil is effective only against young stages of citrophilous mealybug.)

▶ Additional management notes
Cover crops increase orchard humidity and lower orchard temperatures in summer, promoting the build-up of natural enemies.

Longtailed mealybug

Pseudococcus longispinus (Targioni Tozzetti), Hemiptera: Pseudococcidae

Figure 7.3 Importance and distribution of longtailed mealybug.

Plate 7.5 *Longtailed mealybugs on leaf.*

Description

▶ General appearance
The adult female longtailed mealybug is a slow-moving, oval insect 3–4 mm long. It is covered by a thin coating of white, mealy wax which extends into filaments around the edge of the body. At the rear end is a pair of tail filaments which are as long as, or longer than, the mealybug's body. The length of the side filaments is about half the width of the body.

The crawlers are minute, pink and mobile. Later stages are similar in appearance to adult females. Adult males are tiny, delicate, winged insects with long tail filaments.

▶ Distinguishing features
When the longtailed mealybug is squashed, its body fluids are seen to be pale yellow, whereas the body fluids of the citrophilous mealybug are a dark claret-red. The long tail filaments and side filaments are also distinguishing features of this species.

Life cycle

The longtailed mealybug does not produce an egg sac like those of other mealybugs discussed in this book, or lay eggs. The adult female deposits living young under her body, amongst a network of fine waxy threads. About 200 crawlers are produced over 2–3 weeks.

The life cycle takes about 6 weeks in summer and about 12 weeks in winter. The developmental stages are similar to those of citrophilous mealybug. Females moult three times before reaching adulthood, while males moult four times.

Seasonal history

In Queensland and the Northern Territory, there are 4–6 generations per year.

In New South Wales, Victoria and South Australia, there are 3–4 generations per year.

Habits

The habits of the longtailed mealybug are similar to those of citrophilous mealybug. The spring generation of crawlers migrates to young fruit in late November and early December. By the third generation, there is considerable overlap of all stages of development.

Juveniles and adults seek out sheltered sites, such as under the fruit calyx, inside the navel of navel oranges, between touching fruit and leaves, in curled leaves attacked by citrus leafminer, and inside cracks and crevices in bark.

Many adult females reproduce in these sheltered sites, although others migrate down the trunk to reproduce on tree skirts, or on some broadleaf weeds.

Most longtailed mealybugs over-winter as juveniles which reach adulthood by August–September.

Longtailed mealybugs excrete a sticky honeydew that is a good medium for the growth of sooty mould and other fungi.

Hosts

Longtailed mealybug also occurs on grape, pear, mango, various palms, and a wide range of ornamentals.

Origin and distribution

Longtailed mealybug has a worldwide distribution.

Damage

Fruit
Sooty mould.

Leaves
Sooty mould inside curled leaves, and where leaves touch fruit.

Honeydew and sooty mould accumulate under and around the fruit calyx and inside the navel, and on the surface where fruit and foliage touch. Longtailed mealybugs continue to infest the fruit after harvest and can survive to the marketplace. The sticky sooty mould gives the plant, especially the fruit, an unsightly appearance.

Other fungi may also grow on the honeydew and cause rotting and breakdown of fruit.

Varieties attacked

All citrus varieties are attacked, but navel oranges and grapefruit are worst affected.

Plate 7.6 Female of Anagyrus fusciventris, *a wasp parasite of longtailed mealybug.*

Natural enemies

Parasites

In south-eastern Australia the major wasp parasites of longtailed mealybug are *Anagyrus fusciventris*, *Tetracnemoidea sydneyensis* and *Tetracnemoidea peregrina*.

Predators

Predators include lacewing larvae (e.g. *Mallada* spp.), ladybirds, such as the mealybug ladybird (*Cryptolaemus montrouzieri*), *Rhyzobius ruficollis* and *Scymnus* spp. and the predatory larva of the cecid fly *Diadiplosis koebelei*. (See citrus mealybug, pages 83–84, for more information on predators.)

Management

Monitoring

- Longtailed mealybug occurs on all citrus varieties, but is more common on grapefruit and navel oranges.
- Fortnightly monitoring is critically important in November–December. After December, monitor monthly. Monthly monitoring must continue until May.
- Check 5 fruit per tree from randomly selected trees throughout the block. With a ×10 hand lens, examine underneath the calyx for the presence of mealybug. In autumn, also check the navels of navel oranges.
- In January–March, collect 50–100 mealybugs to assess parasitism.

Action level

In navel oranges and grapefruit, action should be taken when the percentage of fruit with mealybugs reaches 10% or more. In other varieties, action should be taken when the percentage

Plate 7.7 The ladybird Diomus notescens, *a predator of longtailed mealybug.*

of fruit with mealybugs reaches 20%. However, some pest scouts prefer to use a lower action level earlier in the season.

If blocks have a history of good biological control, e.g. they are not being sprayed and the parasitism level is 40% or more in January–March, action levels may be raised.

▶ Appropriate action
In late November and early December, control is required if the action level is reached before calyx closure. Spray trees very thoroughly with an appropriate pesticide, or petroleum spray oil (see chapter 27), or a combination of oil and pesticide, as soon as possible. (Note that petroleum spray oil is effective only against young stages of longtailed mealybug.)

▶ Additional management notes
Cover crops increase orchard humidity and decrease orchard temperatures in summer, promoting the build-up of natural enemies.

Citrus mealybug

Planococcus citri (Risso), Hemiptera: Pseudococcidae

Figure 7.4 Importance and distribution of citrus mealybug.

Plate 7.8 Citrus mealybugs on leaf.

Description

▶ General appearance
Adult female citrus mealybugs are white, about 3 mm long, and covered with a white, mealy wax. There are 18 pairs of short wax filaments around the margin of the body. These are shorter near the head end, and lengthen progressively towards the rear end. The last pair are up to one-quarter the length of the body.

The males are short-lived insects. They are similar to the males of armoured scales, with one pair of fragile wings and non-functional mouth parts. They have two long filaments at the rear end.

▶ Distinguishing features
Citrus mealybug is similar to longtailed mealybug and citrophilous mealybug. The long tail filaments of the longtailed mealybug distinguish it from the citrus mealybug.

Citrophilous mealybug is distinguished by its claret-coloured body fluid (observable when the insect is squashed), while citrus mealybug has yellow body fluid.

▶ Life cycle
The pale yellow eggs are laid in an elongated, loose, cottony egg sac extending beneath and behind the female. About 300–600 eggs are laid over 1–2 weeks, and these eggs hatch in about a week. There are three moults for females, and four moults for the male. The complete life cycle takes about 6 weeks during the summer.

▶ Seasonal history
In Queensland and the Northern Territory, there are at least 6 generations per year.

In New South Wales, there are 4–5 generations per year.

In Victoria and South Australia, there are probably 3–4 generations per year.

▶ Habits
During winter, citrus mealybugs shelter in cracks in the branches or trunk, or in leaf axils. Young mealybugs move onto fruit in late spring and usually settle under the calyx or between touching fruit. From late December, they also settle in the navel of navel oranges. Mealybugs produce honeydew, on which sooty mould grows.

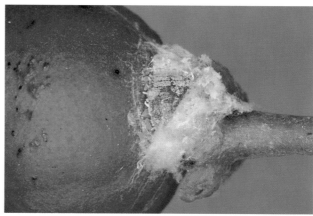

Plate 7.9 Citrus mealybugs and egg masses on fruit stalk.

‣ Hosts
Citrus mealybugs attack a wide range of tropical and subtropical fruits and ornamentals.

‣ Origin and distribution
The citrus mealybug has a worldwide distribution.

Damage

Flowers
Flower drop if infestation is heavy.

Fruit
Sooty mould and navel end rot; deformities and fruit drop if young fruit are heavily infested.

Leaves
Sooty mould.

Twigs
Sooty mould.

Citrus mealybugs produce honeydew, resulting in heavy growths of sooty mould. Sooty mould in the navel of fruit is particularly difficult to remove by washing.

Mealybug infestation inside fruit navels increases the incidence of navel end rots and premature fruit drop. (This is a problem in Queensland from February until harvest in April.)

‣ Varieties attacked
All citrus varieties are susceptible to attack by citrus mealybug, but navel oranges and grapefruit are worst affected.

Natural enemies

The most important natural enemies of citrus mealybug are the leptomastix wasp (*Leptomastix dactylopii*), the mealybug ladybird (*Cryptolaemus montrouzieri*), the lacewings *Oligochrysa lutea* and *Micromus* sp., and the green lacewing (*Mallada signata*).

‣ Parasites
Leptomastix is a small (3 mm long) honey-coloured wasp originating in Brazil. It was introduced to Queensland in 1980 and parasitises third instar and adult mealybugs.

The female deposits its egg inside the mealybug body. After hatching from the egg, the wasp larva eats the mealybug, moulting several times before forming a pupal case from the remnants of the host's skin. These pupal cases are often referred to as 'mummies'.

The pupal case is a honey-coloured cylinder, with a circular lid at one end. At first, the pupa is covered with the white mealy wax of its host but this soon wears off.

The presence of the parasite is usually indicated by clusters of the honey-coloured pupae at sites formerly infested by citrus mealybug.

The complete life cycle of leptomastix takes about 3 weeks.

Another wasp, *Anagyrus* sp., is a parasite of third instars, while *Anagyrus pseudococci* is being considered for introduction from Israel.

Less effective parasites are the wasps *Leptomastidea abnormis*, which parasitises second instar mealybugs, and *Coccidoxenoides peregrina* which parasitises first instars.

‣ Predators
Mealybug ladybird
The mealybug ladybird (*Cryptolaemus montrouzieri*) is 4 mm long, with a black body, and orange-red head and abdominal tip.

Plate 7.11 Adult leptomastix wasp and 'mummies' of parasitised mealybugs.

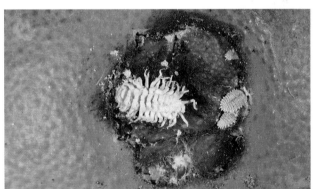

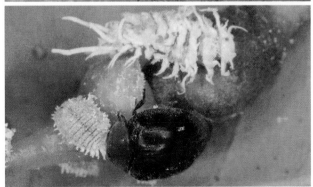

Plate 7.12 Citrus mealybug (top right, bottom left) with predatory mealybug ladybird (adult, bottom; larvae, centre).

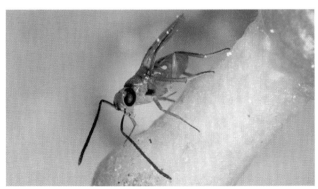

Plate 7.10 Female leptomastix wasp, a parasite of citrus mealybug.

The female deposits its eggs amongst the citrus mealybugs. The eggs hatch into larvae about 3 mm long, superficially similar in appearance to the adult mealybug, being white and mealy. However, the white filaments of the ladybird larva are much longer, and as it develops through five larval instars, the larva becomes much larger (up to 10 mm long) than a mealybug. The complete life cycle takes about 6 weeks.

Both adults and larvae of the mealybug ladybird feed voraciously on mealybugs, particularly on the egg masses. This ladybird is sometimes slow in locating a mealybug infestation, but once it does, numbers build up rapidly and it is an effective predator.

Lacewings

Adult lacewings are slender and delicate insects, often green in colour. Their bodies are about 10 mm long and the wings, held in a tent-like position when at rest, extend back about 10 mm. Fine antennae protrude forward for a similar distance.

Adults feed on nectar and honeydew, but larvae are voracious predators of young scales, mealybugs and mites. They contribute significantly to the control of these pests. Lacewing eggs are laid in clusters of about a dozen (each one on a long stalk) on the surfaces of leaves and fruit.

Young larvae are soft-bodied and spindle-shaped, and have prominent sickle-shaped jaws. After a larva devours its prey,

Plate 7.13 *Adult midge* Diadiplosis koebelei. *Larvae of this midge prey on the eggs of citrus mealybug.*

it attaches remnants of the prey to its back (see plate 8.10, page 92). It soon looks like a fluffy mound 5 mm or more in diameter.

Pupae are grey-white, leathery spheres. Each is attached by a thread to a leaf or fruit.

The complete life cycle takes 3–4 weeks.

Other predators

The larva of the midge *Diadiplosis koebelei* feeds on eggs of the citrus mealybug, but is not regarded as a major predator.

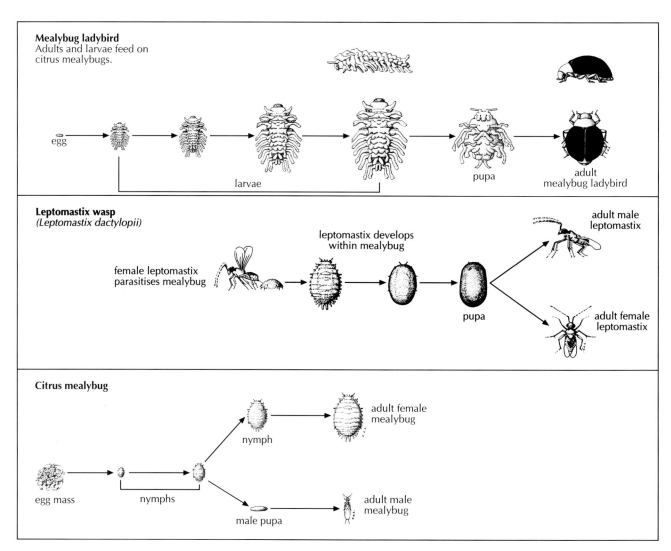

Figure 7.5 Life cycle of citrus mealybug, of the parasitic leptomastix wasp, and of the predatory mealybug ladybird.

Management

▶ Monitoring
- Citrus mealybug occurs on all citrus varieties, but is more common on navel oranges and grapefruit.
- Monitor fortnightly in November–February, and monthly thereafter until May.
- Check 5 fruit per tree from randomly selected trees. Using a ×10 hand lens, look for citrus mealybugs underneath the calyx. Also check the navels of navel oranges.

▶ Action level
The action level is 10% or more of navel oranges or grapefruit infested, and 20% or more of other varieties of fruit infested.

If mealybug ladybirds are present on 25% or more of the infested fruit, spraying should be unnecessary.

▶ Appropriate action
Petroleum spray oil (see chapter 27) gives only partial control, as it is effective only against young stages of citrus mealybug. From spring onwards, young stages and adults are found together on citrus trees.

In most navel orange and grapefruit orchards in Queensland, leptomastix wasps should be released once at the rate of 10 000 adults per hectare between mid-October and late December. (The same should be done in plantings of other varieties if an infestation develops.) If 10% of the fruit are still infested in January and there is little evidence of the parasite, a repeat release can be made before the end of February.

Mealybug ladybirds should be released into the orchard at 800–2000 beetles per hectare if they are not already present on 10% or more of fruit infested with citrus mealybug.

Spherical mealybug

Nipaecoccus viridis (Newstead), Hemiptera: Pseudococcidae

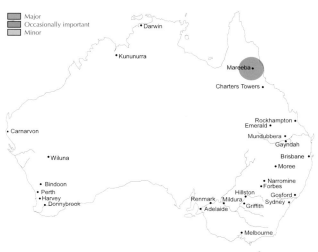

Figure 7.6 Importance and distribution of spherical mealybug.

Description

▶ General appearance
The adult female spherical mealybug is 2.5–4 mm long, oval, slightly flattened in shape and covered with a thick layer of creamy wax. Under the wax, the body is purplish or dark-brown. The body fluid is purple.

The eggs are deposited in a white, striated, hemispherical egg sac. The eggs and crawlers are purple.

▶ Distinguishing features
The mature female is usually hidden by its characteristic large hemispherical egg sac. Unlike other mealybugs, spherical mealybugs commonly cluster in colonies on leaves and shoots.

▶ Life cycle
The female spherical mealybug lays batches of eggs into an egg sac which she has secreted a few days before the start of egg laying. Each female lays a total of about 500 eggs.

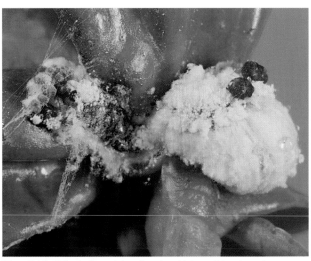

Plate 7.14 Spherical mealybugs under fruit calyx.

Plate 7.15 Spherical mealybug female with her white, domed egg sac (right). Note the two red defensive droplets and the clear droplets of honeydew produced by this female. (Most species of mealybugs produce defensive droplets from excretory glands.) The white covering of the female on the left has been removed to show the purple body and purple eggs.

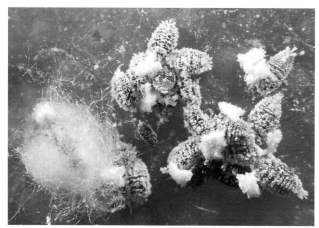

Plate 7.16 Young spherical mealybugs, with their white cast skins, on fruit surface. Note the cottony mass (left), under which there is a young male mealybug, and the crawler (centre).

Female mealybugs pass through three moults before reaching adulthood. Male mealybugs pass through four moults before emerging as a fragile winged adult. The complete life cycle takes about 3 weeks.

▶ Seasonal history
In northern Queensland, there are at least 6 generations per year.

▶ Habits
Spherical mealybug infests twigs, shoots, leaves, flower buds and fruit. Young stages produced by females over-wintering on tree limbs move onto young shoots and fruit in spring. These mealybugs are heavy producers of honeydew, on which sooty mould grows.

▶ Hosts
In Australia, spherical mealybug is currently found only on citrus. In other countries, it infests a wide range of hosts, including mango, date palm, soursop, papaya, avocado, grape and fig.

▶ Origin and distribution
The spherical mealybug occurs in tropical and subtropical areas of Africa, South-East Asia and islands of the western Pacific. So far in Australia, it has been a problem only in north Queensland citrus.

Damage

Flowers
Deformation and flower drop if infestation is heavy.

Fruit
Deformation and fruit drop; sooty mould.

Leaves
Damage to midribs resulting in twisting of leaves.

Twigs
Twisting of shoots if infestation is heavy.

Plate 7.17 Fruit distortion caused by spherical mealybug.

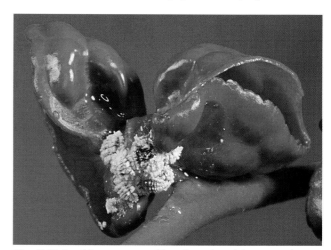

Plate 7.18 Leaf distortion caused by spherical mealybug. Note the mealybugs clustered on the twig.

Plants have a severe reaction to feeding by spherical mealybug. Shoots and leaves become twisted or puckered, and fruit develop grotesque bumps. Severely damaged fruit drops. In addition, spherical mealybug produces a great deal of honeydew, with resulting sooty mould.

▶ Varieties attacked
All varieties of citrus are susceptible.

Natural enemies

▶ Parasites
A small wasp *Anagyrus agraensis* (=*indicus*) is an important parasite of spherical mealybug.

▶ Predators
The mealybug ladybird (*Cryptolaemus montrouzieri*) is an important predator. This ladybird is 4 mm long, with a black body, and orange-red head and abdominal tip.

The female deposits its eggs amongst the spherical mealybugs. The eggs hatch into larvae about 3 mm long, superficially similar in appearance to the adult mealybug, being white and mealy. However, the white filaments of the ladybird larva are much longer, and as it develops through five larval instars, the larva becomes much larger (up to 10 mm long) than a mealybug. The complete life cycle takes about 6 weeks.

Plate 7.19 Female of the wasp Anagyrus agraensis, *a parasite of spherical mealybug.*

Both adults and larvae of the mealybug ladybird feed voraciously on mealybugs, particularly on the egg masses. This ladybird is sometimes slow in locating a mealybug infestation, but once it does, numbers build up rapidly and it is an effective predator.

Management

▶ Monitoring
- Spherical mealybug occurs on all citrus varieties.
- Monitor every 2 weeks from early October until harvest.
- Check 5 fruit per tree from randomly selected trees throughout the block. Using a ×10 hand lens, look for spherical mealybugs underneath the calyx. Also check the navels of navel oranges.

▶ Action level
The action level is 5% of fruit infested.

▶ Appropriate action
Petroleum spray oil (see chapter 27) gives only partial control, as it is effective only against young stages of spherical mealybug. From spring onwards, young stages and adults are found together on citrus trees.

Mealybug ladybirds should be released into the orchard at 800–2000 beetles per hectare if they are not already present on 10% or more of fruit infested with spherical mealybug.

Rastrococcus mealybug

Rastrococcus truncatispinus Williams, Hemiptera: Pseudococcidae

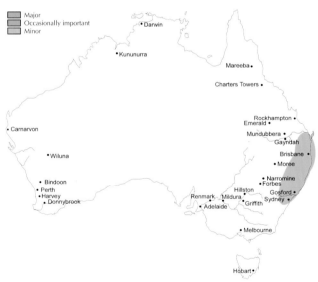

Figure 7.7 Importance and distribution of rastrococcus mealybug.

Plate 7.20 Adult female rastrococcus mealybug. Note the white, recently shed skin next to the mealybug.

▶ Distinguishing features
Rastrococcus is a relatively large mealybug. Nymphs commonly infest leaves. The body colour and long, waxy filaments are characteristic features of rastrococcus.

▶ Life cycle
Little is known about the life cycle of rastrococcus, but it probably takes 6 or more weeks in summer. Each female lays more than 30 eggs in a cottony egg sac. Eggs hatch within 7–14 days. Females moult three times before reaching adulthood, whereas the male moults four times to become a winged adult.

▶ Seasonal history
In central, mid and north coastal New South Wales, and in south-east Queensland, there may be several generations per year.

Description

▶ General appearance
Adult female rastrococcus mealybugs are oval and up to 5–6 mm long. The body colour is a mixture of grey, red and orange. Thin, waxy filaments radiate from the edge of the body, with the longest filaments at the front and the rear.

Young nymphs are pale yellow to orange. They, too, have thin filaments extending from the body.

Plate 7.21 Rastrococcus mealybug nymphs.

▶ Habits

Rastrococcus nymphs feed on leaves, producing honeydew on which sooty mould grows. They can be quite common in late summer and early autumn. It is thought that adults migrate from the outer canopy leaves to older inner canopy leaves and branches.

▶ Hosts

Rastrococcus mealybug also attacks mango, oleander, fig, and eucalypts.

▶ Origin and distribution

Rastrococcus mealybug is native to coastal New South Wales and Queensland.

Damage

 Leaves
Sooty mould, particularly in autumn.

▶ Varieties attacked

All citrus varieties are attacked, but infestations are most common on oranges and mandarins.

Natural enemies

Parasites and predators are commonly observed in orchards, but the species have not been scientifically identified. They cause consistent rapid declines in populations of rastrococcus mealybug in unsprayed orchards in autumn. As many as four unidentified species of wasps parasitise rastrococcus, and eggs of a small ladybird are often seen beside nymphs.

Other *Rastrococcus* spp. are attacked by a range of parasites (e.g. *Anagyrus* sp., *Gyranusoidea* sp. and *Tetrastichus* sp.), predators (e.g. *Scymnus* sp. ladybirds) and fungal pathogens (e.g. *Hirsutella* sp.).

Management

▶ Monitoring

- Rastrococcus mealybug attacks all citrus varieties, but is most common on oranges and mandarins.
- Monitor fortnightly in late summer and early autumn.
- Using a ×10 hand lens, check for the presence or absence of rastrococcus mealybugs on 5 randomly selected green twigs (with 5–10 leaves) per tree.

▶ Action level

Action is required when 25% more of leaves are infested, or if sooty mould threatens the marketing of mature fruit, particularly navel oranges and mandarins.

▶ Appropriate action

Infestations rarely warrant spraying. If necessary, apply petroleum spray oil (1 L oil per 100 L water) (see chapter 27). Ensure thorough coverage of infested leaves.

▶ Additional management notes

Rastrococcus mealybug may be an induced pest, with infestations resulting when natural enemies are destroyed by disruptive pesticides. Minimise the use of such pesticides.

Aphids (Hemiptera)

8

Aphids are soft-bodied insects with characteristic tubular extensions to the abdomen. They feed on plant sap, using their sucking mouthparts. Those which attack citrus are black or greenish and about 2 mm long.

Aphids are commonly found in dense groups on new plant growth. They produce a sugary liquid called honeydew, on which sooty mould grows. Their feeding also distorts shoots, and can transmit plant viruses.

Aphid life cycles are complex. Adults may be winged or wingless, depending on the state of their food supply, and how crowded together they are. Younger stages are all wingless. Some adult females reproduce without mating with a male (i.e. parthenogenetically, or from unfertilised eggs), and others reproduce after mating with a male (i.e. from fertilised eggs).

Monitoring
The following points apply to the monitoring of all aphids and their natural enemies:
- The number of trees from which samples are taken depends on block size (see chapter 25 for details).
- Additional care should be taken when monitoring blocks that have a history of economic damage.

Citrus aphids

Black citrus aphid (also known as brown citrus aphid) *Toxoptera citricida* (Kirkaldy), and black citrus aphid *Toxoptera aurantii* (Boyer de Fonscolombe), Hemiptera: Aphididae

Figure 8.1 Importance and distribution of black citrus aphids.

Plate 8.1 Citrus aphids on shoot. Note the winged female (top), and the wing buds of a last-stage nymph (lower left).

Description

▶ **General appearance**
Adult black citrus aphids are black, 2 mm long and may be winged or wingless. Immature stages are red-brown.

▶ **Distinguishing features**
Both species of black citrus aphid feed in colonies on young growth. Winged adults of *Toxoptera citricida* can be distinguished from *Toxoptera aurantii* by their wholly black third antennal segment, which is succeeded by a pale fourth segment. In *T. aurantii*, the third antennal segment is clear with a black tip.

Citrus aphids can be distinguished from melon aphid and spiraea aphid by their colour: citrus aphids are black or brown, while melon aphid and spiraea aphid are greenish.

▶ **Life cycle**
The life cycle—from birth of young aphids, through numerous moults to the adult form—can take as little as one week. Both winged and wingless adults produce live young. Winged forms develop and disperse as food quality declines, and/or as crowding increases.

▶ **Seasonal history**
There are at least 25–30 generations per year.

▶ **Habits**
Both species of citrus aphid infest citrus blossom and young growth. They produce honeydew, on which sooty mould grows.

Aphids are found on young citrus foliage for most of the year, but are most abundant in spring and autumn. Small colonies over-winter on young shoots inside trees, or in leaves curled by citrus leafminer, and move onto new shoots in late winter and early spring.

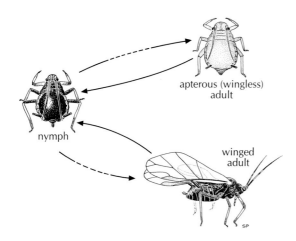

Figure 8.2 Life cycle of black citrus aphid *(Toxoptera citricida)*.

▶ **Hosts**

T. citricida occurs on citrus and some related species, while *T. aurantii* is also known to attack camellia, tea and coffee.

▶ **Origin and distribution**

Both species have a worldwide distribution, but possibly originated in Asia.

Damage

Flowers
Deformation, flower drop, and sooty mould, if infestation is heavy.

Fruit
Sooty mould; reduced fruit set if infestation is heavy.

Leaves
Distortion of young leaves.

Twigs
Possible distortion of shoots.

T. citricida, in particular, and *T. aurantii*, are the most important known vectors of tristeza virus. Sour orange rootstock is very susceptible to tristeza.

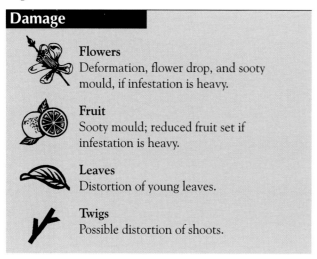

Plate 8.2 *Citrus aphids on shoot and flower stalk. Note the large amounts of honeydew on the largest leaf.*

▶ **Varieties attacked**

All citrus varieties are attacked.

Natural enemies

Natural enemies are important in controlling citrus aphids.

▶ **Parasites**

Wasp parasites include *Aphidius* spp. and *Aphelinus* spp.

▶ **Predators**

A range of predators attack citrus aphids. These include ladybirds (e.g. the transverse ladybird (*Coccinella transversalis*), the common spotted ladybird (*Harmonia conformis*), *Harmonia testudinaria*, the variable ladybird (*Coelophora inaequalis*), and the yellow-shouldered ladybird (*Scymnodes lividigaster*); syrphid flies (e.g. the common hoverfly (*Simosyrphus grandicornis*)); and lacewing larvae.

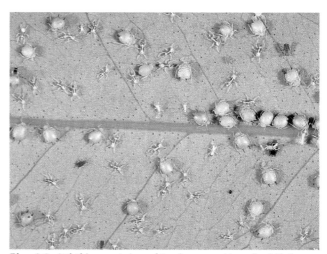

Plate 8.3 *Aphid 'mummies', resulting from parasitism of aphids by a small aphelinid wasp.*

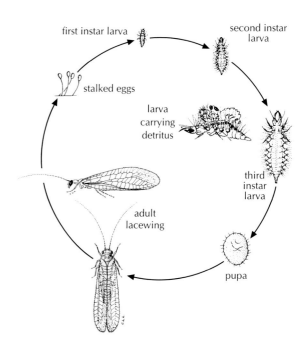

Figure 8.3 Life cycle of the green lacewing (*Mallada signata*).

8 Aphids (Hemiptera)

Plate 8.4 *Ladybirds that feed on citrus aphids: (top left) the variable ladybird (Coelophora inaequalis); (top right) the transverse ladybird (Coccinella transversalis); (bottom right) the common spotted ladybird (Harmonia conformis); (bottom left) Harmonia testudinaria; (centre) the yellow-shouldered ladybird (Scymnodes lividigaster).*

Plate 8.5 *Transverse ladybird (Coccinella transversalis) feeding on aphids. Note the yellow ladybird eggs.*

▸ Pathogens

A fungus *Entomophthora* sp. attacks citrus aphids.

Management

▸ Monitoring

- Citrus aphids occur on all citrus varieties.
- Although aphids are a sporadic problem on young growth, particularly of young trees in the spring, infestations can be heavy and warrant control.
- Check twice between early September and late October to cover the spring flush. If necessary, sample fortnightly during February–April on summer–autumn flushes.
- Check young shoots randomly selected from different sides and heights. In blocks of tall trees (above 3.5 m), one in 10 samples should be taken from the tops of trees. Examine 5–10 immature (flush growth) leaves on each shoot for the presence of aphids, honeydew and sooty mould.

▸ Action level

The action level is 25% or more of leaf flushes infested. If natural enemies (ladybirds, syrphid fly larvae, lacewings) and parasitised aphid mummies are common, e.g. on 25% or more of infested shoots, defer action until next sampling

▸ Appropriate action

When action levels are exceeded, lightly spray young growth with a specific aphicide. This spray may be applied at the same time as a fungicide.

▸ Additional management notes

Ants may protect aphids from their natural enemies, and hence may need to be controlled. (See pages 37–38 on ants.)

It is rarely necessary to spray aphids in mature orchards, as the large numbers of natural enemies usually give satisfactory control.

Extreme heat and frost can kill aphids.

Plate 8.6 *The spotted ladybird (Harmonia conformis) feeding on aphids.*

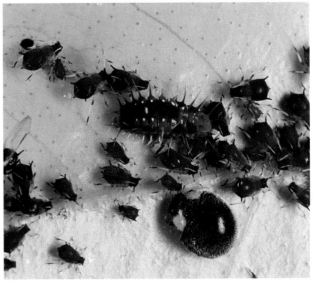

Plate 8.7 *Black citrus aphids, and adult and larva of the predatory yellow-shouldered ladybird (Scymnodes lividigaster).*

8 Aphids (Hemiptera)

Plate 8.8 Adult syrphid fly Melanastoma agrolas *on citrus flower. The larvae of these flies prey on aphids.*

*Plate 8.10 Larva of green lacewing (*Oligochrysa lutea*) feeding on adult female black citrus aphid.*

Plate 8.9 Predatory syrphid fly larva feeding on impaled aphid.

*Plate 8.11 Larva of brown lacewing (*Micromus tasmaniae*), an aphid predator.*

Melon aphid and spiraea aphid

Melon aphid *Aphis gossypii* Glover, spiraea aphid
Aphis spiraecola Patch, Hemiptera: Aphididae

Figure 8.4 Importance and distribution of spiraea aphid.

*Plate 8.12 Spiraea aphids with the predatory larva of a syrphid fly (*Simosyrphus grandicornis*) (top), and a small ladybird larva (lower left).*

8 Aphids (Hemiptera)

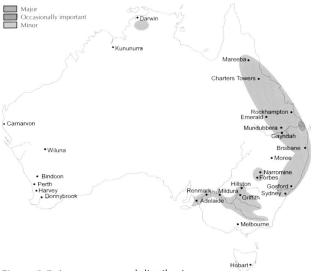

Figure 8.5 Importance and distribution of melon aphid.

Plate 8.13 Shoot infested with spiraea aphids. Note the deformed leaves caused by aphid attack.

Description

▶ General appearance
The adult melon aphid (also known as cotton aphid) is grey-green in colour and about 2 mm long.

The adult spiraea aphid is apple-green with a black tip on the abdomen, and about 2 mm long.

▶ Distinguishing features
Both species of aphid feed in colonies on young growth. They can be distinguished from the two black citrus aphid species by their greenish colour. Melon aphid is the smallest of the four species of aphids that attack citrus. Spiraea aphid is the lightest in colour.

▶ Life cycle
The complete life cycle, from the birth of young aphids, through numerous moults to the adult form, can take less than a week. Both winged and wingless adult females produce up to 60 live young each. Following initial colonisation of part of a plant, infestations of aphids are usually comprised almost solely of wingless stages. Melon aphids produce no males when on citrus.

▶ Seasonal history
There are probably 25 or more generations per year.

▶ Habits
Female winged adult aphids, often migrating from other host plants, colonise young citrus shoots in spring to autumn. Aphids can be found on young growth at most times of the year, but major infestations occur on spring and autumn flushes. Both species also infest blossom. In heavy infestations, large amounts of honeydew are produced.

▶ Hosts
Melon aphid more commonly attacks cucurbits, cotton and ornamentals, rather than citrus. Spiraea aphid attacks rosaceous plants, such as apple trees.

▶ Origin and distribution
Melon aphid and spiraea aphid are distributed throughout the world.

Plate 8.14 A young citrus shoot infested with melon aphids. Note the deformed leaves caused by aphid attack.

Damage

Flowers
Deformation, flower drop, and sooty mould, if aphid infestation is heavy.

Fruit
Sooty mould; reduced fruit set if infestation is heavy.

Leaves
Distortion of young leaves.

Twigs
Possible distortion of shoots.

Heavy infestations result in much honeydew and associated sooty mould. Melon aphid in particular causes severe distortion and twisting of young growth.

▸ Varieties attacked

All varieties of citrus are attacked. In coastal areas, Meyer lemon is the most commonly infested variety.

Natural enemies

Natural enemies are important in controlling melon aphid and spiraea aphid.

▸ Parasites

Known wasp parasites of these aphids include *Aphidius* spp. and *Aphelinus* spp.

▸ Predators

A range of predators attack these aphids. They include ladybirds, e.g. the common spotted ladybird (*Harmonia conformis*), syrphid flies, e.g. the common hoverfly (*Simosyrphus grandicornis*), and lacewing larvae (*Chrysopa* sp.).

Management

▸ Monitoring

- Melon aphid and spiraea aphid occur on all citrus varieties.
- Although aphids are a sporadic problem on young trees in the spring, infestations can be heavy, warranting control.
- It is enough to check 2 samples between early September and late October to cover the spring flush. If necessary, sample fortnightly during February–April on summer–autumn flushes.
- Check young shoots from different sides and heights on the tree. In blocks of tall trees (above 3.5 m), one in 10 samples should be taken from the tops of trees. Examine 5–10 immature (flush growth) leaves on each shoot for the presence of aphids, honeydew and sooty mould.

▸ Action level

The action level is 25% or more of flushes infested. If natural enemies (ladybirds, syrphid fly larvae and lacewing larvae) and parasitised aphid mummies are common, e.g. on 25% or more of infested shoots, defer action until next sampling.

▸ Appropriate action

When action levels are exceeded, lightly spray young growth with a specific aphicide. This spray may be applied at the same time as a fungicide.

▸ Additional management notes

Ants may protect aphids from their natural enemies, and hence may need to be controlled. (See pages 37–38 on ants.)

It is rarely necessary to spray aphids in mature orchards, as the large numbers of natural enemies usually give satisfactory control.

Planthoppers and leafhoppers (Hemiptera)

9

Planthoppers and leafhoppers are small bugs related to aphids, scales and mealybugs. They feed on plant sap, using their sucking mouthparts. Adults usually have wings, but are weak fliers. The younger stages, or nymphs, have waxy tail filaments. All stages produce honeydew, on which sooty mould grows.

Monitoring
The following points apply to the monitoring of all planthoppers and leafhoppers and their natural enemies:
- The number of trees from which samples are taken depends on block size (see chapter 25 for details).
- Additional care should be taken when monitoring blocks that have a history of economic damage.

Planthoppers (flatids)

Citrus planthopper *Colgar peracutum* (Walker), mango planthopper *Colgaroides acuminata* (Walker), green planthopper *Siphanta acuta* (Walker), green planthopper *Siphanta hebes* (Walker), Hemiptera: Flatidae

Figure 9.1 Importance and distribution of citrus planthopper.

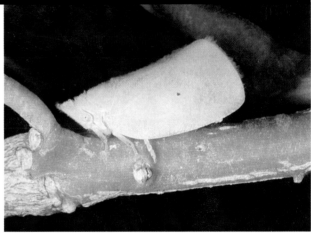

Plate 9.1 Citrus planthopper (*Colgar peracutum*) on fruit stalk.

Figure 9.2 Importance and distribution of mango planthopper.

Plate 9.2 Mango planthopper adult (top). Below the adult are numerous, white cast skins from nymphs that have become adults. At bottom right is the pupal case of an epipyropid moth (a parasite of mango planthopper) on a piece of excised leaf.

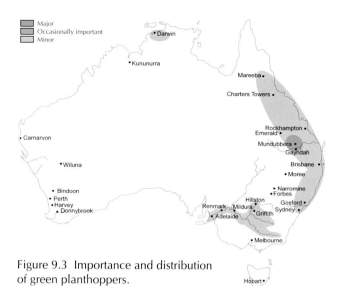

Figure 9.3 Importance and distribution of green planthoppers.

Description

General appearance

Adult citrus planthopper and adult green planthoppers are about 8 mm long, and adult mango planthopper about 13 mm long. When at rest, all these planthoppers hold their wings in a tent-like shape. Viewed from the side they appear triangular.

The citrus and the mango planthopper are pale green to white, with a minute red spot in the middle of each forewing, and usually a red border on the forewings. Green planthoppers are usually green, and covered with small yellow spots.

The young stages of the citrus planthopper and the mango planthopper are pale green to white (sometimes with conspicuous lengthwise brownish stripes), and a clump of feathery, mealy filaments projecting back from the tip of the abdomen.

Young green planthoppers are yellow-green and the feathery tail filaments extend out in two clumps either side of the tip of the abdomen.

Both adult and young planthoppers are mobile, skipping short distances when disturbed. Adults are weak fliers.

Distinguishing features

Citrus and mango planthoppers are pale green to white, while green planthoppers are pale olive-green to grass-green. Mango planthopper is larger than the other planthoppers.

Mango and citrus planthopper egg masses have a white or cream central cap. The eggs are upright and tilted slightly towards the centre. Green planthopper egg masses are flatter with the eggs partly overlapping and facing in one direction. They do not have a central cap.

Plate 9.3 Late instar nymph of citrus planthopper.

Plate 9.4 Green planthopper (Siphanta acuta) adult and nymphs on twig. Note the sooty mould.

Life cycle

The eggs of all species are laid in oval-shaped masses of about 50 eggs. The egg masses are about 5 mm in diameter. At first eggs are white, but they darken near hatching.

The complete life cycle takes 1–2 months. Citrus planthopper is the most common species on citrus in Queensland, while green planthoppers (particularly S. acuta) predominate in southern states.

Plate 9.5 Mango planthopper egg masses. Left, unhatched; right, with newly hatched nymphs.

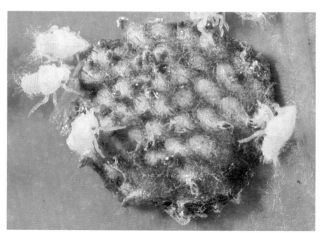

Plate 9.6 Green planthopper egg mass and newly hatched nymphs.

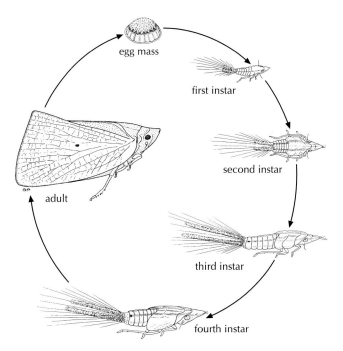

Figure 9.4 Life cycle of citrus planthopper.

▸ Seasonal history
In Queensland and the Northern Territory, there are at least 5–6 generations per year.

In New South Wales, Victoria and South Australia, there are 3–4 generations per year.

▸ Habits
Planthoppers are more common inside the tree canopy than outside. Large numbers of young and adults congregate on the twigs and fruit stalks, producing large amounts of clear honeydew, which supports the growth of sooty mould.

▸ Hosts
Hosts of the citrus planthopper include macadamia, guava, asparagus, papaw, grapes, coral tree, Geraldton wax, potato and bougainvillea.

Mango planthopper attacks mango, as well as citrus.

Hosts of the green planthopper (*S. acuta*) include *Acacia koa*, *Cheirodendron gaudichaudii*, *Suttonia lessertiana*, *Coprosma vontempskyi*, *Eucalyptus* spp., *Metrosideros collina*, *Moraea iridioides*, *Rubus hawaiiensis*, *Styphelia* (*Cyathodes*), *Myrsine* (*Suttonia*), *Tetraplasandra*, coffee, guava, and sumac.

Hosts of the green planthopper (*S. hebes*) include *Kennedia rubicunda*, *Schefflera* and *Cordyline*.

▸ Origin and distribution
These planthoppers are native to Australia. *Siphanta acuta* also occurs in New Zealand and Hawaii.

Damage

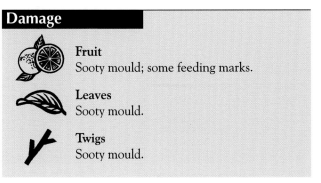

Fruit
Sooty mould; some feeding marks.

Leaves
Sooty mould.

Twigs
Sooty mould.

Plate 9.7 Male of the wasp Achalcerinys *sp., which parasitises the eggs of mango planthopper.*

Plate 9.8 Adult dryinid wasp, an ant-mimicking parasite of green planthopper. Note the disc-like pupal case (left) from which the wasp has emerged.

Plate 9.9 Dryinid wasp larva feeding externally behind the right wing bud of a green planthopper nymph.

Slight damage caused by planthoppers feeding on fruit has been noted in spring and early summer. Damage is in the form of small, circular yellow marks, but as the fruit colours, these marks become less obvious.

▸ Varieties attacked
All citrus varieties are susceptible.

Natural enemies

▸ Parasites
The small wasp *Achalcerinys* sp. parasitises up to 90% of the egg rafts of the citrus and the mango planthopper, and is an important natural enemy of these species. Other wasps which parasitise planthopper eggs include *Ooencyrtus* spp. and *Aphanomerus* spp.

Dryinid wasps and strepsiptera parasitise nymphs and adults of all four species. An epipyropid moth parasitises mango planthopper.

▸ Predators
Spiders and assassin bugs prey on planthoppers.

Management

▶ **Monitoring**
- Planthoppers occur on all citrus varieties, but particularly mandarins.
- In Queensland, monitor fortnightly from mid-December until March. In South Australia and Victoria, monitor at 3–4 weekly intervals from early February to late May.
- From each tree selected for checking, examine 5 randomly selected green twigs, with fruit attached. Two samples should be taken from the centre of trees.
- To assess parasitism levels, 10–20 planthopper egg rafts may be collected and placed in vials. Wait for any wasps to hatch, and count the numbers of egg rafts from which numerous parasites have emerged.

▶ **Action level**
The action level is 20% or more of green twigs infested. If more than half of the sampled egg rafts produce parasites, the planthopper population can be expected to decline.

▶ **Appropriate action**
Spray once with a selective pesticide.

Passionvine hopper

Scolypopa australis (Walker), Homoptera: Ricaniidae

Figure 9.5 Importance and distribution of passionvine hopper.

Plate 9.10 Passionvine hopper adult.

Description

▶ **General appearance**
Adult passionvine hoppers are about 8 mm long and superficially resemble moths. From the side they appear triangular. Their bodies are brown, and the wings are mottled with brown and clear areas. The wings extend beyond the body, and are often spread flat over the body during feeding.

Nymphs are squat, and mottled brown and white (sometimes with a greenish tinge in the later instars). They have white waxy filaments, which resemble tail feathers, at the end of the abdomen.

▶ **Distinguishing features**
The adult passionvine hopper has a very distinctive moth-like appearance. Nymphs have feathery tail filaments. Adults and nymphs are very alert. When disturbed, both hop readily and the adults fly.

▶ **Life cycle**
Females lay their elongated oval eggs in slits cut into the bark of thin shoots and twigs. Nymphs emerge and develop from mid-spring to early summer. There are five nymphal instars. Adults are active in summer and early autumn.

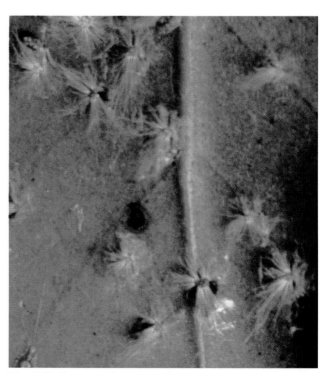

Plate 9.11 Passionvine hopper nymphs.

Seasonal history
There may be more than one generation per year.

Habits
Passionvine hopper is common in spring, occasionally in very large numbers. As many as 5–6 nymphs have been seen on almost every fruit stalk in an orchard. Generally, adults and nymphs cluster on twigs, fruit stalks, and along the mid-veins of leaves, sucking sap and producing honeydew on which sooty mould grows.

Hosts
Hosts include many cultivated plants and weeds.

Origin and distribution
Passionvine hopper is native to the east coast of Australia, where it is common in spring. It also occurs in New Zealand.

Damage

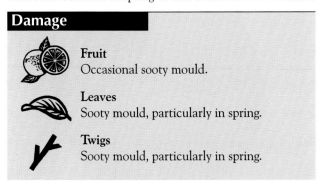

Fruit
Occasional sooty mould.

Leaves
Sooty mould, particularly in spring.

Twigs
Sooty mould, particularly in spring.

Although occasionally affected by sooty mould, fruit are generally undamaged and yield unaffected, even if infestations are heavy. There can be heavy sooty mould growth on leaves and twigs if infestations are heavy.

Varieties attacked
All citrus varieties are attacked, but infestations are most common on oranges and mandarins.

Natural enemies
The levels of passionvine hopper infestations vary from season to season. Natural enemies may be important in controlling infestations, but little is known about them.

Parasites
The small wasps *Centrodora scolypopae* and an unidentified scelionid are egg parasites. *C. scolypopae*, in turn, is parasitised by an unidentified species of *Ablerus*.

Predators
Lacewings prey on young nymphs. Spiders and birds also prey on passionvine hoppers.

Management

Monitoring
- Passionvine hopper occurs on all citrus varieties, but most commonly on oranges and mandarins.
- Monitor fortnightly from mid-spring until autumn.
- Check 5 randomly selected green twigs (with fruit present) per tree.

Action level
Action is required when 20% or more of green twigs are infested, or if sooty mould associated with passionvine hopper threatens the marketing of mature fruit.

Appropriate action
Passionvine hopper is a minor pest. Sprays are rarely required and may disrupt the control of more important pests.

If spraying is required, spray once with a selective pesticide. However, control is not easy to achieve by spraying, as the insects are so active. It is necessary to spray very thoroughly to ensure effective coverage of infested trees and the ground underneath them.

Citrus leafhopper (citrus jassid)

Empoasca smithi (Fletcher and Donaldson),
Hemiptera: Cicadellidae

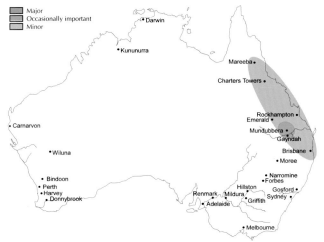

Figure 9.6 Importance and distribution of citrus leafhopper.

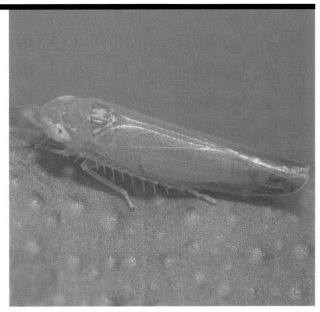

Plate 9.12 *Adult citrus leafhopper (also known as citrus jassid).*

Plate 9.13 Last instar nymph of citrus leafhopper. Note the well-developed wing buds.

Description

▶ **General appearance**

The adult citrus leafhopper is 3–4 mm long, light green, with bristle-like antennae and prominent rows of spines along its hind legs. The broadest part is the head, with the body tapering towards the tail.

▶ **Distinguishing features**

Citrus leafhopper (or citrus jassid) is not easily confused with other insects which occur on citrus. It hops readily when disturbed.

▶ **Life cycle**

The eggs are laid singly in the midrib of young leaves, and take about a week to hatch. The nymphs look like smaller wingless versions of the adult. They progress through five growth stages before reaching adulthood. The life cycle takes 4–5 weeks to complete.

▶ **Seasonal history**

There are at least 6 generations per year.

▶ **Habits**

Both adults and young are very active, jumping readily when disturbed, or shifting position quickly to the opposite side of a leaf or fruit.

The citrus leafhopper feeds on shoots during winter–spring, and moves onto ripening fruit in February–April.

▶ **Hosts**

Citrus leafhopper has also been recorded on castor oil plant.

▶ **Origin and distribution**

This leafhopper is native to Australia.

Damage

Fruit
Disfiguring oleocellosis-like brown spots, 3–10 mm in diameter on colouring fruit, with up to 100 spots per fruit.

Leaves
Young shoots become hardened and twisted; leaf drop.

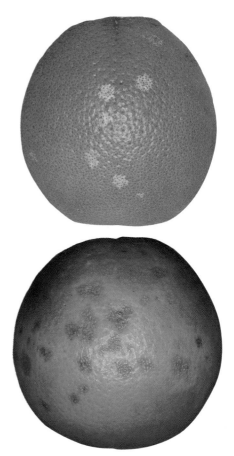

Plate 9.14 Navel oranges damaged by citrus leafhopper. Top, immature fruit; bottom, mature fruit.

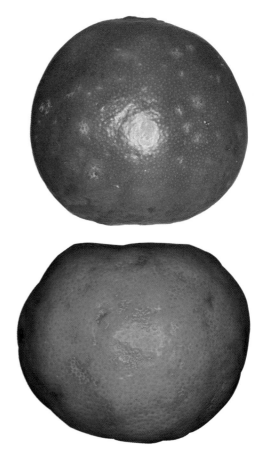

Plate 9.15 Imperial mandarins damaged by citrus leafhopper. Top, immature fruit; bottom, mature fruit.

In Queensland, adults and nymphs feed on the young foliage or on the rind of maturing fruit. Their feeding has a toxic effect on the leaves and shoots, causing twisting, hardening and leaf fall, even when numbers of leafhoppers are relatively low. In other areas, fruit damage and toxic symptoms are rare.

On fruit nearing maturity (beginning to colour), discoloured spots are produced, and may at first glance be confused with oleocellosis. Oleocellosis is caused by damage to rind tissue by oil released from damaged oil cells, whereas leafhopper damage is caused after the very fine sucking mouthparts of the leafhopper puncture surface cells. This results in death and sinking of affected tissue, leaving the oil cells standing out prominently.

Up to 50% of a crop may be affected. Levels of damage are usually highest following rain.

▶ Varieties attacked

All citrus varieties are attacked, but the worst affected are Imperial mandarin, navel orange and grapefruit.

Natural enemies

Natural enemies (including mymarid wasp egg parasites) have been observed, and are suspected to be important in controlling this pest.

Management

▶ Monitoring

- Citrus leafhopper occurs on all varieties, but is most damaging on Imperial mandarin, navels and grapefruit.
- Sample fortnightly from mid-January to early June.
- Check 5 young shoots from the tree centre for the presence or absence of citrus leafhoppers. From early March, also check 5 randomly selected fruit per tree, including fruit from inside the tree canopy, for the presence of leafhoppers and/or for fresh rind damage.
- Additional care should be taken on blocks that have leafhoppers present on young shoots before February.

Plate 9.16 Mymarid wasp (Anagrus sp.), a parasite of citrus leafhopper eggs.

▶ Action level

The action level is 20% or more of flushes, or 5% or more of the fruit, infested and/or damaged.

▶ Appropriate action

Apply a selective pesticide when infestations exceed the action level.

Additional management notes

Citrus leafhopper infestations are often worst in blocks sprayed early in the season or in the previous season with an organophosphate such as methidathion or chlorpyrifos. The most likely reason for this is that important egg parasites, which normally keep the leafhopper at insignificant levels, are killed. The best way of controlling this pest is by minimising the use of pesticides disruptive to natural enemies, and by following sound IPM practices.

10 Cicadas (Hemiptera)

Cicadas are large insects up to 50 mm long. They develop as nymphs for 2–5 years before reaching adulthood. Nymphs have large, strong front legs for digging in the soil, where they feed on plant roots. Adult male cicadas have highly developed musical organs with which they make their characteristic shrill or piercing noise.

Bladder cicada

Cytosoma schmeltzi Distant, Hemiptera: Cicadidae

Figure 10.1 Importance and distribution of bladder cicada.

Plate 10.1 Adult bladder cicada.

Description

▶ **General appearance**

Adult bladder cicadas are 30–40 mm long, with an inflated green abdomen and opaque, green, leaf-like wings.

Nymphs (up to 30 mm long) occur in the soil beneath trees, including citrus trees. They have strong burrowing forelegs.

Both nymphs and adults have strong sucking mouthparts.

▶ **Distinguishing features**

Bladder cicada is unlikely to be mistaken for any other citrus pest. The appearance of both nymphs and adults is distinctive: large, soft, green adults in the tree; nymphs with burrowing forelegs in the soil; both adults and nymphs with strong, sucking mouthparts.

▶ **Life cycle**

The adult females slit the bark of young twigs with their sword-like ovipositor to insert rows of quite large eggs. On hatching, the young nymphs enter the soil beneath the tree, where they spend the next 2–3 years.

Fully grown nymphs burrow upwards, and climb the tree trunk. When adults emerge, they leave behind the empty shell so characteristic of cicadas.

▶ **Seasonal history**

One generation takes about 2–3 years, but there may be 2–3 overlapping generations of different-sized nymphs present at any one time.

▶ **Habits**

The males have well-developed sound-producing organs and they begin their chorus around dusk during summer.

The numbers of nymphs feeding on the roots can be up to hundreds per tree.

▶ **Hosts**

Bladder cicada occurs on many tree species.

▶ **Origin and distribution**

Bladder cicada is native to Australia.

Damage

 Fruit
Drop of maturing fruit, caused by egg laying in the fruit stem.

 Twigs
Splitting due to egg laying; gummosis, leaf drop and dieback if heavily attacked.

Most damage is caused by the insertion of eggs beneath the bark. If the twigs damaged by egg laying are carrying fruit, they tend to break when the fruit reach full size. Young trees are occasionally heavily attacked, which can result in severe dieback.

While nymphs feed on roots, their impact is not known. Citrus trees seem to be able to support appreciable numbers without showing serious symptoms.

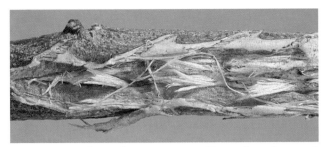

Plate 10.2 Twig showing burst tissues, caused by cicada egg laying and subsequent emergence of nymphs.

▶ Varieties attacked
All citrus varieties are attacked.

Natural enemies
Little is known about any parasites or pathogens of this cicada. Birds are known predators.

Management

▶ Monitoring
- If cicadas have previously caused fruit drop, estimate the numbers of adults per tree, or the numbers of nymphs beneath trees. Check once in December–January.

▶ Action level
If numbers of cicadas are high enough to cause fruit drop, action should be taken.

▶ Appropriate action
Use an appropriate selective pesticide against adult bladder cicada. However, sprays are rarely required, and may disrupt control of other pests.

Plate 10.3 Damage to shoots of a young tree caused by cicada egg laying. Leaves have dropped from affected twigs.

11 Whiteflies (Hemiptera)

Whiteflies are small, sap-sucking insects about 1.5–2 mm long, related to scales, aphids and mealybugs. They often congregate on the undersides of leaves. The white, mealy adults resemble small moths, and readily fly when disturbed. The nymphs, or younger stages, are a flattened oval shape and wingless. Both adults and nymphs produce sticky honeydew.

Most whitefly pest species feed on a range of plants. They damage crops directly by their feeding, by producing honeydew on which sooty mould grows, and by transmitting viral diseases.

Monitoring
The following points apply to the monitoring of all whiteflies and their natural enemies:

- The number of trees from which samples are taken depends on block size (see chapter 25 for details).
- Additional care should be taken when monitoring blocks that have a history of economic damage.

Australian citrus whitefly

Orchamoplatus citri (Takahashi), Hemiptera: Aleyrodidae

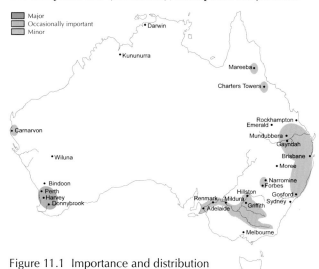

Figure 11.1 Importance and distribution of Australian citrus whitefly.

Plate 11.1 Australian citrus whitefly. Note the moth-like adults, and circular egg laying pattern.

Description

▶ General appearance
The adult Australian citrus whitefly is a delicate insect, 2.5 mm long with white powdery wings. Although they look like small moths, whiteflies are related to scale insects.

▶ Distinguishing features
This whitefly has a distinctive appearance, and lays its eggs in a circular pattern on the undersides of young leaves. Adults will fly out in swarms if foliage is disturbed.

▶ Life cycle
The eggs are yellow and oval-shaped, and appear to have been dusted with white powdery wax.

The larvae settle in groups on the undersides of leaves. At first, they are flat and resemble scale crawlers, feeding and developing beneath a protective waxy covering.

Three growth stages are followed by a pupa, and then the winged adult. The pupae are often found on leaf midribs, and mistaken for young soft brown scale.

Most individuals that survive over winter are immature stages. During winter, their development is delayed, and then it speeds up in the spring.

▶ Seasonal history
In Queensland, there are at least 5–6 generations per year, with spring, summer and autumn growth flushes becoming infested.

In New South Wales, Victoria and South Australia, there are 4–5 generations per year.

▶ Habits
Australian citrus whitefly infests the undersides of mature young leaves on the inside of the tree canopy, particularly in the spring and autumn. They produce honeydew, on which sooty mould grows.

Plate 11.2 Leaf infested with Australian citrus whitefly nymphs.

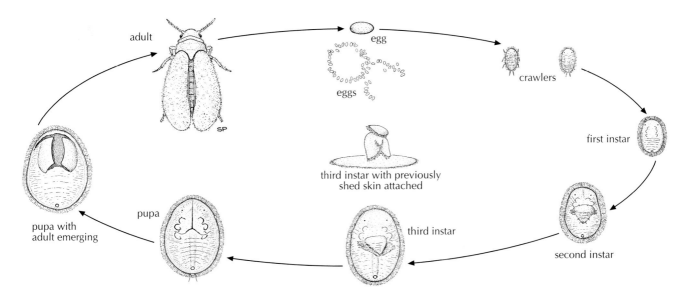

Figure 11.2 Life cycle of Australian citrus whitefly.

▶ Hosts
Citrus is the only known host, but there are probably also other plant hosts native to Australia.

▶ Origin and distribution
This whitefly is found only in Australia.

Damage

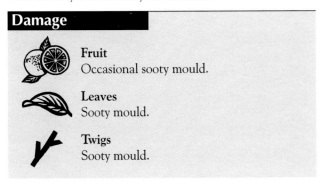

Fruit
Occasional sooty mould.

Leaves
Sooty mould.

Twigs
Sooty mould.

▶ Varieties attacked
All citrus varieties are attacked, especially lemons, which produce new leaf flushes regularly.

Natural enemies

▶ Parasites
Two small wasps (*Encarsia* sp. and *Prospaltella* sp.) are important parasites. The predominant parasite in South Australia is the wasp *Cales noacki*.

▶ Predators
Ladybird and lacewing larvae are important predators of Australian citrus whitefly. A tiny nitidulid beetle (*Cybocephalus aleyrodiphagus*) appears to be the major natural enemy in inland New South Wales, Victoria and South Australia, and provides good control. The predatory mite *Euseius victoriensis* will feed on small whitefly nymphs. Syrphid flies feed on the eggs, and predatory thrips may also be an important predator of eggs and nymphs.

▶ Pathogens
The fungus *Verticillium lecanii* infects whitefly juveniles during prolonged wet weather in coastal New South Wales and Queensland. Inland irrigated areas are usually too dry for this fungus to be a major control agent.

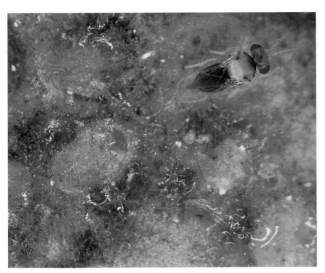

Plate 11.3 *Encarsia sp. wasp (top right), a parasite of Australian citrus whitefly, with parasitised nymphs. The dark matter is sooty mould.*

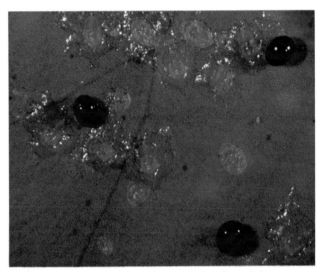

Plate 11.4 *The predatory beetle* Cybocephalus aleyrodiphagus *feeding on young Australian citrus whitefly.*

Management

▶ Monitoring

- Australian citrus whitefly occurs on all citrus varieties, but is more common on lemons.
- If the pest is troublesome, monitor young maturing leaves in spring and autumn, e.g. once in October and once or twice during February–May.
- Randomly select 5 shoots with newly expanded leaves from different sides and heights of the tree.
- Examine 5 young mature leaves on each shoot for the presence of whitefly, honeydew and sooty mould. Also check for the presence of predators and parasites.

▶ Action level

The action level is 25% or more of shoots infested with whitefly, with no evidence of the presence of natural enemies.

▶ Appropriate action

Thoroughly apply petroleum spray oil (500 mL – 1 L in each 100 L of water) (see chapter 27), at times and volumes appropriate for control of other pests such as scales.

▶ Additional management notes

Action rarely needs to be taken to control Australian citrus whitefly, particularly in inland areas, and in fact the low levels of honeydew it produces on the undersides of leaves may be an important source of food for predatory mites.

Aleurocanthus whitefly

Aleurocanthus sp., Hemiptera: Aleyrodidae

Figure 11.3 Importance and distribution of aleurocanthus whitefly.

Description

▶ General appearance

Adult aleurocanthus whitefly are delicate, cream-coloured, moth-like insects, about 2.5 mm long, with white powdery wings. Nymphs secrete an oval-shaped, black and white patterned, scaly covering.

▶ Distinguishing features

Yellow eggs are laid in spiral patterns on the leaf surface. The nymphs hatching from the eggs settle nearby. Their black and white patterned covering is distinctive.

▶ Life cycle

Adult females lay about 100 eggs, mostly on the undersides of fully expanded leaves. Yellow crawlers emerge, and after settling, start feeding by inserting their long, fine mouthparts into leaf tissues. The next two nymphal stages stay in one place to feed. After these stages, a pupa is formed, from which the adult emerges. The complete life cycle takes about 4–6 weeks in summer.

Plate 11.5 Aleurocanthus whitefly adults (white) and nymphs (mostly black) on citrus leaf.

Plate 11.6 Black and white patterned aleurocanthus nymphs. Note the eggs at bottom left.

Seasonal history
There are at least 6 generations per year.

Habits
Aleurocanthus whitefly mostly infests the undersides of mature young leaves on the inside of the canopy. Honeydew is produced on which sooty mould grows.

Hosts
Citrus is the only known host.

Origin and distribution
Aleurocanthus whitefly is probably native to Australia.

Damage

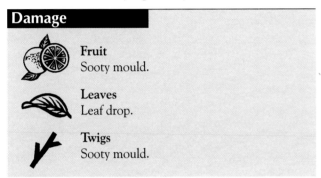

Fruit
Sooty mould.

Leaves
Leaf drop.

Twigs
Sooty mould.

Varieties attacked
All citrus varieties are attacked.

Natural enemies
The predatory mite *Euseius victoriensis* feeds on crawlers and small nymphs of aleurocanthus whitefly. The larva of the green lacewing feeds on nymphs.

Management

Monitoring
- Aleurocanthus whitefly occurs on all varieties.

Plate 11.7 Green lacewing larva feeding on aleurocanthus whitefly nymphs. Note the dead prey used as camouflage on the back of the lacewing.

- If the pest is troublesome, monitor young maturing leaves in spring and autumn, e.g. once in October–November and once or twice during February–May.
- Randomly select 5 shoots with newly expanded leaves from different sides and heights of the tree.
- Examine 5 young mature leaves on each shoot for the presence of whitefly, honeydew and sooty mould. Also check for the presence of predators and parasites.

Action level
Action is required if 25% or more of shoots with newly matured leaves are infested, and natural enemies are absent.

Appropriate action
Thoroughly apply petroleum spray oil (500 mL – 1 L in each 100 L of water) (see chapter 27), at times and volumes appropriate for control of other pests such as scales.

12 True bugs (Hemiptera)

True bugs are a large group that includes both plant-sucking and predatory insects, usually ranging in size from 3–20 mm long. They are active, and the adults are strong fliers. Nymphs resemble adults but have dorsal stink glands, and lack wings. Both nymphs and adults of plant-feeding species cause plant and fruit damage with their sucking mouthparts.

Monitoring
The following points apply to the monitoring of all true bugs and their natural enemies:
- The number of trees from which samples are taken depends on block size (see chapter 25 for details).
- Additional care should be taken when monitoring blocks that have a history of economic damage.

Spined citrus bug

Biprorulus bibax Breddin, Hemiptera: Pentatomidae

Figure 12.1 Importance and distribution of spined citrus bug.

Plate 12.1 Spined citrus bug: adult, nymphs and eggs.

Description

▶ General appearance

The spined citrus bug is a green shield or stink bug, 15–20 mm long with a prominent spine on each shoulder of the thorax.

▶ Distinguishing features

The paired spines and large size are characteristic of this species. Some smaller, green stinkbugs, which also have spines on the thorax, are of only minor importance.

▶ Life cycle

The life cycle takes 4–6 weeks in summer, but may take as long as 10–12 weeks in spring to early summer in the southern states.

In southern Australia, egg laying starts in October and continues until late March, with peaks normally in October–November and February–March.

Eggs are laid on leaves, fruit or twigs in batches (or rafts) of 4–36. The average number is 14–16. Females lay a total of 100–200 eggs over several weeks. The eggs are about 1 mm

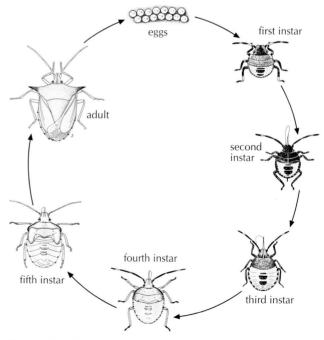

Figure 12.2 Life cycle of spined citrus bug.

in diameter and initially white, but become marked with black and red as they develop.

During summer, eggs hatch in 3–4 days. There are five nymphal stages, with early developmental stages marked with black, green, yellow, white and orange. Late stages are mainly green with black markings.

The first-stage nymphs congregate on the empty egg shells and do not feed. Second to fifth nymphal stages disperse and develop over a period of 3–10 weeks, depending on temperature.

Nymphs are found in all months except July to September. Adult females begin laying after 2–3 weeks, and can live for up to 18 months.

▶ Seasonal history

In Queensland, there are 4 generations per year—in spring, early summer, midsummer and autumn. Overlapping of generations occurs progressively during the season. Reproduction stops in autumn and non-reproductive adults survive over the winter.

In New South Wales, Victoria, and South Australia, there are 3 generations per year—in spring, summer and autumn.

▶ Habits

Adult bugs over-winter in aggregations on non-lemon citrus that are near lemons, or on their native host the desert lime *Eremocitrus glauca*. They disperse in spring to nearby lemon blocks and begin egg laying.

Lemons are the favoured host and damage can occur from October to April. In Queensland, spined citrus bug damages fruit of Lisbon lemon, Imperial and Glen Retreat mandarins, and Ellendale tangor.

Mandarins (particularly Imperial) are often attacked from December to April in all areas. Murcott mandarin is occasionally damaged in Queensland. Oranges are increasingly being damaged by spined citrus bug in the southern states. There is some variability among strains of the bug in their adaptation to oranges.

Heavy infestations of spined citrus bug are common on home-garden lemons and mandarins on the central coast of New South Wales, but the bug is rare in commercial orchards in this region.

▶ Hosts

Hosts include citrus and native Rutaceae, e.g. *Eremocitrus glauca* and *Microcitrus australasica*. Spined citrus bug will also feed on many other kinds of plants. However, breeding occurs only on rutaceous plants.

▶ Origin and distribution

This insect is native to Australia.

Damage

Fruit
Rind lesions; brown staining and gumming in the albedo; drying out of fruit segments and fruit drop; seed damage in rootstock trees.

Bug populations are largest and damage most prevalent during late summer to autumn in southern states. In Queensland, numbers and damage are greatest in spring and late summer.

In lemons, the bugs pierce the rind of fruit at any stage, causing drying and brown staining of segments, gumming and fruit drop. Gum often exudes from fruit at the site of attack, and immature fruit colours prematurely. In ripe fruit, damage is often not evident until the fruit is cut open. Very young fruit fall readily after attack. In the southern states, spined citrus bugs can cause total crop loss.

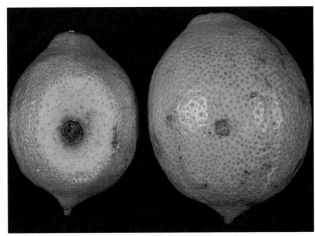
Plate 12.3 Lisbon lemons damaged by spined citrus bug. The lemon on the left has been sliced to show internal damage.

Plate 12.2 Closeup of spined citrus bug eggs and nymphs.

Plate 12.4 Fruit drop (of Lisbon lemons) caused by spined citrus bug.

Plate 12.5 Imperial mandarins damaged by spined citrus bug.

Plate 12.6 Spined citrus bug eggs parasitised by the wasp Trissolcus *sp. Note the emerging wasp (top centre), eggs from which wasps have emerged, and a wasp chewing its way out of an egg (bottom right). The white egg with a black mark was unparasitised, and the spined citrus bug nymph has emerged.*

Plate 12.7 Recently emerged wasp parasite Anastatus biproruli. *This wasp parasitises spined citrus bug eggs.*

Plate 12.8 Wasp parasite Centrodora darwini. *This wasp parasitises spined citrus bug eggs. Note the hole through which the wasp emerged, and the parasitised egg (right).*

Similar damage is caused to mandarins, but these fruit fall even more readily than lemons, particularly when immature. Mature oranges attacked by spined citrus bug often show little external or internal damage; however, immature fruit colours prematurely and falls.

Spined citrus bug attack can also cause darkening and shrivelling of seeds. This can be important on trifoliate orange and other varieties grown for seed production in citrus propagation programs.

▶ Varieties attacked

All citrus varieties are attacked, but the bugs prefer lemons, mandarins and rootstock trees being grown for seed. In the southern states, there are increasing levels of damage to Valencia oranges during spring to autumn, and to navels in autumn.

Natural enemies

▶ Parasites

Small wasps 1–2 mm long, which parasitise eggs, are important in controlling spined citrus bug. There are at least 12 wasp species, but the main ones are *Trissolcus oenone*, *Anastatus biproruli*, *Trissolcus ogyges*, *Acroclisoides tectacorisi* and *Centrodora darwini*. The first two species are the dominant parasites in all areas.

Wasps parasitise 60–100% of eggs, particularly during spring to early summer. Parasitised eggs are easily recognised by their uniform black or brown colour.

An unidentified tachinid fly parasitises adult spined citrus bugs in Queensland.

▶ Predators

Several predators attack spined citrus bug. These include spiders, predatory bugs, praying mantises and the assassin bug (*Pristhesancus plagipennis*). They all feed on bug nymphs and adults. Ants and lacewing larvae consume significant numbers of bug eggs.

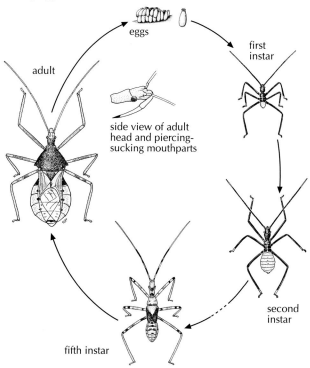

Figure 12.3 Life cycle of assassin bug.

Plate 12.9 Assassin bug preying on a spined citrus bug nymph.

Management

▶ **Monitoring**

- Spined citrus bug occurs on all citrus varieties, but is more common on lemons and mandarins.
- Monitor at fortnightly to monthly intervals from September to May.
- Conduct a search for spined citrus bug while sampling for other pests. At each sampling site, check an additional 4–5 neighbouring trees for the presence of adults and nymphs.
- Collect 5–10 egg batches, and store in separate muslin-covered containers to check for the emergence of bug nymphs or parasitic wasps.
- Also examine the first 2 rows of non-lemon citrus (oranges, mandarins, grapefruit) next to lemons during winter (June–August) for bug aggregations. Search trees intensively on a sunny day.

▶ **Action level**

The action level is 10% or more trees infested with one or more bugs, in susceptible varieties. (Some pest scouts prefer to use 5%.)

Plate 12.10 Courting assassin bugs.

Do not spray if high levels of parasitism (greater than 50%) occur in spring, and bugs are not causing damage. Spot-spray or remove by hand any over-wintering aggregations.

▶ **Appropriate action**

Because of the importance of natural enemies in controlling spined citrus bug, use only a selective pesticide at the recommended rate for integrated pest management.

▶ **Additional management notes**

Integrated pest management of spined citrus bug based on removal of over-wintering aggregations, biological control by parasites and predators, and judicious use of a low rate of a selective pesticide, is effective and easily achieved.

Valencia orange orchards in the southern states should be monitored closely for damaging levels of spined citrus bug.

Bronze orange bug

Musgraveia sulciventris (Stål), Hemiptera: Tessaratomidae

Figure 12.4 Importance and distribution of bronze orange bug.

Plate 12.11 Adult bronze orange bug feeding on shoot.

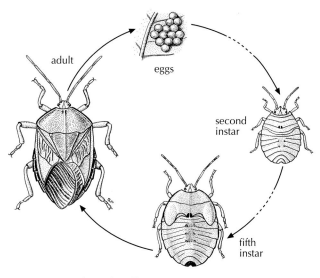

Figure 12.5 Life cycle of bronze orange bug.

Plate 12.12 Second, third and fourth instar bronze orange bug nymphs.

Plate 12.13 Fifth instar bronze orange bug nymph.

Description

) **General appearance**

Adult bronze orange bugs are shield-shaped, 25 mm long, and bronze to nearly black in colour.

Early stage nymphs are thin, flat and greenish, while later stage nymphs are salmon-coloured (see plates 12.12, 12.13).

) **Distinguishing features**

These bugs are the only ones of their size and colour to infest citrus.

) **Life cycle**

Female bronze orange bugs lay several batches of eggs during spring to early summer in Queensland, and in midsummer to autumn in New South Wales. Eggs are laid in clusters of about a dozen on the undersides of leaves. They are apple-green and about 3 mm in diameter.

In 8–14 days after laying, nymphs hatch from the eggs. These are the first of five nymphal stages, or instars. The first-stage nymphs remain near the egg shells and do not feed. Second-stage nymphs are thin and flat and over-winter in groups on the undersides of leaves.

In August or September, the second-stage nymphs start feeding and growing. Within two months they have passed through the third, fourth and fifth stages. The fourth and fifth stages are salmon-coloured, and possess wing buds.

Adulthood is reached during September–October in Queensland, and November–December in New South Wales.

) **Seasonal history**

There is one generation per year.

) **Habits**

Bronze orange bug occurs in coastal areas, and is more likely to be found in home gardens than in commercial citrus orchards.

Older nymphs and adults cluster on the shoots and young fruit. When disturbed or threatened, they spray a pungent liquid which burns plant foliage, and people's eyes and skin. Care must be taken to avoid coming into contact with this secretion.

) **Hosts**

Only citrus is attacked by bronze orange bug.

) **Origin and distribution**

This insect is native to Australia.

Damage

 Flowers
Damage to flower stalks, causing flowers to fall.

 Fruit
Damage to stalks of young fruit, causing fruit drop.

 Leaves
Wilting of shoots; leaves scalded by bugs' secretions (a minor problem).

) **Varieties attacked**

All citrus varieties are attacked, but particularly rough lemon, and also oranges.

Natural enemies

) **Parasites**

The parasitic wasp *Anastatus biproruli* parasitises eggs of bronze orange bug.

Plate 12.14 The wasp Anastatus biproruli, *which parasitises the eggs of bronze orange bug.*

Plate 12.15 Unparasitised (left) and parasitised (right) eggs of the bronze orange bug.

Predators

The assassin bug *(Pristhesancus plagipennis)* preys on bronze orange bug, and can cause high localised mortality. The pentatomid bug *Amyotea hamatus* is another common predator. Some bird species prey on the pest during spring and summer.

Management

Monitoring

- Bronze orange bug occurs on all citrus varieties, but is more common on lemons and oranges.
- Monitoring is not usually necessary. Pickers usually notice infestations of young nymphs in navel oranges, and of late-stage nymphs and adults in other varieties.
- If monitoring is required, thoroughly examine one-quarter of each randomly selected tree for the presence or absence of the bug. Do this once in late winter or in early summer.
- The young first-stage and second-stage nymphs are less obvious than older nymphs and adults.

Action level

The action level is 25% or more of trees infested with one or more bugs.

Appropriate action

Young stages can be controlled in winter by applications of a non-disruptive soft-soap spray. For late-stage nymphs and adult bugs, apply a selective pesticide once only. This is more effective on mild days, when bugs inhabit outer canopies.

Oil sprays applied thoroughly in autumn for the control of other pests, such as scales and citrus leafminer, will also kill first-stage and second-stage nymphs.

Bronze orange bug is rarely a problem in commercial orchards. Minor infestations should be spot-sprayed or squashed by hand. If removing the bugs by hand, remember to wear gloves and eye protection.

Green vegetable bug

Nezara viridula (Linnaeus), Hemiptera: Pentatomidae

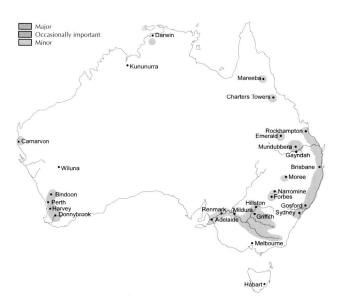

Figure 12.6 Importance and distribution of green vegetable bug.

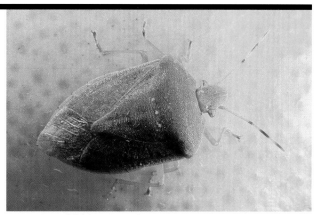

Plate 12.16 Adult green vegetable bug on Imperial mandarin.

Description

General appearance

Adult green vegetable bugs are oval, shield-shaped, 15 mm long and generally green in colour. In winter, they may be more brownish in colour. The eyes are dark, and alternate antennal segments are dark and light.

The nymphs are black, or green and black, and spotted with red, green, and orange.

Distinguishing features

Adult green vegetable bugs can be distinguished from adult spined citrus bugs by their lack of shoulder spines.

The nymphs of green vegetable bug and spined citrus bug are similar. Early stage nymphs of spined citrus bug are completely black on top of the thorax, and yellow and green with black and white markings on the abdomen. Similar stages of green vegetable bug also have yellow markings on the thorax.

Later stage nymphs of spined citrus bug are predominantly yellow-green with black markings. Similar stages of green vegetable bug are mainly black, or green and black, and spotted with red, green, and orange.

Nymphs of both species are easily distinguished from those of bronze orange bug, which are noticeably flattened in profile.

Life cycle

The adult female lays a total of about four clusters of 40–80 eggs in parallel rows. The eggs are pale yellow, becoming pink as they age, and are 1.25 mm long. They hatch after about a week.

There are five nymphal stages before the final moult to the adult.

The complete life cycle takes 5–8 weeks.

Seasonal history

There are at least 3–4 generations per year.

Habits

The green vegetable bug over-winters as an adult and starts laying eggs in the spring. Populations start increasing in the spring in warmer coastal areas, and later in summer–autumn in inland areas.

Green vegetable bug normally does not develop on citrus. When an infestation does occur on citrus, it is usually the result of adult migration into the orchard from an adjacent host, e.g. from a recently mowed field of lucerne.

The adult bugs feed on citrus fruit, piercing the rind with their proboscis.

Hosts

Hosts preferred over citrus include legumes, tomato, maize, passionfruit and cucurbits.

Origin and distribution

Green vegetable bug is found throughout the world.

Damage

Fruit
Small dark scars on the rind and discoloured dry areas in the albedo caused by bugs feeding; significant fruit drop if the bugs are numerous.

Varieties attacked

All citrus varieties are attacked.

Natural enemies

Parasites

A small wasp, *Trissolcus basalis*, parasitises the eggs of green vegetable bug, usually giving a good level of control.

Tachinid flies of the genus *Trichopoda* are important natural enemies in other countries. Attempts have been made to establish one species in Australia without success. Introductions of a second species are in progress.

Plate 12.18 *Fruit drop caused by green vegetable bug.*

Plate 12.17 *Green vegetable bugs: adult (top), late instar nymph (bottom left) and younger nymphs.*

Plate 12.19 *The wasp* Trissolcus basalis, *a parasite of green vegetable bug eggs.*

Predators
The assassin bug (*Pristhesancus plagipennis*) is a common predator.

Management

Monitoring
- Green vegetable bug occurs on all citrus varieties.
- Monitor at fortnightly to monthly intervals from September to May.
- On a sunny morning, conduct a 20–30 minute intensive search of each block for green vegetable bugs. Collect 5–10 egg batches, and store in separate muslin-covered containers to check for the emergence of bug nymphs or parasitic wasps.

Action level
Take action if more than 25% of trees are infested. Action is not needed if there are high levels of parasitism (60–100% of all eggs), and the bugs are not causing damage.

Appropriate action
Because of the importance of natural enemies in controlling green vegetable bug, use only a selective pesticide at the recommended, integrated pest management rate.

Additional management notes
Integrated management of green vegetable bug based on biological control provided by parasites and predators, and judicious use of a low rate of a selective pesticide, is effective and easily achieved.

Citrus blossom bug

Austropeplus sp., Hemiptera: Miridae

Figure 12.7 Importance and distribution of citrus blossom bug.

Description

General appearance
The adult citrus blossom bug is 5 mm long, mainly green on the underside and dark above, with yellow on the thorax and red spots on the wings, and a distinctive green V-shape on the back.

Distinguishing features
Citrus blossom bug is a small, distinctively coloured and patterned bug that feeds on leaf and flower shoots in early spring.

Life cycle
Little is known of the life history of this bug. Eggs are laid on citrus shoots, and hatch into nymphs, which develop through five instars, or stages. The complete life cycle takes 6–8 weeks in summer.

Seasonal history
There are at least 6 generations per year.

Plate 12.20 Citrus blossom bug.

Habits
Adults and young bugs feed on young shoots, including small flower shoots, up to 12 mm long.

Hosts
No hosts other than citrus are known.

Origin and distribution
Citrus blossom bug is native to eastern Australia.

Damage

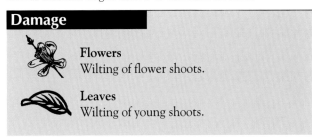

Flowers
Wilting of flower shoots.

Leaves
Wilting of young shoots.

Even light infestations of citrus blossom bug in September can destroy a good proportion of the blossom shoots. This can be a significant loss if the trees attacked have a light blossoming.

Varieties attacked
All citrus varieties are susceptible.

Natural enemies

Natural enemies of this bug have not been identified.

Management

▶ Monitoring

- Citrus blossom bug occurs on all citrus varieties.
- In coastal New South Wales, from mid-September to late October, monitor fortnightly on oranges and mandarins during flowering and fruit set.
- From each tree, randomly select 5 young shoots, with flowers and young fruit present. Examine 5 flowers and/or young fruits on each shoot for bugs.

▶ Action level

The action level is 25% or more of shoots infested.

▶ Appropriate action

This pest is usually not significant unless the blossoming is light. If blossoming is light and the action level reached, apply a selective pesticide.

Crusader bug

Mictis profana (Fabricius), Hemiptera: Coreidae

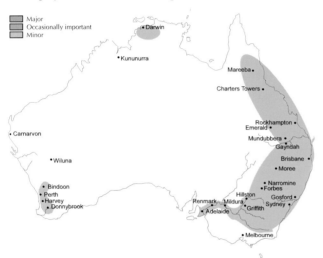

Figure 12.8 Importance and distribution of crusader bug.

Plate 12.21 Crusader bugs: nymph (lower left), adult female (upper left) and male (right).

Description

▶ General appearance

The adult crusader bug is 20–25 mm long, 7–10 mm wide, and brown with a yellow X-shaped cross (St Andrew's cross) on the back.

▶ Distinguishing features

The yellow St Andrew's cross on the back of this bug is a distinguishing characteristic.

▶ Life cycle

Eggs are laid in rows or clusters on the leaves, twigs or fruit. There are five nymphal stages. First-stage nymphs have a red abdomen. Later stages have a pair of orange spots in the middle of the upper surface of the abdomen. The complete life cycle takes about 8 weeks in summer.

▶ Seasonal history

There are 3–4 generations per year.

▶ Habits

Nymphs and adults feed on young plant growth, including flowering shoots. Adults over-winter and attack new growth in the spring. The numbers of crusader bugs are usually greatest in late summer and autumn.

▶ Hosts

Crusader bug also feeds on native wattles, cassias, eucalypts, grapes and ornamentals.

▶ Origin and distribution

Crusader bug is native to Australia.

Damage

Leaves
Wilting of young shoots above the point of feeding.

Damage is occasionally important on young trees in inland areas.

▶ Varieties attacked

All citrus varieties are susceptible.

Natural enemies

▶ Parasites

Wasps that parasitise eggs are the main natural enemies of crusader bug.

▶ Predators
The assassin bug (*Pristhesancus plagipennis*), praying mantises and birds are known predators of crusader bug.

Management

▶ Monitoring
- Crusader bug occurs on all citrus varieties.
- Monitor once or twice from early October to late April (depending on the development of infestations).
- Check 5 randomly selected young shoots on each tree.

▶ Action level
The action level is 25% or more shoots infested on young trees.

▶ Appropriate action
Action is rarely required. If it is, spot-spray with a selective pesticide. This is generally required only on young trees.

Plate 12.22 Wilted shoot caused by crusader bug feeding.

13 Moths and butterflies (Lepidoptera)

Many moths and some butterflies are significant plant pests. It is the larval stage (or caterpillar) which usually causes plant damage. However, in the case of fruitpiercing moths, adults cause problems by feeding on fruit.

Adult moths and butterflies are covered with fine scales, and, except for a few wingless species, have two pairs of wings. The scales of many butterflies, and of some moths, are brightly coloured.

Adults have well-developed eyes, long antennae, and well-developed legs. The adult mouthparts are modified to form a long, coiled tube, designed to suck up liquid food.

Monitoring
The following points apply to the monitoring of all moths and butterflies and their natural enemies:
- The number of trees from which samples are taken depends on block size (see chapter 25 for details).
- Additional care should be taken when monitoring blocks that have a history of economic damage.

Citrus leafminer

Phyllocnistis citrella Stainton, Lepidoptera: Gracillariidae

Figure 13.1 Importance and distribution of citrus leafminer.

Plate 13.1 Adult citrus leafminer.

Life cycle
Under favourable summer and autumn conditions, the complete life cycle takes 14–17 days. In late autumn, winter and spring it can take two or three times longer.

Adults emerge from pupae during the early morning hours. Mating occurs at dusk and in the early evening.

Most adults live for less than a week, but some can live for up to 160 days.

Female moths start laying eggs about 24 hours after mating. Eggs are laid at night. A female can lay more than 50 eggs during her life, and as many as 20 per night.

The flat, slightly oval eggs are about 0.3 mm long. They are translucent, and look like tiny water droplets on the leaves.

Eggs are deposited singly, on the undersides of leaves near the midrib, usually at the base of the leaf. Young leaves 10–20 mm in length are preferred sites for egg laying.

Hatching can occur within a day in summer, and the young larva immediately burrows under the leaf surface.

There are three larval stages, a pre-pupa and a pupa. Larvae feed on sap from epidermal cells ruptured by their blade-like, finely-toothed mouthparts.

Each pale-green larva tunnels a characteristic, sinuous, silvery mine in the leaf, with a raised parchment-like skin lined centrally with dark excreta. Larvae never leave their mines to form other mines or move between lower and upper sides of leaves.

Description

General appearance
Adult citrus leafminers are small, delicate moths with narrow paired forewings, and hindwings fringed with long hairs. At rest, the moths are about 2 mm long. In flight, their wingspan is about 4 mm.

Because adults are nocturnal, they are rarely seen, except in heavy infestations. They may also be seen flying out when foliage is disturbed.

Distinguishing features
Distorted foliage and silver mining trails are the most obvious signs of citrus leafminer activity.

Citrus leafminer is the only leafminer that attacks citrus in Australia.

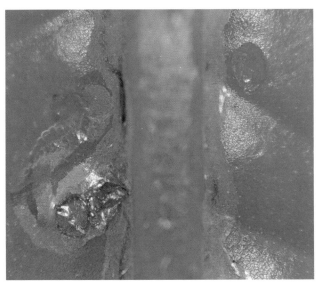

Plate 13.2 *The egg of citrus leafminer resembles a small drop of water (top right). A newly hatched transparent larva can be seen (middle left), and also the brownish empty egg shell from which the larva has hatched.*

Plate 13.3 *Second and third instar larvae of citrus leafminer.*

Plate 13.4 *Pupa of citrus leafminer.*

Development of the first three stages takes about 5–6 days in summer. Mature third instar larvae are about 3 mm long.

The fourth stage (the pre-pupa) is yellowish brown and resembles the third-stage larva, but it does not feed. This stage lasts for about one day in summer.

The pre-pupa uses silk produced from its mouthparts to form a pupal chamber, which is usually sheltered within the rolled-over edge of the leaf. Transformation into a pupa occurs within the chamber.

The yellowish brown pupae are about 2.5 mm long. After about 6 days, adults emerge. The cast pupal skins are usually left protruding from the chambers.

▶ Seasonal history
There are up to 15 generations per year.

▶ Habits
Severe infestations usually occur only in late summer and autumn, and are often related to low levels of activity of natural enemies. Severe infestations do not normally occur in spring because leafminer numbers are low after winter.

In Queensland, where citrus leafminer has been established for 30 years, significant activity begins in early October after the major spring growth flush has hardened. In southern states, significant activity begins a few weeks later.

Leafminer activity increases during December–January on early summer growth flushes, and peaks in February–March. Populations drop off rapidly from late April, due to a lack of suitable new growth, to increased parasitism and to the cooler conditions. A few active mines can usually be found where there is a growth flush, even in June–July, but the pest is very scarce during August–September.

▶ Hosts
Citrus is the only known host of citrus leafminer in Australia.

▶ Origin and distribution
Originally from South-East Asia, citrus leafminer is now found almost throughout the world, following its rapid spread since 1990.

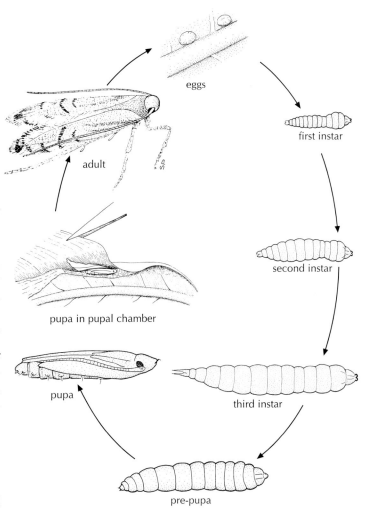

Figure 13.2 Life cycle of citrus leafminer.

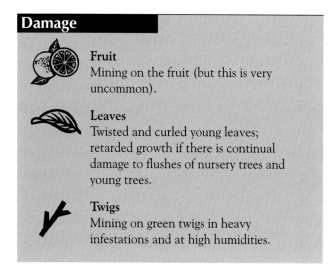

Damage

Fruit
Mining on the fruit (but this is very uncommon).

Leaves
Twisted and curled young leaves; retarded growth if there is continual damage to flushes of nursery trees and young trees.

Twigs
Mining on green twigs in heavy infestations and at high humidities.

Damage is caused by the larvae as they mine immature foliage. Twisted and curled leaves are generally the first symptoms noticed.

Severe infestations (an average of 2 or more mines per leaf) can retard the growth of nursery and newly planted trees, but their effect on trees older than 5 years is usually insignificant.

▸ **Varieties attacked**

All citrus varieties are attacked.

Plate 13.5 *Autumn leaf flush damaged by citrus leafminer.*

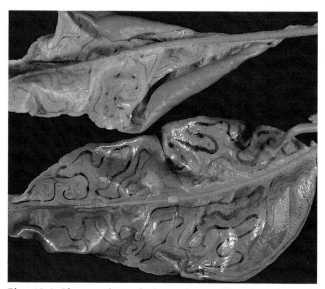

Plate 13.6 *Close-up of citrus leafminer damage. Note the curling and distortion of leaves, the mines with their dark excrement trail, the larva within a mine (lower left), and brown pupae within folds near the edges of the leaves.*

Natural enemies

▸ **Parasites**

The most effective wasp parasites of citrus leafminer are *Ageniaspis citricola* and *Cirrospilus quadristriatus* (both introduced from South-East Asia in 1990–92) and *Semielacher petiolatus*, an indigenous species. Parasitism by other wasps native to Australia (*Cirrospilus* near *ingenuus*, *Sympiesis* sp., and *Zaommomentedon brevipetiolatus*) has also been observed.

In the Murrumbidgee Irrigation Area, the Sunraysia and the Riverland, *S. petiolatus* is the major parasite. Parasitism levels are generally below 20% of all larvae in January–February, but build up to 50% or more in March–May.

In Queensland, the levels of parasitism of larvae by *A. citricola* reach 90% by February–March, but *S. petiolatus* and *C. quadristriatus* are also important.

▸ **Predators**

Predators include lacewings, which are generally associated with heavy infestations.

Plate 13.7 *The wasp* Ageniaspis citricola, *a parasite of citrus leafminer: bottom, pupae; top, adult.*

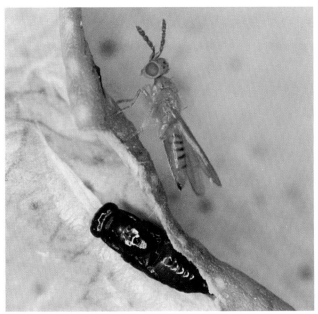

Plate 13.8 The wasp Cirrospilus quadristriatus, *a parasite of citrus leafminer: bottom, pupa; top, adult.*

Plate 13.10 The wasp Semielacher petiolatus, *a parasite of citrus leafminer: bottom, pupa; top, adult.*

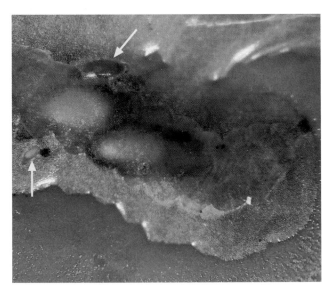

Plate 13.9 *Larva of citrus leafminer within its mine. It is being attacked by the wasps* Ageniaspis citricola *(see the two barrel-shaped pupae, centre left, inside the leafminer larva) and* Cirrospilus quadristriatus *(the larva of which (see top arrow) feeds externally on the citrus leafminer larva). There is also a parasite egg at middle left (bottom arrow).*

Management

◗ Monitoring

- Citrus leafminer occurs on all varieties.
- From January to mid-March, monitor production of significant growth flushes (more than 25% of trees flushing) on blocks of young trees less than 5 years old. It is important to detect the flush at a very early stage, when shoots are less than 20 mm long, especially if planning to use a petroleum spray oil to control leafminer. Examine at least 20 advanced shoots (i.e. shoots that have emerged ahead of the main flush) for active mines.
- Recently hedged trees may also need monitoring.
- Because spring growth on which fruit is borne is not attacked to any extent, and because summer and autumn flushing on mature trees is not usually vigorous, monitoring of mature trees is generally not warranted.

◗ Action level

Action is based on plant flushing cycles as well as infestation levels, because the initial stages of leafminer infestations are very difficult to detect, and infestations develop extremely quickly.

There are two action levels. The action level for larvicides is 25% of the block flushing, and 50% of the more advanced flushes infested with active leafminer.

The action level for petroleum spray oil is 25% of the block flushing, and 10% of the more advanced flushes infested.

◗ Appropriate action

Apply a larvicide or petroleum spray oil (250–500 mL oil per 100 L water, see chapter 27). Petroleum spray oil reduces the number of eggs laid by adult female leafminers because they do not like laying their eggs on sprayed surfaces. Petroleum spray oil is therefore preventative rather than curative.

Apply oil every 6–10 days until the youngest leaves on the majority of flushes within each cycle are 40 mm long. Approximately 2–4 sprays per flush are required.

◗ Additional management notes

Effective control of citrus leafminer using larvicides can be difficult because larvae and pupae are protected within the leaf. Some larvicides, e.g. synthetic pyrethroids, are effective, but disrupt the activity of natural enemies of leafminer and other citrus pests, and are not recommended.

Because infestations are restricted to flush growth, particularly in late summer and autumn, their severity can be reduced by:

- fertilising in late winter to promote flush growth in spring when the pest is either absent or relatively scarce
- limiting flush growth in late summer and autumn by not fertilising and irrigating during summer and autumn in excess of the amount needed for normal growth
- not hedging trees in summer and autumn.

Lightbrown apple moth

Epiphyas postvittana (Walker), Lepidoptera: Tortricidae

Figure 13.3 Importance and distribution of lightbrown apple moth.

Plate 13.11 Larva of lightbrown apple moth feeding beneath the calyx. Larvae often move onto fruit from leaves that are touching the fruit. Note the webbing in the young leaf at right.

Description

▶ General appearance
The adult lightbrown apple moth is a small, pale-brown moth, with a wingspan of about 18 mm. When viewed from above, the adult is bell-shaped.

▶ Distinguishing features
The caterpillars form silken feeding shelters under the calyx, in flowers, and between young leaves. They wriggle vigorously backwards when disturbed and may drop off the plant to hang from a silken thread.

▶ Life cycle
The eggs are pale green and laid in flat overlapping masses that look like fish scales. Eggs are laid on leaves, stems and fruit.

After hatching, the caterpillars pass through about six developmental stages, increasing in size from one to 18 mm in length. Young caterpillars are pale yellow-green, while mature caterpillars are pale green with a darker green central stripe and a brown head.

The pupa is 10–12 mm long and enclosed in silken webbing. Pupation takes place in a rolled-up leaf or in flower debris.

The life cycle is completed in 2–3 months.

▶ Seasonal history
There are normally 4 generations per year in inland irrigated districts. In cooler districts, there are normally 3 generations per year.

The two early generations (September–October and December–January) cause most damage to young and mature fruit. Eggs from the last generation hatch to become the overwintering caterpillar population.

Plate 13.12 Adult female lightbrown apple moth.

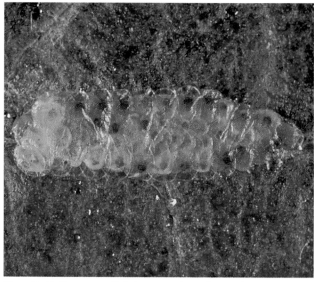

Plate 13.13 Egg mass of lightbrown apple moth. Note how the eggs overlap, resembling fish scales.

Adults developing from the over-wintering caterpillar population emerge in mid-spring. They lay eggs that hatch into the spring generation of caterpillars.

▶ Habits
Moths are weak fliers and do not disperse far. However, in early summer they can move into orchards from native plants or weeds. Juvenile stages can be transported in nursery trees.

When disturbed, young larvae wriggle backwards or drop on silken threads.

In inland districts, hot dry summers reduce populations of lightbrown apple moth.

▶ Hosts
Lightbrown apple moth is a commonly occurring pest in orchards and grapevines. It also infests a wide range of native plants, ornamentals, vegetables, pasture crops and weeds.

▶ Origin and distribution
This insect is native to Australia, and has also become established in New Zealand, the United Kingdom, New Caledonia, and Hawaii.

Damage

 Flowers
Flowers webbed into clumps; some may fall off or be eaten.

 Fruit
Circular scar around the calyx (halo damage); fruit drop if badly damaged; shallow scarring of older fruit.

 Leaves
Young leaves webbed or rolled together.

Halo damage is caused by caterpillars feeding under the calyx of very young fruit. Badly damaged fruit can fall. Caterpillars also feed on older fruit at points of contact with leaves and other fruit, causing shallow scarring. During autumn and spring, caterpillars may bore into ripe fruit causing fruit to drop.

Some export markets do not tolerate the presence of any lightbrown apple moth in fruit. Detection of the pest in packing sheds may result in the rejection of fruit for export.

▶ Varieties attacked
All citrus varieties are attacked by lightbrown apple moth, but particularly oranges. Levels of mature fruit drop are highest in Valencia and navel oranges, and lowest in lemons and mandarins.

Natural enemies

▶ Parasites
Tiny *Trichogramma* wasps parasitise the eggs of lightbrown apple moth.

Parasites of the caterpillars include the wasps *Dolichogenidea arisanus* and *Xanthopimpla* spp., and the parasitic flies *Voriella* spp., *Goniozus* spp. and *Zosteromyia* spp.

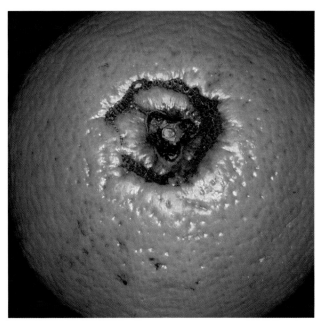

Plate 13.14 Characteristic circular scarring (halo damage) around the fruit calyx, caused by larvae of lightbrown apple moth. The scar is the result of larval feeding on the fruit when it was at a much younger stage. Sooty mould has darkened the scar.

▶ Predators
The most important predators of lightbrown apple moth caterpillars are the predatory bug *Oechalia schellembergii*, lacewing larvae (*Micromus* spp.) and spiders.

▶ Pathogens
Lightbrown apple moth caterpillars can be affected by a nuclear polyhedrosis virus (NPV), and are susceptible to the bacterial pathogen *Bacillus thuringiensis*.

Management

▶ Monitoring
- Lightbrown apple moth attacks all varieties.
- Monitor fortnightly, checking flowers, fruitlets and maturing fruit.
- The most critical times for monitoring are mid-September to mid-February for young fruitlets, and June to November for mature fruit.
- Pheromone traps and wine lures, both of which attract moths, can assist with monitoring.
- It can also be helpful to collect temperature data, which can be used to predict development times and seasonal activity.
- Fruit for export markets require careful monitoring during March–June.

▶ Action level
In navel oranges, the action level is 5% or more flowers, fruitlets or fruit infested in spring. The action level is 10% for other varieties.

▶ Appropriate action
Sprays of *Bacillus thuringiensis* should be timed to coincide with first-stage caterpillars. Larger caterpillars in sheltered feeding sites are difficult to control.

Orange fruitborer

Isotenes miserana (Walker), Lepidoptera: Tortricidae

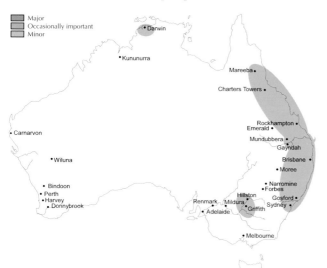

Figure 13.4 Importance and distribution of orange fruitborer.

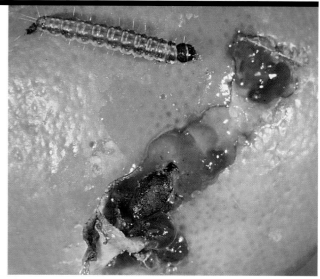

Plate 13.15 Orange fruitborer larva and the damage it has caused by feeding on fruit. Larvae often feed in the sheltered area between fruit that are touching.

Description

▶ **General appearance**

The adult orange fruitborer is a pale-grey moth, speckled with small brown marks, and with a wingspan of about 15 mm.

Very young larvae are green, but older larvae are cream-coloured with three red-brown longitudinal bands and a dark brown head capsule.

▶ **Distinguishing features**

Damage by this species could be confused with damage caused by blastobasid fruitborers and other fruitborers. However, the orange fruitborer larva has three distinct red-brown stripes, which larvae of the other fruitborers do not have.

▶ **Life cycle**

The female lays pale-green eggs in overlapping clusters that look like fish scales. Small green caterpillars hatch out of the eggs. When disturbed, the caterpillars wriggle backwards vigorously and drop off the fruit. Mature larvae are about 20 mm long.

Pupation takes place either in a dead, rolled-up leaf or in a mass of flower debris or webbed foliage.

During the summer, the life cycle takes about one month. Young larvae hatching out in late autumn feed through the winter and pupate in early spring.

▶ **Seasonal history**

There are 5–6 generations per year.

▶ **Habits**

Orange fruitborer larvae bore into the fruit, often at the calyx end or between touching fruit.

When disturbed, larvae wriggle backwards vigorously and drop off the fruit.

▶ **Hosts**

Orange fruitborer attacks a wide range of tropical and subtropical fruit and ornamentals.

Plate 13.16 Orange fruitborer moth.

▶ **Origin and distribution**

Orange fruitborer is native to Australia.

Damage

 Fruit
Holes bored through the rind into the pulp; cavities eaten out between touching fruit, or where a leaf touches fruit; fruit drop.

Leaves
Slight damage (webbing).

On maturing fruit, larvae bore round holes about 3 mm in diameter into the rind (often where fruit touch, and into the navel on navel oranges). Examination usually reveals a shallow excavation under the hole. On young green fruit, the larvae make shallow pits in the rind, and gum may exude from the wounds. Premature colouring of fruit often indicates attack. Attack on fruit in clusters is common.

Varieties attacked
All citrus varieties are susceptible to attack, but particularly navel and Valencia oranges, and Meyer and Lisbon lemons.

Natural enemies
The ichneumonid wasp *Phytodietus* sp. parasitises orange fruitborer.

Management

Monitoring
- Orange fruitborer attacks all varieties, but is more common on navel oranges, and Meyer and Lisbon lemons. Navel oranges are worst affected from February to June.
- This is a difficult pest to monitor. Check trees fortnightly from February to harvest. An important sign of orange fruitborer activity is the presence of fallen fruit. If there is an average of 3 or more freshly fallen fruit per tree, slice 20–50 fruit open to confirm whether there is orange fruitborer damage. The larva is often found in these fruit.
- If the pest scout wishes, also sample 5 randomly selected fruit on each tree.

Action level
The action level is 10% or more of fallen fruit infested, or 5% or more of fruit infested on the tree.

Appropriate action
Because fruit drop is often the first indication of damage, decisions to apply sprays need to be based on careful monitoring in addition to the frequency of damage in previous years. If action is needed, a selective pesticide should be applied in autumn to control orange fruitborer.

Additional management notes
Orange fruitborer is usually not a major problem, but may cause frequent damage in particular orange orchards. Some damage can be done to young and half-grown fruit, particularly Meyer and Lisbon lemons in spring to summer.

Orchard hygiene is important in controlling this pest. Fallen infested fruit should be destroyed, if practicable.

Citrus rindborer

Adoxophyes sp., Lepidoptera: Tortricidae

Figure 13.5 Importance and distribution of citrus rindborer.

Plate 13.17 Citrus rindborer caterpillar and damage caused to fruit. Note the webbing on the leaf, and the round hole in the fruit.

Plate 13.18 Adult female citrus rindborer.

Description

General appearance
The adult citrus rindborer is a light-brown moth with darker brown markings. The wingspan is about 10 mm. The larva is pale yellow with a light yellow-brown head.

Distinguishing features
The uniform pale yellow colour of the larva distinguishes it from the orange fruitborer larva, which has three red-brown longitudinal stripes, and the lightbrown apple moth larva, which is yellow-green.

Life cycle
The female moth lays pale-green eggs on young shoots or fruit. Larvae pass through about five larval stages before pupating in webbing in a folded leaf. The complete life cycle takes about 4 weeks in summer.

Seasonal history
There are about 7 generations per year.

Habits
The larvae feed on young foliage and bore into fruit at a point where leaves touch fruit.

Hosts
Citrus is the only known host.

Origin and distribution
This pest appears to be native to Australia.

Damage

Fruit
Holes in the rind of young and mature fruit; fruit drop.

Leaves
Large holes eaten out of young leaves.

Larvae bore round holes about 3 mm in diameter in the rind of maturing fruit. They also make unsightly shallow excavations in the rind. Gum exudes from damaged areas of young fruit.

Varieties attacked
All citrus varieties are susceptible to citrus rindborer, but particularly Imperial mandarin.

Natural enemies
No parasites or predators of this pest have yet been identified. The pathogenic bacterium *Bacillus thuringiensis* shows promise as a control agent.

Management

Monitoring
- Citrus rindborer attacks all varieties, but is more common on Imperial mandarin.
- Monitor fortnightly from October to April. A sign of citrus rindborer activity is the presence of fallen fruit. If there is an average of 3 or more freshly fallen fruit per tree, slice 20–50 fruit open to confirm whether there is citrus rindborer damage. The larva is often found in these fruit.
- If the pest scout wishes, also sample 5 randomly selected fruit on each tree.

Action level
The action level is 10% or more of fallen fruit infested, or 5% or more of fruit infested on the tree.

Appropriate action
Apply a selective pesticide.

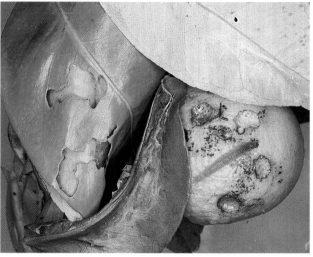
Plate 13.19 Leaves and fruit damaged by citrus rindborer. Note the holes eaten in the leaves, and the webbing which ties the leaves together. There is a brown pupa in the webbing and a larva on the fruit (right).

Other fruitborers

Sorghum head caterpillar *Cryptoblabes adoceta* Turner, yellow peach moth *Conogethes punctiferalis* (Guenée), Lepidoptera: Pyralidae, and *Blastobasis* spp., Lepidoptera: Blastobasidae

Figure 13.6 Importance and distribution of sorghum head caterpillar.

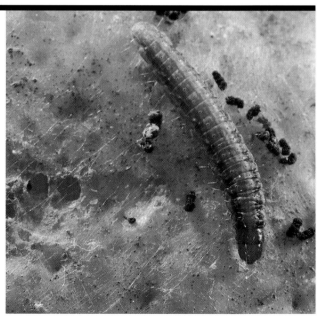
Plate 13.20 Larva of sorghum head caterpillar feeding between navel oranges. Note the black faecal pellets beside the larva, the fine webbing over the fruit surface, and the damage to the fruit.

13 Moths and butterflies (Lepidoptera)

Figure 13.7 Importance and distribution of yellow peach moth.

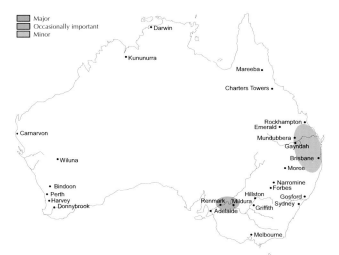

Figure 13.8 Importance and distribution of blastobasid fruitborers.

Plate 13.21 Larva of yellow peach moth and damage caused to navel orange. Note the feeding tube (opened to reveal the larva) made of frass webbed together.

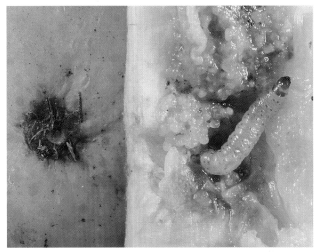

Plate 13.22 Larva of Blastobasis *sp.* and damage caused to navel orange. The entry hole can be seen on the rind on the left. The rind has been cut to reveal the feeding larva and its sago-like frass (right).

Plate 13.23 Moth of sorghum head caterpillar.

Plate 13.24 Yellow peach moth.

Description

▶ General appearance

The adult of sorghum head caterpillar is a nondescript brown-grey moth. The larva is about 7 mm long, tapering towards the dark-brown head, and has two green longitudinal stripes.

The yellow peach moth adult is an attractive yellow moth with black spots and a wingspan of 25 mm. The larva is pinkish, with grey-brown rectangular spots extending along the back.

Blastobasis spp. are grey moths, with a wingspan of 10 mm. The larvae are fat pink-white grubs, with dark heads.

▶ Distinguishing features

The appearance of the larvae of these species is distinctive, and they are unlikely to be confused with each other or with the larvae of other species (see under 'General appearance' above).

▶ Life cycle

The life cycle is outlined for yellow peach moth only, as the life cycles of all these species are similar.

The eggs of yellow peach moth are about 1 mm long, oval, like fish scales in appearance, and are deposited singly on the fruit surface. Newly hatched larvae bore into the fruit. After three weeks of feeding and growth (mostly inside fruit, between touching fruit, or under frass), the last-stage larvae emerge onto the surface of the fruit and pupate in the shelter of webbed frass. Adults later emerge from the pupae. The complete life cycle takes 3–5 weeks during summer.

▶ Seasonal history
There are at least 5–6 generations per year.

▶ Habits
Yellow peach moth is uncommon in citrus. Larvae bore deeply into maturing oranges in autumn.

Sorghum head caterpillar feeds between touching fruit in the autumn. Often the fruit are infested with citrus mealybug, and the moth seems attracted to lay eggs on fruit with honeydew and sooty mould on them.

Blastobasids lay their eggs on maturing fruit in autumn, and young larvae bore through the rind into the pulp (mainly of navel oranges). The point of entry may be the navel end or through the side.

▶ Hosts
Yellow peach moth infests a wide range of fruits and nuts such as custard apple, macadamia, stonefruit, papaw and mango, as well as field crops such as cotton and maize.

Sorghum head caterpillar also occurs on a range of hosts, including sorghum, maize, hibiscus, mango, grape, persimmon and castor-oil plant.

Other hosts of blastobasids have not yet been recorded.

▶ Origin and distribution
Yellow peach moth also occurs in South-East Asia, Japan and Indonesia. Sorghum head caterpillar and blastobasid fruitborers are native to Australia. Other blastobasid species occur in the USA.

Damage

Fruit
Fruit drop (all species); shallow but disfiguring gouging of the rind between maturing fruit (sorghum head caterpillar); holes in the rind (yellow peach moth); larvae in the fruit pulp (blastobasid fruitborers and yellow peach moth).

Because the larvae of blastobasid fruitborers are very small when they enter the fruit, the entry holes are often not obvious, being marked only by a prematurely coloured spot. Cutting open the fruit will reveal the pink-white larvae feeding in the pulp. Damaged fruit drop, and in severe infestations, up to 100 fruit per tree can be lost.

Damage caused by blastobasid fruitborer should not be confused with damage caused by orange fruitborer. Blastobasid fruitborer larvae bore deeply into the fruit, while orange fruitborers make only shallow excavations.

▶ Varieties attacked
These fruitborers attack all citrus varieties. However, yellow peach moth prefers navel and Joppa oranges, blastobasid fruitborers prefer navel oranges, and sorghum head caterpillar prefers mainly Siletta and Joppa oranges.

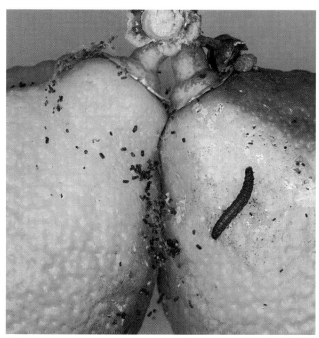
Plate 13.25 Sorghum head caterpillar feeding between Meyer lemons.

Natural enemies

Natural enemies, particularly larval parasites, are important in regulating populations of fruitborers.

▶ Parasites
The main parasite of yellow peach moth is the tachinid fly *Argyrophylax proclinata*. The adult is about the same size as a housefly. It attaches its eggs to the host larvae.

The parasite larva develops inside its host, killing the host larva at or near pupation. The fly pupa is brown, cylindrical and about 8 mm long. Up to 40% of all larvae may be parasitised.

Another two species of parasitic tachinid flies have been recorded from yellow peach moth, but not identified.

Parasites of sorghum head caterpillar and blastobasid fruitborers are as yet unknown.

▶ Predators
The assassin bug (*Pristhesancus plagipennis*) is a predator of all these species of fruitborers.

Plate 13.26 Tachinid fly *Argyrophylax proclinata*, a parasite of yellow peach moth. At left, the silken cocoon within which the yellow peach moth larva pupated has been cut open to reveal the barrel-shaped fly pupa (top), which was formed after the tachinid fly larva emerged from the yellow peach moth pupa (bottom).

Management

> **Monitoring**
- These fruitborers occur on all varieties, especially navel, Joppa and Siletta oranges.
- These are difficult pests to monitor. An important sign of fruitborer activity is the presence of fallen fruit. If there is an average of 3 or more freshly fallen fruit per tree, slice open 20–50 fruit to confirm whether there is fruitborer damage. Larvae are often found in these fruit.
- Check trees fortnightly from February to harvest. If the pest scout wishes, 5 randomly selected fruit on each tree sampled can also be checked. If the presence of sorghum head caterpillar is suspected, it is important to sample between touching fruit.

> **Action level**

The action level is 10% or more of fallen fruit infested, or 5% or more of fruit infested on the tree.

> **Appropriate action**

Apply an appropriate selective insecticide.

> **Additional management notes**

Orchard hygiene is important in controlling these pests. Fallen infested fruit should be destroyed, if practicable.

Fruitpiercing moths

Othreis fullonia (Clerck), *Othreis materna* (L.), and *Eudocima salaminia* (Cramer), Lepidoptera: Noctuidae

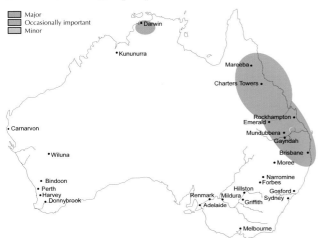

Figure 13.9 Importance and distribution of fruitpiercing moths.

Plate 13.28 Female fruitpiercing moth (*Othreis materna*) feeding on an orange. Males of this species have brown forewings. Both sexes have a prominent black spot in the centre of each hindwing.

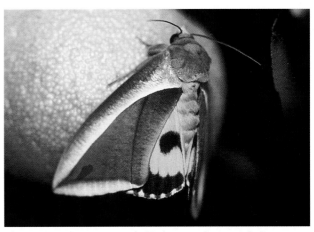

Plate 13.29 Fruitpiercing moth (*Eudocima salaminia*) feeding on an orange. Note the characteristic forewing pattern of this species, which mimics a curled leaf.

Plate 13.27 Fruitpiercing moth (*Othreis fullonia*) feeding on an orange. Note the large size of the moth, the brown forewings, and orange hindwings with dark patches.

Description

> **General appearance**

Adult fruitpiercing moths are large and stout-bodied, with a wingspan of 100 mm. Hindwings are yellow-orange, with black patches or spots. Larvae are velvety-black.

> **Distinguishing features**

All three species are large colourful moths, with mainly brown, cream or green forewings, and yellow and black hindwings. The larvae of *Othreis* have two large spots (mainly white with black centres) on either side of body segments 6 and 7.

> **Life cycle**

The larvae feed on native vines for about 3 weeks, progressing through five stages, or instars, before forming a dark-brown pupa within a delicate silk cocoon between webbed leaves. After 3 weeks, adults emerge from the pupa.

Plate 13.30 Larva of fruitpiercing moth feeding on the leaves of a forest vine (family Menispermaceae).

▶ Seasonal history

There are 3–4 generations per year north of Rockhampton and in the Northern Territory, and probably 2 generations in summer–autumn in southern Queensland and northern New South Wales.

▶ Habits

The larvae feed on native vines of the plant family Menispermaceae, which are usually found growing in moist forested areas.

The adult moths fly to orchards during the night to suck juice from ripening fruit. Long, sucking mouthparts (the proboscis) are used to pierce the fruit rind. Moths rest outside the orchard by day.

Migration of moths from areas north of Rockhampton into south Queensland may occur in late summer and early autumn. Moths are capable of over-wintering in south Queensland when conditions are mild.

▶ Hosts

The moths also feed on carambola, banana, fig, guava, kiwifruit, lychee, mango, stonefruit, persimmon and papaw.

Larval hosts include native vines of the family Menispermaceae (of which there are 22 species in north Queensland). The preferred species are *Tinospora smilacina* and *Stephania japonica* var. *timoriensis*.

▶ Origin and distribution

These three species of fruitpiercing moths are widely distributed through the western Pacific, South-East Asia and parts of Africa.

Damage

Fruit
Fruit pulp sucked out; rot develops; fruit drop.

Damage to fruit might be confused with that caused by Queensland fruit fly. The fruit fly oviposition hole is much smaller, and fly larvae can usually be seen in the fruit. A fruitpiercing moth's proboscis leaves a hole 2 mm in diameter.

The feeding hole may also occasionally be mistaken for damage caused by an orange fruitborer larva. However, decay and premature colouring sets in quickly around the hole and fruit drop follows.

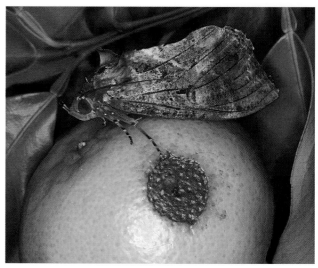

Plate 13.31 Fruitpiercing moth (Othreis fullonia) resting on a damaged Imperial mandarin. (The forewings of this moth have lost their decorative scales.)

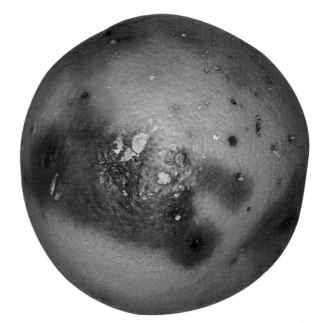

Plate 13.32 Navel orange damaged by fruitpiercing moths. Note the numerous feeding punctures with associated white sugarine crystals, and discolouration caused by secondary rots.

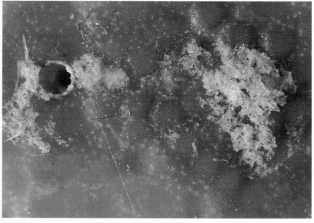

Plate 13.33 Closeup of feeding puncture made by fruitpiercing moth. The white material is sugarine crystals resulting from the evaporation of juice which has escaped through the wound.

Plate 13.34 *Fruit drop caused by fruitpiercing moths feeding on Imperial mandarins.*

▶ **Varieties attacked**

All citrus varieties are attacked, but early-maturing varieties (i.e. maturing in February–April) are most susceptible, e.g. navel oranges, early mandarins and grapefruit.

Natural enemies

▶ **Parasites**

Egg parasites include the wasps *Trichogramma* spp., *Telenomus* spp. and *Ooencyrtus* spp. Larval parasites include the tachinid fly *Exoristae sorbillans*.

▶ **Predators**

A lygaeid bug preys on eggs.

Management

▶ **Monitoring**

- Early-maturing varieties are attacked, especially navel oranges, mandarins (particularly Imperial), and grapefruit.
- Various bait traps, e.g. ripening bananas in an inverted sack, are useful in trapping moths.
- Monitor fortnightly in February and weekly in March–April (heaviest attack occurs during 3–4 weeks in March–April). Concentrate on the edge rows, in the direction of nightly moth invasion (known from previous observation).
- Sample 5 randomly selected maturing fruit per tree. Look for fruit prematurely ripening on one side, pierced with 2 mm diameter holes with a sucked-out area beneath.

▶ **Action level**

Action is required when 5% or more of fruit are affected.

▶ **Appropriate action**

Chemical control is difficult. Best results have been obtained by spraying edge rows in the direction of moth invasion, within 1 hour before dusk. Nearby vegetation, in which moths rest during the day, can be also be sprayed in late afternoon. Repeat 1 week later. Use a non-residual, quick-knockdown chemical.

▶ **Additional management notes**

It is advisable to limit the planting of early-maturing varieties, such as navel oranges, grapefruit, and early mandarins, in tropical Queensland, especially in orchards near rainforest.

Orchard hygiene is important in controlling fruitpiercing moths. Rotting and fallen fruit should be removed.

Where these moths are troublesome, harvest fruit early.

Lemon bud moth and citrus flower moth

Lemon bud moth, *Prays parilis* Turner and citrus flower moth, *Prays nephelomima* Meyrick, Lepidoptera: Yponomeutidae

Figure 13.10 Importance and distribution of lemon bud moth and citrus flower moth.

Plate 13.35 *Lemon bud moth: mature larva (right) near feeding hole on lemon bud; pupa within silken mesh cocoon (left). Buds are webbed together with silk. Note the larval frass amongst the buds.*

Description

General appearance
The young larvae of lemon bud moth are yellow, and later stages (which grow up to 10 mm in length) are green and reddish-brown. Citrus flower moth larvae are lemon-yellow throughout their development.

Adult moths of both species have a wingspan of 12 mm. Wings are grey with brown markings.

Distinguishing features
The small larvae are found within buds and flowers. Extensive webbing is associated with their presence.

Plate 13.36 Larva of citrus flower moth on lemon flower. There is also a pupa within a silken cocoon to the left of the larva.

Plate 13.37 Lemon bud moth.

Plate 13.38 Citrus flower moth and pupa in silken cocoon (top left).

Life cycle
The life cycles of the two species are generally similar, but more detail is known about the life cycle of lemon bud moth. Female lemon bud moths lay up to 20 white hemispherical eggs on the fruit buds, although only one larva survives in each bud. A fortnight after hatching, the larva forms a lace-like cocoon, often at the edge of a curled leaf or on the flowers. The complete life cycle takes about 3 weeks in summer.

Seasonal history
There are 6–7 generations each year.

Habits
Larvae hatch from the eggs laid on the flower buds, bore into the blossom and eat the developing fruit. Groups of flowers are often webbed by several larvae feeding simultaneously.

The pest occurs in summer–autumn flowers, and can destroy 50% or more of them. However, it is uncommon in the main spring blossoming.

Hosts
Citrus is the only recorded host.

Origin and distribution
Lemon bud moth also occurs in South-East Asia and the Philippines. Little is known about the origin and distribution of citrus flower moth.

Damage

Flowers
Damaged by larval feeding, and covered with webbing.

Fruit
Very young fruit eaten by larvae.

Varieties attacked
All citrus varieties are attacked, but most commonly lemons, usually during 'out of season' flowerings.

Natural enemies

The natural enemies of these moths have not yet been identified. However, lemon bud moth is known to be parasitised by at least one small wasp.

Plate 13.39 Unidentified wasp parasites recently emerged from a lemon bud moth pupa. Note the emergence holes in the pupa.

Management

▶ Monitoring
- Lemon bud moth and citrus flower moth attack all varieties of citrus trees, but are more commonly found on lemons.
- Monitor fortnightly during flowering, especially during summer–autumn, assessing 5 randomly selected flowering shoots per tree. A sign of activity is the presence of webbing on buds and flowers.

▶ Action level
The action level is 50% or more of flower terminals, or flowering shoots, infested. Reduce the action level if flowering is poor, and raise it if flowering is abundant.

▶ Appropriate action
A selective pesticide should be applied once or twice.

Plate 13.40 Unidentified wasp, a parasite of lemon bud moth, resting on flower stamens.

▶ Additional management notes
Lemon bud moth and citrus flower moth are not normally a problem in the main crop. They can, however, be a serious threat to fruit set between late summer and early winter. This second-crop fruit is difficult to keep free of pests.

Citrus leafroller

Psorosticha zizyphi (Stainton), Lepidoptera: Oecophoridae

Figure 13.11 Importance and distribution of citrus leafroller.

Plate 13.41 Young citrus leafroller larva beneath silken webbing. Later stage larvae roll young leaves lengthwise.

Description

▶ General appearance
Citrus leafroller larvae have a dark head and yellow-orange body. The adult is a small, greyish-brown moth with black spots on the wings.

▶ Distinguishing features
Early-stage larvae spin a thin, silken gallery on the leaf surface, and feed within this. Later stages roll young leaves lengthwise to form tubes.

▶ Life cycle
Adult moths shelter in the tree in daytime, and are active at night. A total of about 300 eggs are laid during 6 weeks.

The pale-green eggs are deposited singly on either side of young leaves, usually near the midrib. They hatch in 3–4 days. Larvae develop through five stages over about 2 weeks.

Plate 13.42 Citrus leafroller moth.

Fully grown larvae spin a silken cocoon inside the leaf roll, before changing into a pupa. Adults emerge after another 2 weeks.

The complete life cycle takes about 24 days during summer.

Plate 13.43 Damage to young leaves caused by citrus leafroller. (There is also citrus leafminer damage on the lower leaves.)

▸ Seasonal history
There are about 10 generations per year.

▸ Habits
This insect is characterised by its leafrolling habits. It attacks the young growth, mainly in November to March.

▸ Hosts
In South-East Asia, citrus leafroller also attacks curry-leaf (*Murraya koenigii*) and jujube (*Zizyphus jujuba*).

▸ Origin and distribution
This pest is probably South-East Asian in origin.

Damage

 Leaves
Young leaves rolled, and with a ragged appearance.

▸ Varieties attacked
All citrus varieties are attacked, but particularly lemons.

Natural enemies

▸ Parasites
An ichneumonid wasp is known to parasitise citrus leafroller.

▸ Predators
The assassin bug (*Pristhesancus plagipennis*) is a known predator.

Management

▸ Monitoring
- Citrus leafroller occurs on all varieties, especially lemons.
- Monitor fortnightly, checking 5 randomly selected shoots per tree from November to March.

▸ Action level
Take action if 50% or more of the shoots are infested.

▸ Appropriate action
Apply a selective pesticide.

Corn earworm and native budworm

Corn earworm, *Helicoverpa armigera* (Hübner) and native budworm, *Helicoverpa punctigera* (Wallengren), Lepidoptera: Noctuidae

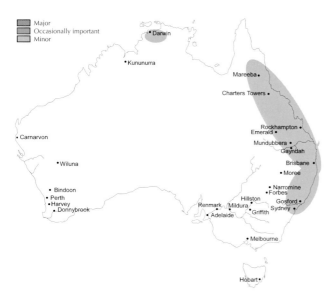

Figure 13.12 Importance and distribution of corn earworm and native budworm.

Plate 13.44 Corn earworm larva feeding on young fruit.

Plate 13.45 Mature corn earworm larva on young shoot.

Description

▸ General appearance
Both corn earworm and native budworm are also commonly called heliothis moths.

The young larvae are grey-white to brown, but later stages are yellow, tan or reddish-brown, with black or brown, and thin white, longitudinal stripes along the back. They have noticeable body hairs. Mature larvae are 40–50 mm long.

Adult moths are stout-bodied, with a wingspan of about 40 mm. They are tan, reddish-brown or grey, with darker markings on the forewings, and a dark area at the outer margin of the pale-coloured hindwings.

▸ Distinguishing features
The appearance of the larvae is distinctive, making them unlikely to be confused with other caterpillars on citrus.

▸ Life cycle
Females deposit their eggs singly on buds and flowers. The eggs are cream or white when newly laid, dome-shaped and finely striped. About 1000 eggs are laid during 2 weeks.

Larvae develop through five to seven (usually six) instars, or stages, before pupation. Pupation takes place in the soil at

Plate 13.46 Corn earworm moth.

a depth of about 100 mm. The adults may emerge a fortnight later, or many months later, depending on the temperature and wetness of the soil.

The complete life cycle in summer takes as little as 30 days.

▸ Seasonal history
In the warmer parts of Queensland, there may be 10 or more generations per year.

In cooler regions, there are fewer generations. The pupae enter diapause (a resting state) in autumn, and adult moths emerge in spring.

▸ Habits
Adults are active at night, and begin to fly at dusk.

Both species occasionally infest flowers and very young fruit from mid-September to mid-October. Larvae feed on

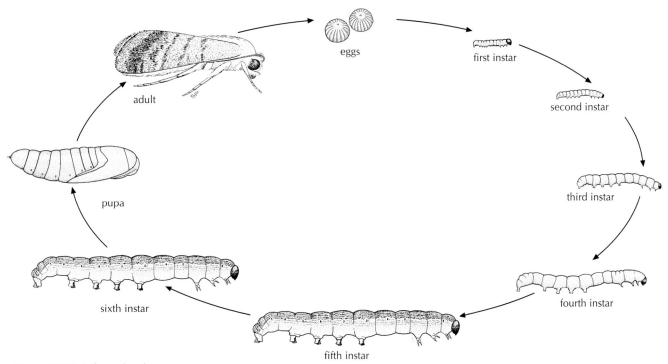

Figure 13.13 Life cycle of corn earworm.

the blossom and the newly set fruit. Oranges are worst affected. Sometimes larger fruit are also damaged.

Corn earworm occurs mostly in southern coastal areas, while native budworm is more common in subcoastal citrus.

▶ Hosts
Both species infest a very wide range of hosts, including commercial crops and ornamentals in the home garden.

▶ Origin and distribution
Corn earworm occurs in many countries. Native budworm is native to Australia.

Damage

Flowers
Damaged by larval feeding, and fall.

Fruit
Very young fruit are consumed; large, irregular holes eaten in young fruit up to 40 mm diameter; damaged fruit fall.

Leaves
Young leaves sometimes eaten.

▶ Varieties attacked
All citrus varieties are attacked.

Natural enemies

▶ Parasites
The very small wasps *Trichogramma* spp., *Trichogrammatoidea* spp. and *Telenomus* spp. are important parasites of corn earworm and native budworm eggs. The wasp *Microplitis demolitor* is an important internal parasite of larvae.

There are also tachinid fly parasites and ichneumonid wasp parasites, which emerge during the pupal stage of the host.

▶ Predators
Predators include predatory bugs, beetles, lacewings and spiders. Important predatory bugs belong to the genera *Geocoris*, *Oechalia*, *Deraeocoris*, *Nabis*, *Orius* and *Pristhesancus*.

▶ Pathogens
Larvae are moderately susceptible to *Bacillus thuringiensis*. They are attacked by nuclear polyhedrosis and granulosis viruses, and the fungi *Entomophthora* spp. and *Nomuraea rileyi*.

Management

▶ Monitoring
- Corn earworm and native budworm attack all varieties.
- Monitor fortnightly from mid-September to mid-October, checking 5 randomly selected flowering shoots per tree.
- Corn earworm and native budworm are usually minor pests.

▶ Action level
The action level is 25% or more of flowering shoots infested.

▶ Appropriate action
A single application of a selective pesticide (preferably low-volume sprays of less than 2000 L/hectare for 3–4 m high trees) is recommended.

Petroleum oil sprays (500 mL oil per 100 L water) (see chapter 27) reduce egg laying by both moths. Egg laying by native budworm, however, is reduced more by these sprays than the egg laying of corn earworm.

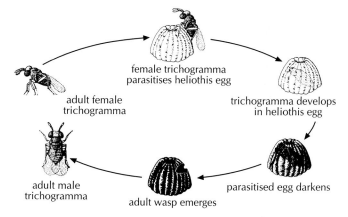
Figure 13.14 Life cycle of trichogramma wasp.

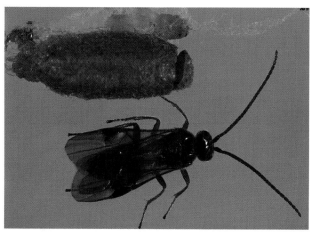
Plate 13.47 The wasp parasite *Microplitis demolitor* and its cocoon. The larva of this wasp is an internal parasite of the corn earworm larva and the native budworm larva.

Plate 13.48 Predatory pentatomid bug *Oechalia schellembergii* preying on a corn earworm caterpillar. Note the bug's piercing–sucking mouthparts, on which the caterpillar is impaled.

Banana fruit caterpillar

Tiracola plagiata (Walker), Lepidoptera: Noctuidae

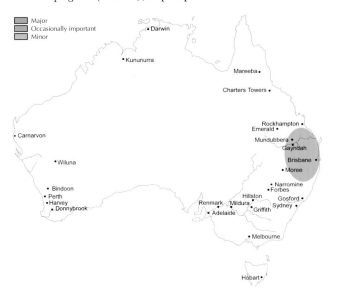

Figure 13.15 Importance and distribution of banana fruit caterpillar.

Plate 13.49 Banana fruit caterpillars feeding on young citrus fruit.

Description

▶ General appearance
Mature larvae, or caterpillars, are about 60 mm long and khaki-coloured, with two pairs of black markings on the upper surface of the body near the rear, and a narrow, broken, yellow-white band on either side of the body.

The adult moth of banana fruit caterpillar has a wingspan of about 50 mm. The wings are grey or chocolate brown in the female, and fawn or reddish-brown in the male. Each wing has a dark V-shaped marking on the margin.

▶ Distinguishing features
The size and colour of late instar larvae (see above) distinguishes them from corn earworm and native budworm larvae.

▶ Life cycle
There are six larval stages which develop over a period of 6–7 weeks. Pupation takes place in the soil or in leaf litter under the tree, and development of the adult moth within the pupa takes approximately 4 weeks. The complete life cycle takes about 3 months.

▶ Seasonal history
There are 3–4 generations per year.

▶ Habits
Eggs are laid on citrus flowers, young fruit and young foliage in the spring. The larvae feed on young fruit (up to 30 mm in diameter).

▶ Hosts
Banana fruit caterpillar infests a wide range of hosts including pumpkin, passionfruit, papaw, banana and maize.

▶ Origin and occurrence
This insect is probably native to South-East Asia.

Damage

Flowers
Some flowers eaten.

Fruit
Large holes gouged out of young fruit (up to 30 mm diameter); many damaged fruit drop.

Leaves
Holes eaten in some leaves.

▶ Varieties attacked
All citrus varieties are susceptible, but especially oranges.

Natural enemies

▶ Parasites
The tachinid flies *Palexoristus solemis* and *Sturmia* spp. and the ichneumonid wasps *Lissopimpla semipunctata* and *Paniscus testaceous* are important parasites of the banana fruit caterpillar (larval stages). The eulophid wasp *Euplectrus kurandaensis* also parasitises the larvae.

▶ Predators
Predators include assassin bugs, predatory pentatomid bugs and praying mantises.

Management

▶ Monitoring
- Banana fruit caterpillar attacks all citrus varieties, but particularly oranges.
- Monitor fortnightly from mid-September to mid-October, checking 5 randomly selected flowering shoots per tree.
- This pest is rare. Monitoring for corn earworm and native budworm will also detect banana fruit caterpillar.

▶ Action level
Action is required if 25% or more of shoots are infested.

▶ Appropriate action
Apply a low-volume spray of a selective pesticide, e.g. less than 2000 L/ha for trees 3–4 metres high.

Citrus butterflies

Small citrus butterfly, *Eleppone anactus* (W. S. Macleay), and large citrus butterfly, *Princeps aegeus* (Donovan), Lepidoptera: Papilionidae

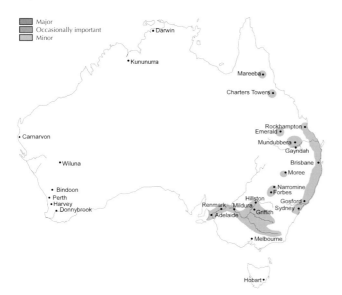

Figure 13.16 Importance and distribution of citrus butterflies.

Plate 13.50 Three different-sized larvae of the large citrus butterfly (left), showing the colour variation that occurs during growth, and three different-sized larvae of the small citrus butterfly (right).

Description

▶ General appearance

The fully grown larva of the large citrus butterfly is 70 mm long, and brown to olive green in colour, with a reddish band near the front. The fully grown larva of the small citrus butterfly larva is 50 mm long, and brown to black with three rows of orange-yellow spots along its body.

Larvae of both species have rows of small fleshy spines on the body. They also protrude a red, fleshy, forked process from the thorax when disturbed, and emit a strong defensive odour.

The small citrus butterfly has a wingspan of about 75 mm. Males and females are similar in colour. The forewings are black with grey and white markings, and hindwings have white, orange-red and blue markings.

The large citrus butterfly has a wing span of about 130 mm. Wings of the male are black with white markings, while the hindwings of the female are brightly marked with white, orange and blue.

▶ Distinguishing features

The appearance of the larvae is distinctive (see above).

▶ Life cycle

Butterflies emerge in early spring from over-wintering pupae, and lay their eggs singly on the tips of young leaves. The eggs of both species are yellow, spherical and 2–2.5 mm in diameter.

After hatching, larvae pass through five instars, or stages, before pupating in an upright position attached to a citrus twig by a silken pad at the tail end, and a fine silken girdle at the waist.

The complete life cycle takes 2–3 months.

▶ Seasonal history

There are at least 3 generations per year.

Plate 13.51 Small citrus butterfly adult, with larvae and pupa. This species is more common than the large citrus butterfly in southern Australia.

Plate 13.52 Male (top) and female large citrus butterflies.

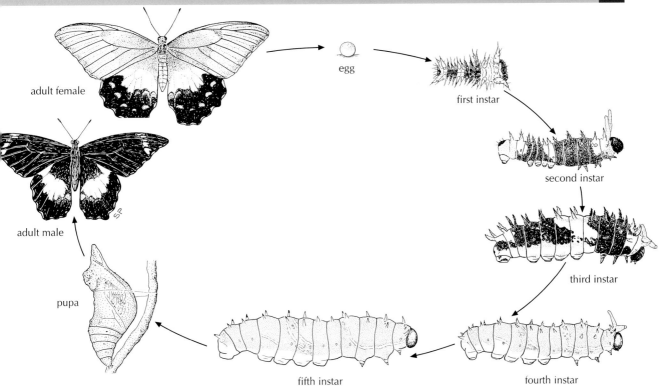

Figure 13.17 Life cycle of large citrus butterfly.

▶ Habits
Caterpillars feed voraciously on young foliage. The butterflies can usually be seen flying in the orchard during spring and autumn.

▶ Hosts
Citrus butterflies also attack other members of the plant family Rutaceae.

▶ Origin and distribution
Citrus butterflies are native to Australia.

Damage

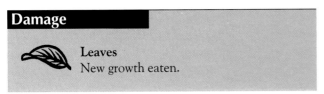

Damage is usually not important, except in young trees, which can be defoliated.

▶ Varieties attacked
All citrus varieties are susceptible.

Natural enemies

▶ Parasites
Parasitic wasps, including *Pachyneuron kingsleyi* and *Pteromalus puparum*, and parasitic tachinid flies (*Blepharipa* spp.) are important natural enemies that attack larvae. Wasps of *Trichogramma* spp. are known to parasitise eggs.

▶ Predators
The assassin bug (*Pristhesancus plagipennis*) is a common predator of citrus butterfly larvae in Queensland. Shield bugs (*Oechalia schellembergii* and *Cermatulus nasalis*) commonly prey on larvae in southern inland areas. Birds also feed on larvae, while praying mantises feed on larvae and on adult butterflies.

Plate 13.53 Tachinid fly Blepharipa fulviventris, *a parasite of small citrus butterfly. The fly larva emerges from, and kills, the butterfly pupa (left), then pupates within a barrel-shaped puparium (lower right), from which the adult fly emerges at a later date.*

Plate 13.54 *A minute trichogramma wasp parasitising the egg of a large citrus butterfly.*

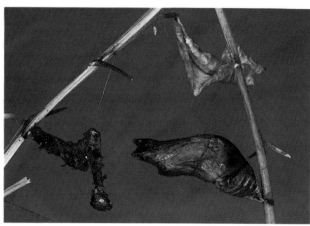

Plate 13.55 Large citrus butterfly pre-pupa (left) and pupa (right) killed by a virus.

▶ Pathogens
A number of viral and bacterial diseases affect caterpillars of both species.

Management

▶ Monitoring
- Citrus butterflies occur on all varieties.
- Monitor once or twice on young trees from October to April, checking 5 randomly selected young shoots per tree.
- Butterflies flying in the orchard also indicate that the pests are active.

▶ Action level
The action level is 25% or more of shoots of young trees infested.

▶ Appropriate action
Both species very occasionally defoliate young flushing trees (less than 2 years old). However, normally they are kept at low levels by natural enemies.

If a troublesome infestation occurs on very young trees, a single spray of a selective pesticide can be applied. *Bacillus thuringiensis* bacterial sprays can also be effective.

Beetles (Coleoptera)

Beetles are insects with chewing mouthparts. They vary greatly in size, shape and colouring.

Adult beetles usually have two pairs of wings. The outer pair is hard and shell-like, and folds together to form a tough cover over the abdomen. The inner pair is membranous, and used for flying.

Larvae normally have distinct heads, antennae and three pairs of thoracic legs. However, some beetle larvae, e.g. larvae of some weevils, do not have legs.

The diagram on page 84 shows an example of the life cycle of a beetle, in this case the mealybug ladybird, a beneficial organism.

The beetle pests of citrus in Australia include weevils, longicorns and leaf beetles. Depending on the species, either the larvae, or the adults, or occasionally both, cause damage.

Monitoring

The following points apply to the monitoring of most beetles and their natural enemies:

- The number of trees from which samples are taken depends on block size (see chapter 25 for details).
- Additional care should be taken when monitoring blocks that have a history of economic damage.

Fuller's rose weevil

Asynonychus cervinus (Boheman), Coleoptera: Curculionidae

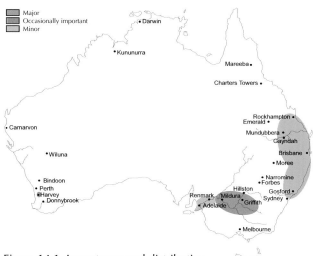

Figure 14.1 Importance and distribution of Fuller's rose weevil.

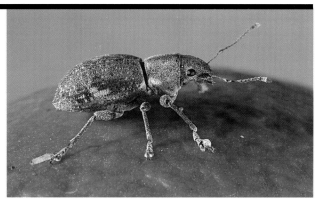

Plate 14.1 Fuller's rose weevil on fruit.

Description

▶ **General appearance**

Adults are wingless, grey-brown weevils about 8 mm long, with a faint crescent-shaped white mark on each wing cover.

Mature larvae, or grubs, are yellow, legless, C-shaped and about 6 mm long.

▶ **Distinguishing features**

The size of Fuller's rose weevil and the crescent-shaped markings on the wing covers are identifying features.

▶ **Life cycle**

Females produce eggs without mating. Eggs are yellow and laid in masses of 20–30. Early in the life cycle, 1–2 masses are laid per week. About 160 eggs are laid during a life span of about 10 weeks.

The eggs are laid under fruit calyces, in the navel of oranges, in splits in the bark, or in microsprinklers under the tree. After hatching, the yellow grubs drop to the soil and feed on citrus roots.

The grubs pupate in the soil. Adults emerge from February to July, often after rain.

▶ **Seasonal history**

There are 1–2 generations per year.

▶ **Habits**

Adult weevils cannot fly. They climb the trunk, or branches in contact with the ground, to reach the canopy. They feed on foliage, chewing the leaf margins and leaving a coarsely serrated edge. The weevils drop off foliage when disturbed, and are readily transported on bulk bins and fruit boxes.

▶ **Hosts**

Fuller's rose weevil occurs on a wide range of broadleaf plants, including ornamentals and weeds. Commercial tree crops affected include stonefruit, avocado and walnut.

▶ **Origin and distribution**

This weevil is widely distributed throughout citrus growing areas of the world.

Damage

 Fruit
Eggs and scurfing around the calyx.

 Leaves
Chewed, coarsely serrated edges.

 Roots
Root fibres nipped off by larvae and deep grooves eaten in larger roots.

The presence of eggs under the calyx may cause a problem in some export markets.

When common, this weevil also poses a problem by blocking up microsprinklers.

▶ **Varieties attacked**
All citrus varieties are susceptible to attack.

Natural enemies

▶ **Parasites**
The wasp *Fidiobia citri* is a common parasite of weevil eggs.

▶ **Predators**
The assassin bug (*Pristhesancus plagipennis*) and praying mantises feed on Fuller's rose weevil.

▶ **Pathogens**
Several soil-inhabiting fungi, including *Metarhizium* sp. and *Beauveria* sp. attack larvae. Parasitic nematodes such as *Heterorhabditis* sp. also play a major role in controlling weevil populations.

Management

▶ **Monitoring**
- Fuller's rose weevil attacks all citrus varieties.

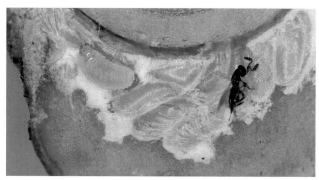

Plate 14.2 Eggs of Fuller's rose weevil under a fruit calyx. At left, a newly hatched larva and, at right, the wasp parasite Fidiobia citri.

- Fortnightly monitoring is necessary from early August to late October, and again from February to late June.
- Examine 5 low-hanging fruit per tree near the trunk. Carefully lift the fruit calyx and look for egg masses, using a ×10 hand lens.
- Adult weevils can also be monitored by beating tree limbs, and collecting fallen weevils on a tray or cloth. Beat one limb on each of 10–20 trees per block.

▶ **Action level**
Fuller's rose weevil is of significance only on trees producing fruit for export to Japan, as the presence of any egg masses on fruit for this market is not tolerated.

▶ **Appropriate action**
Control is rarely required. On blocks where fruit are grown for export, prune the skirts of the trees 1 metre off the ground at the end of each season and control weeds.

A sticky polybutene trunk band can prevent the adult weevils from climbing the trunk. Wrap a piece of plastic around the trunk before applying a sticky band.

Trunks can also be chemically treated with a recommended pesticide. There is no advantage in treating the trunks with a combination of sticky bands *and* pesticide.

Citrus leafeating weevil

Eutinophaea bicristata Lea, Coleoptera: Curculionidae

Figure 14.2 Importance and distribution of citrus leafeating weevil.

Plate 14.3 Mating citrus leafeating weevils. Note the two ridges on the male's back.

Description

▶ **General appearance**

The adult is a small, mottled, grey to tan weevil about 3 mm long. The larvae are small, white grubs up to 3 mm long.

▶ **Distinguishing features**

Citrus leafeating weevil is similar to the spinelegged citrus weevil (dicky rice weevil). However the male citrus leafeating weevil does not have spines on the forelegs.

▶ **Life cycle**

The larvae feed on roots and pupate in the soil. The adults emerge in the spring and climb onto the lower part of the tree. Adults are capable of flight.

▶ **Seasonal history**

There is one generation per year.

▶ **Habits**

The larvae of citrus leafeating weevil develop on the tree roots. Adults first appear in early September, and persist until late summer. They feed on the foliage, especially in the lower third of the tree.

▶ **Hosts**

Citrus is the only known host.

▶ **Origin and distribution**

Citrus leafeating weevil is native to Australia.

Damage

 Leaves
Irregularly shaped patches about 5 mm across chewed out of both leaf surfaces; possible leaf drop.

Damage to foliage is worse in the lower third of the tree.

The level of feeding on fruit is insignificant.

Although larvae feed on the tree roots, there are no obvious signs of damage.

▶ **Varieties attacked**

All varieties of citrus are susceptible.

Natural enemies

No natural enemies have yet been identified.

Management

▶ **Monitoring**

- Citrus leafeating weevil attacks all citrus varieties.
- From October onwards, monitor weevil numbers at fortnightly intervals by beating 10 randomly selected small branches over a white, expanded-polystyrene box (570 mm × 380 mm).

▶ **Action level**

Action is required if 50 or more citrus leafeating weevils are collected from beating 10 small branches in the way described above.

▶ **Appropriate action**

Apply an appropriate insecticide when weevil numbers exceed the action level.

On blocks with a history of weevil damage, a pre-emergence soil application of insecticide at a higher rate will prevent the emergence of adult weevils.

Plate 14.4 Leaf damage caused by citrus leafeating weevils. Yellow areas show where the tissues have been eaten.

Plate 14.5 Close-up of leafeating weevil and damage caused to a leaf.

Elephant weevil

Orthorhinus cylindrirostris (Fabricius), Coleoptera: Curculionidae

Figure 14.3 Importance and distribution of elephant weevil.

Plate 14.6 Adult elephant weevil.

Description

▶ General appearance
The adults are large weevils (up to 20 mm in length), and mostly grey with darker markings over the body.

▶ Distinguishing features
Elephant weevil is a large weevil, with a 5–7 mm long proboscis, front legs longer than the other two pairs, and wing covers with four distinct protuberances near the wing tips.

▶ Life cycle
Eggs are laid underneath the bark at the tree butt, in October–November. Larvae are creamy in colour and do not have legs. They tunnel within branches, trunk and upper roots for about 10 months. Pupation takes place in the trunk just underneath the bark surface, and adults emerge 2–3 months later, usually in September–October. The complete life cycle takes about 12 months.

▶ Seasonal history
There is one generation per year.

▶ Habits
Adult elephant weevils drill into tree trunks with their proboscis to make small holes in which to lay eggs. They also cause minor damage by chewing angular pieces of green bark from twigs and younger branches.

Larvae tunnel through the trunk and woody roots, initially downwards. Later in the life cycle, they turn around and bore upwards.

Pupation usually takes place within the trunk up to 1 metre above ground level. Characteristic holes approximately 6 mm in diameter are made in the trunk and lower branches by emerging adults. These holes are very neatly made, and look as if they have been machine-drilled.

Weevils are attracted to weak and unhealthy trees that are suffering from waterlogging, root disease or drought.

▶ Hosts
Elephant weevil attacks a range of species, including eucalypts, grape, apple and stonefruit.

▶ Origin and distribution
Elephant weevil is native to Australia.

Damage

 Twigs
Angular pieces of bark chewed off by adults; a minor problem.

 Branches
Angular pieces of bark chewed off by adults; a minor problem.

 Main trunk
Tunnelling by larvae throughout lower trunk and branches.

 Roots
Tunnelling by larvae throughout woody roots.

Larvae cause more damage than adults. Heavy or prolonged infestations will seriously weaken or kill mature trees.

▶ Varieties attacked
Lemons are most susceptible to attack. Other citrus varieties are also attacked, but damage is usually minor.

Natural enemies

No natural enemies have been identified.

Plate 14.7 *Twigs damaged by adult elephant weevil.*

Management

▶ Monitoring

- Elephant weevil occurs on all citrus varieties, but is more common on lemons.
- Check blocks generally for signs of damage, particularly on unhealthy or less vigorous trees. Unhealthy trees should be checked for adult emergence holes in the lower trunk and branches, indicating previous damage. To confirm this, branches can be crosscut to show tunnels containing tightly packed sawdust.
- Adults are not easily found, but freshly chewed bark on twigs is an indication of their presence.
- There is no set number of trees recommended for sampling.

▶ Action level
Treat any affected trees.

▶ Appropriate action
If trees are suffering from over-watering, high water tables, root disease or water stress, try to remedy these problems.

Spraying the trunks and lower branches of affected trees with a residual broad-spectrum insecticide during September–November may help reduce reinfestation.

Citrus fruit weevil

Neomerimnetes sobrinus Lea, Coleoptera: Curculionidae

Figure 14.4 Importance and distribution of citrus fruit weevil.

Plate 14.8 *Citrus fruit weevil feeding on young fruit.*

Description

▶ General appearance
The adult is a tan-coloured weevil, about 7 mm long.

▶ Distinguishing features
The citrus fruit weevil is larger than the spinelegged citrus weevil and the citrus leafeating weevil, but smaller than Fuller's rose weevil.

▶ Life cycle
The larvae feed on tree roots, and the adults emerge and climb onto the tree during September–October.

▶ Seasonal history
There is one generation per year.

▶ Habits
The citrus fruit weevil is a localised pest of citrus in the Howard area, north of Maryborough in Queensland. During October–November, adults chew holes in fruit up to 20 mm in diameter. Up to half the fruit on individual trees can be affected.

Hosts
Citrus is the only known host.

Origin and distribution
The citrus fruit weevil is native to Australia.

Damage

Fruit
Holes chewed out of young fruit, many of which drop.

Although larvae feed on tree roots, there is no apparent damage.

Varieties attacked
All citrus varieties are attacked, but particularly oranges.

Natural enemies
No natural enemies have yet been identified.

Management

Monitoring
- Citrus fruit weevil occurs on all varieties, but is more common on oranges.
- In the Howard area of south-east Queensland, check 5 randomly selected young fruit per tree. Fruit should be sampled once or twice, starting in early October.

Action level
Action is required when 10% or more of fruit are damaged.

Appropriate action
Apply a spray as required.

Spinelegged citrus weevil

Maleuterpes spinipes Blackburn, Coleoptera: Curculionidae

Figure 14.5 Importance and distribution of spinelegged citrus weevil.

Plate 14.9 Adult spinelegged citrus weevil (also known as dicky rice weevil).

Description

General appearance
The adult spinelegged citrus weevil (also known as the dicky rice weevil) is 3 mm long, and brown with grey-white markings on the back and legs. The adults are flightless.

Distinguishing features
The spinelegged citrus weevil is similar to the citrus leafeating weevil, but the male has a long curved spine at the middle of each front leg.

Life cycle
Larvae of the spinelegged citrus weevil feed on roots, and adults emerge during October–November.

Seasonal history
There are 1–2 generations per year.

Habits
Adult spinelegged citrus weevils feed on the rind of young fruit and on young foliage, mainly in late spring.

Hosts
Spinelegged citrus weevil also attacks some other members of the Rutaceae family.

Origin and distribution
This weevil is probably native to Australia.

Damage

Fruit
Rind scarring on young fruit.

Leaves
Saw-toothed pattern eaten out of leaf margins.

Fruit damaged by spinelegged citrus weevil have a network of irregular furrows on the rind.

Although larvae feed on tree roots, there is no apparent damage.

Plate 14.10 Damage to Valencia oranges caused by spinelegged citrus weevil.

Plate 14.11 Damage to leaves caused by spinelegged citrus weevil.

Varieties attacked
All citrus varieties are susceptible.

Natural enemies
No natural enemies have yet been identified.

Management

Monitoring
- Spinelegged citrus weevil occurs on all citrus varieties.
- Monitor once or twice from early October to late November.

- In blocks with a history of attack, check 5 young fruit per tree on randomly selected trees. Look for the presence or absence of adults and/or fresh damage.

Action level
The action level is 10% or more of fruit or 25% or more of leaves damaged.

Appropriate action
Control is rarely required.

Apple weevil

Otiorhynchus cribricollis Gyllenhal, Coleoptera: Curculionidae

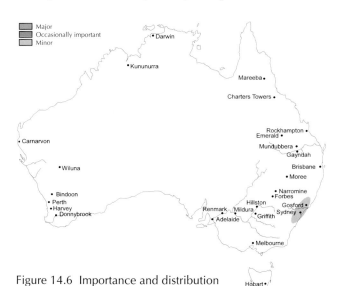

Figure 14.6 Importance and distribution of apple weevil.

Plate 14.12 Adult apple weevil.

Life cycle
Larvae of the apple weevil feed on roots, and adults emerge in early December and January.

Seasonal history
There are 1–2 generations per year.

Habits
Adult apple weevils feed on foliage at night, and shelter in soil by day.

Hosts
The apple weevil also attacks pome fruit, roses and some ornamentals.

Description

General appearance
The adult apple weevil is 9 mm long, shiny and dark brown.

Distinguishing features
The apple weevil is distinguished from spinelegged citrus weevil by its larger size, together with its brown colour.

Origin and distribution
The apple weevil is probably native to Australia.

Damage

Leaves
Defoliation of young trees if infestation is heavy.

Apple weevil is present in South Australia and Victoria, but is not known to cause significant damage in those states.

Although larvae feed on tree roots, there is no apparent damage.

Varieties attacked
All citrus varieties are susceptible.

Natural enemies
No natural enemies have yet been identified.

Management

Monitoring
- Apple weevil occurs on all citrus varieties.
- Monitor once or twice in December–January.
- In blocks with a history of attack, check 5 fully expanded leaves per tree on randomly selected trees. Look for the presence or absence of adults and/or fresh damage.

Action level
The action level is 25% or more of leaves damaged.

Appropriate action
Control is rarely required.

Citrus branch borer

Uracanthus cryptophagus Olliff, Coleoptera: Cerambycidae

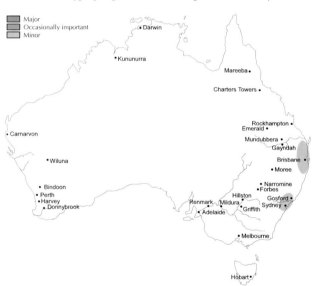

Figure 14.7 Importance and distribution of citrus branch borer.

Plate 14.13 Adult and larva of citrus branch borer, and larval gallery filled with frass.

Description

General appearance
The adult citrus branch borer is a narrow, reddish brown beetle about 40 mm long, with antennae as long as the body.

The larva is white, with a concealed, small, dark-brown or black head, and a club-shaped body tapering from head to tail.

Distinguishing features
Larvae grow to 40 mm in length. They burrow in limbs and branches from 10–100 mm in diameter.

Life cycle
Adult females lay eggs in the cracks of twigs and small branches. The eggs hatch in about 10 days and the young larvae immediately bore into the wood. Development continues through summer, autumn and winter.

Pupation can occur as early as September, and adults start to emerge during October–November. Emergence continues through the summer.

Mating usually occurs shortly after adult emergence. Egg laying then begins and continues throughout most of the summer months.

Seasonal history
There is one generation per year.

Habits
Larvae work their way from the twigs and small branches towards the trunk. Periodically they construct a dome-shaped cavity with a small opening in the bark. This is used for disposal of waste from the main tunnel. Often their feeding ringbarks branches, and wilting occurs beyond this point.

Hosts
Plants in the citrus family are known hosts. A native host is the finger lime (*Microcitrus australasica*).

Origin and distribution
The citrus branch borer is found only in Australia.

Damage

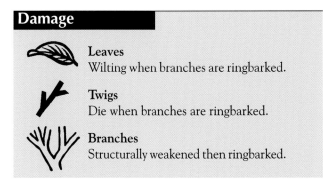

Leaves
Wilting when branches are ringbarked.

Twigs
Die when branches are ringbarked.

Branches
Structurally weakened then ringbarked.

Large larvae feed in branches up to 100 mm in diameter, and can ringbark and kill large sections of the tree. Sometimes damaged branches break off under their own weight.

▶ **Varieties attacked**
All citrus varieties are attacked, but particularly oranges.

Natural enemies

The natural enemies of citrus branch borer have not yet been identified.

Management

▶ **Monitoring**
- Citrus branch borer occurs on all varieties, but is more common on oranges.
- In orchards susceptible to branch borer, trees should be checked regularly for early signs of wilting. This checking usually occurs during monitoring for other pests, and during normal orchard operations.
- There is no set number of trees to be monitored.

▶ **Action level**
The first signs of wilting caused by this pest indicate that action is necessary.

Plate 14.14 *Wilted Valencia orange branch, caused by citrus branch borer. Note the proximity of the orange trees to native forest.*

▶ **Appropriate action**
There is no effective chemical control for citrus branch borer. It is important to stop the development of larvae by pruning twigs and small branches before larvae tunnel into larger branches. Regular pruning and burning of infested small branches will not only prevent loss of large sections of trees, but minimise build-up of the pest within the orchard.

▶ **Additional management notes**
It is inadvisable to grow citrus commercially in former rainforest areas which are very prone to attack by citrus branch borer, e.g. in the Blackall Range in south-east Queensland.

Trees in poor condition are more susceptible to attack than vigorous trees. Good orchard management will maintain vigorous trees, and lessen the chances of attack.

Evidence from other countries suggests that excessive use of fungicides may result in increased damage caused by borers. This is because the fungicides kill fungi that attack the borers.

Fig, pittosporum and citrus longicorns

Fig longicorn *Acalolepta vastator* (Newman), pittosporum longicorn *Strongylurus thoracicus* (Pascoe), and citrus longicorn *Skeletodes tetrops* Newman, Coleoptera: Cerambycidae

Figure 14.8 Importance and distribution of fig, pittosporum and citrus longicorns.

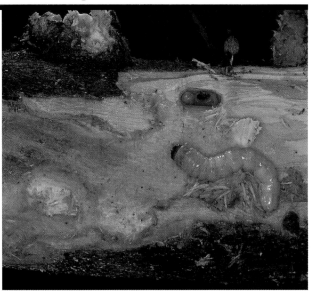

Plate 14.15 *Fig longicorn larvae and damage to a Tahitian lime. The bark has been removed to reveal the larvae, their feeding galleries and frass.*

Plate 14.16 Fig longicorn adult, resting near its emergence hole.

Plate 14.17 Pittosporum longicorn.

Plate 14.18 Citrus longicorn.

Description

▶ General appearance

The larvae of all three of these longicorns are white grubs with a concealed, dark-brown or black head, and a club-shaped body, tapering from head to tail. They grow up to 40 mm long.

The adults of all three of these longicorns, particularly the fig longicorn, have very long antennae. (The name 'longicorn', meaning 'longhorned', refers to these antennae.) The fig longicorn adult is 30 mm long, speckled grey-brown, with a prominent spine on each side of the thorax. The pittosporum longicorn adult is 30 mm long, light brown and has a characteristic row of white spots on each side of the thorax. The citrus longicorn is sometimes called the spider longicorn because of its long legs. The adult is 12 mm long and brownish-grey.

▶ Distinguishing features

The spines on the side of the thorax of fig longicorn, the white spots on the thorax of pittosporum longicorn, and the long legs and smaller size of the citrus longicorn are the main distinguishing features. The citrus branch borer is reddish brown, and has a longer, narrower body than these three species.

▶ Life cycle

These three longicorns have a life cycle similar to that of the citrus branch borer. Adult females lay eggs in the cracks of twigs and small branches. The eggs hatch in about 10 days and the young larvae immediately bore into the wood. Development continues through summer, autumn and winter.

Pupation can occur as early as September, and adults start to emerge during October–November. Emergence continues through the summer.

Mating usually occurs shortly after adult emergence. Egg laying then begins and continues throughout most of the summer months.

▶ Seasonal history

There is one generation per year.

▶ Habits

The fig longicorn larva attacks the tree trunks or heavier limbs, feeding beneath the bark and boring deeply into the wood.

The pittosporum longicorn larva feeds along the twigs and limbs (similar to the citrus branch borer).

The citrus longicorn larva feeds under the bark (similar to the speckled longicorn). The female lays her eggs in dead bark or where other borers have inflicted damage.

Pittosporum and citrus longicorns are attracted to older trees lacking vigour or to trees affected by disease. However, the fig longicorn attacks young, healthy trees, as well as older trees. It has been observed to be more prevalent after pruning, especially in limes.

▶ Hosts

The full range of plant hosts is unknown. The fig longicorn and pittosporum longicorn are named after their major hosts. Fig longicorn is also a serious pest of grapevines in the lower Hunter Valley of New South Wales.

▶ Origin and distribution

These species are native to Australia.

Damage

 Twigs
Death of twigs and small branches (caused by pittosporum longicorn only).

 Branches
Holes, tunnelling, frass; branches can snap off.

 Main trunk
Holes, tunnelling, frass; secondary rots.

Plate 14.19 Severe damage to the lower trunk of a Tahitian lime caused by fig longicorn. Note the frass beneath the tree.

Of these three species, fig longicorn causes the most severe damage.

Moisture and disease-causing fungi can enter through damaged areas of the trunk and major limbs, resulting in secondary rots.

▸ Varieties attacked
All citrus varieties are susceptible, particularly oranges.

Natural enemies

▸ Parasites
Parasitic wasps attack longicorn larvae and pupae, but these have not been identified.

▸ Predators
Predatory beetles of the family Cleridae attack the larvae of these borers.

Management

▸ Monitoring
- These longicorn beetles attack all citrus varieties.
- Monitor blocks that have a history of borer problems.
- Rate the damage, particularly for fig and citrus longicorns, on the following scale:
    - **0** no damage
    - **1** small patch on the trunk or a limb
    - **2** trunk and a limb with two or three patches
    - **3** most lower limbs and trunk with serious patches
    - **4** extensive damage, serious dieback and secondary rot.

▸ Action level
Take action when the average rating is 2 or more.

▸ Appropriate action
These longicorns cannot be controlled by spraying foliage with chemicals. The best way of controlling fig longicorn is by treating individual trees as follows. Scrape away diseased and eaten bark, and apply a low-pressure spray containing a mixture of fungicide and insecticide to the affected areas on the trunk and lower limbs.

Prevent development of longicorns within the orchard. Control larvae by pruning wilted branches and scraping away affected tissue. Regular pruning and burning of infested small branches will prevent loss of large sections of trees and minimise build-up of the pests within the orchard.

▸ Additional management notes
Trees in poor condition are more susceptible to attack than vigorous trees. Good orchard management will maintain vigorous trees, and lessen the chances of attack by longicorns.

Evidence from other countries suggests that excessive use of fungicides may result in increased damage caused by borers. This is because the fungicides kill fungi that attack the borers.

Speckled longicorn

Paradisterna plumifera (Pascoe), Coleoptera: Cerambycidae

Figure 14.9 Importance and distribution of speckled longicorn.

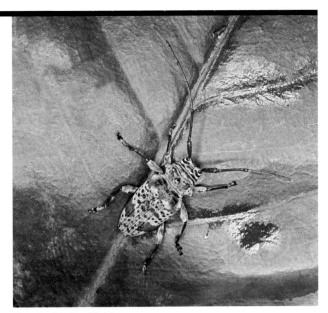

Plate 14.20 Speckled longicorn.

Description

▶ **Appearance**

The larvae are white grubs with a concealed, dark-brown or black head, and a club-shaped body, tapering from head to tail. They grow up to 30 mm long.

The adult beetle is 12 mm long, light or dark grey, and speckled with brown marks. The antennae are longer than the body. (The name 'longicorn', meaning 'longhorned', refers to these antennae.)

▶ **Distinguishing features**

The size and appearance of the adult beetle are distinctive. The speckled longicorn tends to feed just under the bark on the trunk and larger limbs of trees.

▶ **Life cycle**

The life cycle of the speckled longicorn is similar to that of other longicorns and the citrus branch borer. Adult females lay eggs on the bark of the trunk, and lower sections of the main limbs. The eggs hatch in about 10 days and the young larvae immediately bore into the wood. Development continues through summer, autumn and winter.

Pupation can occur as early as September, and adults start to emerge during October–November. Emergence continues through the summer.

Mating usually occurs shortly after adult emergence. Egg laying then begins and continues throughout most of the summer months.

▶ **Seasonal history**

There is one generation per year.

▶ **Habits**

Speckled longicorn seems to be attracted to older trees (particularly if they are lacking in vigour), phytophthora-affected trees, or those with dead or diseased areas in the trunk or main limbs. Often the larvae feed where the bark is already diseased, but extend their activity into adjacent healthy bark. The larvae bore under the bark and produce large amounts of sawdust-like frass, which extrudes from the tunnels, and falls to the bottom of the trunk.

▶ **Hosts**

Speckled longicorn also attacks radiata pine.

▶ **Origin and distribution**

The speckled longicorn is native to Australia.

Damage

Branches
Large patches of affected bark on lower sections of main limbs; holes, tunnelling, frass; main limbs structurally weakened.

Main trunk
Holes, tunnelling, frass; secondary rots.

Moisture and disease-causing fungi can enter through damaged areas of the trunk and major limbs, resulting in secondary rots.

Plate 14.21 Damage caused by speckled longicorn to the lower trunk of a navel orange tree.

▶ **Varieties attacked**

All citrus varieties, particularly navel oranges, are attacked. Usually trees more than 15 years old are worst affected.

Natural enemies

Nothing is known about the natural enemies of speckled longicorn.

Management

▶ **Monitoring**

- Specked longicorn attacks all citrus varieties, but is more common on navel oranges.
- Monitor trees in older blocks that have a history of borer problems.
- Rate the damage on the following scale:
  - 0 no damage
  - 1 small patch on the trunk or a limb
  - 2 trunk and a limb with two or three patches
  - 3 most lower limbs and trunk with serious patches
  - 4 tree extensively damaged with serious dieback and secondary rot.

▶ **Action level**

An average rating of 2 or more indicates a serious infestation, control of which should be attempted.

Appropriate action
Attempts to control speckled longicorn by spraying foliage with chemicals are unlikely to succeed. Systemic pesticides can be expensive and unreliable in large, old trees.

The best chance of controlling infestations is by treating individual trees. Scrape away diseased and eaten bark, and apply a low-pressure spray containing fungicide and insecticide to the affected areas on the trunk and lower limbs.

Ultimately, old blocks of trees affected by borers and disease should be replaced.

Additional management notes
Older trees and trees in poor condition (particularly phytophthora-affected older trees) are more susceptible to attack than vigorous trees. Good orchard management will maintain vigorous trees, and lessen the chances of attack by speckled longicorn.

Evidence from other countries suggests that excessive use of fungicides may result in increased damage caused by borers. This is because the fungicides kill fungi that normally attack the borers.

Monolepta beetle and rhyparida beetle

Monolepta australis (Jacoby) and *Rhyparida* spp., Coleoptera: Chrysomelidae

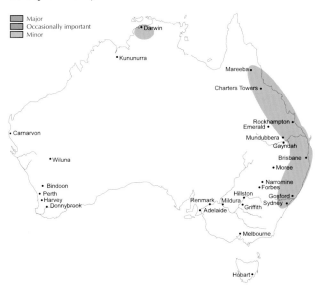

Figure 14.10 Importance and distribution of monolepta beetle.

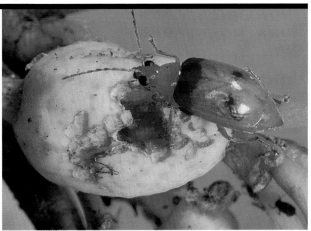

Plate 14.22 Monolepta beetle feeding on a citrus flower bud.

Plate 14.23 Rhyparida beetle.

Figure 14.11 Importance and distribution of rhyparida beetle.

Description

General appearance
Monolepta beetle (also known as redshouldered leaf beetle) is about 6 mm long, and yellow with red markings on each wing cover.

Rhyparida beetle is 7 mm long, and metallic brown in colour.

Distinguishing features
The red markings on the wing covers are characteristic of monolepta beetle, and the metallic brown colour characteristic of rhyparida beetle.

Plate 14.24 Monolepta beetles and the damage they have caused to leaves and buds.

▶ Life cycle
Monolepta eggs are laid under the soil surface, mainly in pastures, and larvae (or grubs) feed on grass and clover roots. The mature larva is about 5 mm long and pupates in the soil.

Adults usually emerge after good rains following a dry period, and swarms develop and migrate into tree crops during spring, summer and autumn. The life cycle takes about 2 months during summer. Adult beetles live about 10 weeks.

The rhyparida beetle has a similar life cycle.

▶ Seasonal history
In Queensland and the Northern Territory, there are 3–4 generations per year.

In New South Wales, there are 2 generations during the warmer months (in spring to autumn), and possibly one generation during winter.

▶ Habits
Swarms of these beetles attack citrus blossom, buds and young fruit. Such swarms cause significant damage within 1–2 days.

▶ Hosts
These beetles have a wide host range. The commercial crops most affected are stonefruit, avocado, macadamia, carambola, lychee, cashew and mango. *Eucalyptus torelliana* is also commonly attacked.

▶ Origin and distribution
These beetles are native to Australia.

Damage

Flowers
Flowers eaten by swarms of beetles.

Fruit
Young fruit eaten by swarms of beetles.

Leaves
Leaves eaten by swarms of beetles; foliage has a scorched appearance.

Twigs
Twig dieback after leaves eaten.

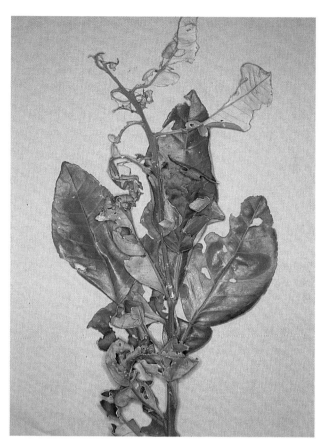

Plate 14.25 Damage caused to pummelo leaves by rhyparida beetle.

▶ Varieties attacked
All citrus varieties are attacked.

Natural enemies

▶ Parasites
The tachinid fly *Monoleptophaga caldwelli* parasitises adult monolepta beetles, killing up to 5% of the adult population.

Management

▶ Monitoring
- Monolepta and rhyparida beetles attack all citrus varieties.
- Swarms of beetles are very common from spring to autumn. Vigilance in the orchard during normal cultural operations, and while monitoring other pests, should result in swarms being observed when they occur.
- It is usual for swarms to be found on the same trees or in the same sections of the orchard year after year, and swarms can be detected early if these 'hot spots' are regularly monitored.
- Some windbreak trees (e.g. *Eucalyptus torelliana*) are favoured hosts, and are often the first trees infested in an orchard. They can be used as an early warning system indicating the threat of infestation.

▶ Action level
If swarms of beetles are present, action is required. However, ignore small numbers of beetles.

▶ Appropriate action
Trees affected by the swarm should be spot-treated with a selective insecticide.

Katydids, crickets and grasshoppers (Orthoptera)

Katydids, crickets and grasshoppers are insects with chewing mouthparts and large hind legs, often used for jumping. While some are wingless, most have wings that are used for flight.

Many katydids are green in colour and have antennae that are as long as or longer than their bodies. Crickets and grasshoppers range in colour from green to brown to black. Most have long antennae, but some grasshoppers, in particular, have short antennae.

The young stages of katydids, crickets and grasshoppers are called 'nymphs'. They are wingless, but otherwise similar to the adults, although smaller.

These insects cause damage to citrus by chewing holes in leaves and young fruit.

Monitoring
The following points apply to the monitoring of katydids and crickets:
- The number of trees from which samples are taken depends on block size (see chapter 25 for details).
- Additional care should be taken when monitoring blocks that have a history of economic damage.

Katydids

Citrus katydids *Caedicia* spp., inland katydid *Caedicia simplex* (Walker) and spotted katydid *Ephippitytha trigintiduoguttata* (Serville), Orthoptera: Tettigoniidae

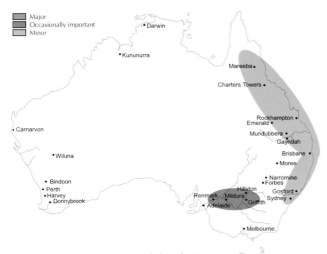

Figure 15.1 Importance and distribution of citrus katydids.

Plate 15.1 *Katydid on lemon blossom. Note the damage to the stigma and style of the young fruit (near the katydid's hind leg).*

Description

▶ General appearance
Katydids are 'longhorned' grasshoppers, i.e. grasshoppers with long antennae.

Adult citrus katydids are green, about 40 mm long, with strong hind legs for jumping, and fat rounded abdomens with purple undersides. The forewings are narrow and opaque, the hindwings fan-like, transparent and pale green. The female has a short, upwardly curving ovipositor.

The adult inland katydid is about 50 mm long. It has a yellow band on the front margin of the wings.

The adult spotted katydid is about 50 mm long, olive green and brown, with dark-brown markings on the wings and body. It is particularly common in inland citrus areas.

▶ Distinguishing features
Katydids are similar to other grasshoppers, but have much longer antennae. The citrus leafeating cricket also has long antennae, but is light brown.

▶ Life cycle
The life cycles of all these katydids are similar. Eggs are laid mostly from January to April. They are disk-shaped and 5 mm long. Eggs are deposited on limbs or tree trunks, and cemented together in parallel rows of about a dozen.

Figure 15.2 Importance and distribution of inland katydid.

Figure 15.3 Importance and distribution of spotted katydid.

Plate 15.2 Adult inland katydid.

Plate 15.3 Adult spotted katydid.

Plate 15.4 Inland katydid nymph.

The nymphs hatch in early spring, and begin feeding on the newly set fruit in October–November. They are a similar colour to the adults, and develop through five instars, or stages. The adult stage is reached during December–January.

▶ Seasonal history
There is one generation per year.

▶ Habits
Katydid nymphs and adults feed on young foliage, flowers, and young fruit.

The adults fly strongly from tree to tree, and infestations tend to be patchy.

During summer and early autumn, the adult male katydid makes a metallic clicking 'chirrup' sound in dull weather and in the early evening.

▶ Hosts
Little is known about hosts other than citrus, but in the southern states the worst affected orchards are usually next to bushland. This suggests that native plants are important hosts. Blackberries are also a host of citrus katydid.

▶ Origin and distribution
These species of katydids are native to Australia.

Damage

 Flowers
Petals and other floral parts eaten.

 Fruit
Deep gouges in very young fruit, which heal to give disfiguring chalky-white scars.

 Leaves
Holes chewed out of leaves, especially on young flushes.

Katydids eat large pieces out of young fruit. Older nymphs and adults make deep gouges, with resulting chalk-white scars on the sides and bases of fruit, particularly navel oranges. The damage can be confused with the damage caused by other grasshoppers.

Slightly damaged fruit heal and tend to stay on the tree, but half-eaten fruit drop. On mature fruit damaged when young, the scar may expand to the size of a 50 cent piece, and develop a chalky appearance.

▶ Varieties attacked
These katydids are more common on oranges, especially navels. Damage is rarely seen on lemons, grapefruit and mandarins.

Natural enemies

▶ Parasites
Citrus katydids are parasitised by an unidentified species of tachinid fly, and unidentified species of wasps.

▶ Predators
The assassin bug (*Pristhesancus plagipennis*), praying mantises, sphecid wasps, and birds prey on citrus katydids.

Management

▶ Monitoring
- Citrus katydids attack oranges, particularly navels, more commonly than other citrus.
- Check 5 randomly selected young fruit per tree every 2 weeks from mid-September to late November. Sticky trunk bands are an additional tool for monitoring the numbers of emerging nymphs.

▶ Action level
The action level is 5% or more of fruit showing fresh injury. This level may be raised if there is a very heavy fruit set.

▶ Appropriate action
Use a selective pesticide. Infestations are usually patchy, and spraying whole orchards is rarely warranted.

Plate 15.5 Damage to young navel oranges caused by citrus katydid. Deep gouges have been made in the fruit, and large amounts eaten.

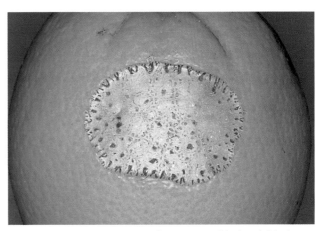

Plate 15.6 Scar on a mature navel orange caused by katydid feeding. The original damage was done when the fruit was very young.

Citrus leafeating cricket

Citrus leafeating cricket *Tamborina australis* (Walker), Orthoptera: Gryllidae

Figure 15.4 Importance and distribution of citrus leafeating cricket.

Plate 15.7 Male leafeating cricket (top), female (bottom). The male uses his strongly veined forewings to make chirping sounds that attract females.

Plate 15.8 Two types of damage caused by leafeating cricket: chewing through leaf blades and chewing holes in the leaf margins (left); and grazing on the leaf surface, which results in browning and leaf curling (right).

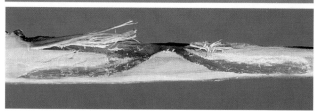

Plate 15.9 Twigs damaged by egg laying of the citrus leafeating cricket. Top, view from above of damaged area; bottom, longitudinal section through the twig to show where the eggs have been inserted, 6–7 each side of the central wound.

Description

▶ General appearance

Adult citrus leafeating crickets are light brown, 35–40 mm long, with long, thin antennae (twice the body length). There is a prominent pair of cerci (sensory appendages) projecting from the end of the abdomen. The curled and folded hindwings also extend beyond the abdomen.

Females are slightly larger than males, with a long ovipositor (egg-laying apparatus) extending to the tip of their folded wings. The forewings of females have many long indistinct, almost parallel veins, whereas male forewings are modified to produce sound, and have short, prominent, curved veins.

▶ Distinguishing features

Citrus leafeating cricket has long antennae and cerci, which distinguish it from most grasshoppers. Katydids have long antennae, but they are usually green (or olive-green and brown, in the case of spotted katydid), while this cricket is light brown.

▶ Life cycle

Development through the complete life cycle takes place in the tree. Eggs are laid in green twigs (especially in water shoots). Nymphs develop through five instars, or stages, before reaching the adult stage in October–November. Adults live for several months, and breeding continues up to late autumn.

▶ Seasonal history

There are 4–5 generations per year.

▶ Habits

Adults and nymphs feed on old and young leaves. They may chew holes through leaves, chew pieces out of the edges of leaves, or graze on the surface of leaves. During the day, adults and nymphs rest under leaves; they feed mostly at night.

When disturbed, they jump to another part of the tree and hide beneath leaves. Adults, when disturbed, also readily fly for short distances.

▶ Hosts

This cricket has been observed only on citrus.

▶ Origin and distribution

The citrus leafeating cricket is native to Australia, and is found mostly around Gayndah and Mundubbera in southern Queensland.

Damage

 Leaves
Grazed leaves brown and curl inwards; holes and chewed edges.

Badly affected trees look tatty, with up to 20% of mature leaves damaged in mid-summer.

▶ Varieties attacked

All varieties may be attacked, but particularly navel oranges, and Murcott and Imperial mandarins.

Natural enemies

Wasps are known to parasitise citrus leafeating cricket, particularly the eggs, but they have not yet been identified.

Management

▶ Monitoring

- Citrus leafeating cricket attacks all citrus varieties, but especially navel oranges, and Murcott and Imperial mandarins.
- Monitor fortnightly during October–November.
- Check for the presence or absence of the pest on 5 randomly selected shoots per tree.

▶ Action level

The action level is 25% or more of shoots showing fresh injury.

▶ Appropriate action

Use a selective pesticide.

Giant grasshopper

Valanga irregularis (Walker), Orthoptera: Acrididae

Figure 15.5 Importance and distribution of giant grasshopper.

Plate 15.11 Adult giant grasshopper.

Description

▶ General appearance
The giant grasshopper is probably the biggest of its type in Australia, with adults growing up to 90 mm long. Adults are light brown, often with a narrow, white to green stripe running from head to tail. Nymphs are pale green.

▶ Distinguishing features
Both adult and nymphal giant grasshoppers have enlarged hind legs used for jumping, and short antennae. These features, and the size of the adult, make this species fairly easy to recognise.

▶ Life cycle
Eggs are deposited in a pod beneath the soil surface in October–November. There are seven nymphal stages before adulthood is reached.

Nymphs are present from September to March, and adults from April to November. No breeding takes place during winter.

▶ Seasonal history
There is one generation per year.

▶ Habits
Giant grasshoppers feed mainly on foliage and sometimes on very young citrus fruit. They are commonly found in home gardens, but are less common in commercial orchards.

▶ Hosts
The giant grasshopper feeds on a wide range of shrubs, trees and herbaceous plants.

▶ Origin and distribution
This grasshopper is native to Australia.

Damage

 Fruit
Deep gouges on very young fruit, which heal to disfiguring chalky-white scars; damage is uncommon.

 Leaves
Edges of leaves chewed.

▶ Varieties attacked
All citrus varieties are attacked.

Natural enemies
The small wasp *Scelio flavicornis* parasitises eggs.

Management

▶ Monitoring
When monitoring for other pests, check for damaged leaves, or for the presence of grasshopper adults and nymphs.

▶ Action level
Action is required when 25% or more of young shoots are damaged by giant grasshopper.

▶ Appropriate action
In commercial orchards, spot-spray infested trees with a selective pesticide. In home gardens, grasshoppers can be removed from plants by hand, especially in the early morning, and destroyed.

Plate 15.10 Giant grasshopper nymph. Note the large holes chewed out of the edges of the leaves.

16 Fruit flies and midges (Diptera)

Flies are insects with one pair of membranous forewings. Instead of hindwings, they have a pair of slender, knobbed balancing organs, called 'halteres'. They have mouthparts that are either for piercing and sucking, or for rasping and lapping. Larvae, often called maggots, do not have legs, and their heads are often reduced and retracted.

Fruit flies are the main representatives of this order of insects which cause damage to citrus. Fruit flies are about the size of the common housefly. Adults lay eggs in ripening and/or overripe fruit. Damage is caused by the feeding of the developing larvae on the fruit pulp, and by secondary rots.

Queensland fruit fly

Bactrocera tryoni (Froggatt), Diptera: Tephritidae

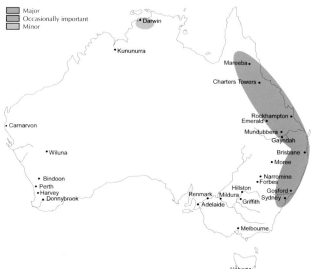

Figure 16.1 Importance and distribution of Queensland fruit fly.

Plate 16.1 Female Queensland fruit fly.

Description

▶ **General appearance**

Adult Queensland fruit flies are about the size of houseflies (7 mm long). They are reddish brown with yellow markings on the thorax. The pointed ovipositor is clearly visible at the end of the female's abdomen. The male fly is distinguished by a row of spines on either side of the abdomen.

The larvae are cream to white maggots, tapering towards the front end. They do not have legs. The last-stage larvae are about 9 mm long.

▶ **Distinguishing features**

Queensland fruit fly is the major fruit fly pest in eastern Australia. It is similar to other *Bactrocera* spp., e.g. lesser Queensland fruit fly (*Bactrocera neohumeralis*). *B. tryoni* has yellow shoulder patches on the thorax, immediately behind the head, whereas *B. neohumeralis* has no shoulder patches.

B. tryoni can easily be distinguished from the recently introduced papaya fruit fly (*Bactrocera papayae*), which has a black T-shaped mark on the abdomen. The papaya fruit fly is currently the target of an eradication campaign, and only occurs in far north Queensland.

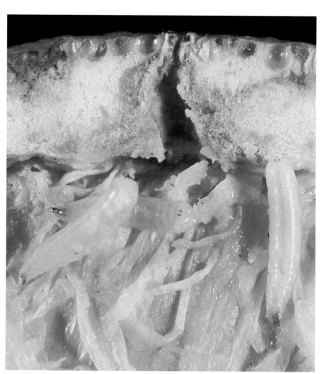

Plate 16.2 Queensland fruit fly larvae in pulp of fruit.

16 Fruit flies and midges (Diptera) 161

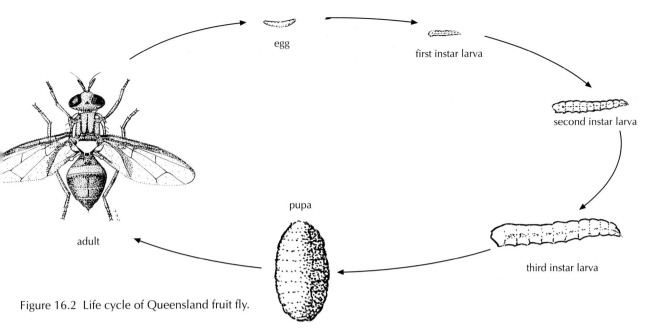

Figure 16.2 Life cycle of Queensland fruit fly.

Smaller ferment flies (*Drosophila* spp.), whose larvae are found only in fruit that are already damaged or rotting, can often be seen flying around rotting fruit, and should not be confused with primary fruit flies (i.e. important pest species that cause initial damage to fruit).

▶ Life cycle

The adult female fly feeds for up to a week on protein, e.g. on bacteria growing on fruit and plant surfaces, and on sugars, e.g. in honeydew and nectar, before laying eggs.

Eggs are white, banana-shaped and 1 mm long. They are deposited in batches of 10–12 into the spongy albedo of the fruit. The rind puncture is not visible at first, but later a yellow area develops around the 'sting' site.

Eggs hatch in 2–3 days and the larvae burrow into the fruit pulp. The number of larvae per fruit varies from one to 12 or more.

In summer, larvae complete their development in about 10 days. They then drop to the ground to find a suitable place to pupate. On the ground they can move up to 150 mm at a time by 'skipping', or flicking themselves into the air.

Pupation occurs in the soil, and during summer can take as little as 9 days. Pupae are brown, barrel-shaped and about 4–5 mm long.

▶ Seasonal history

In Queensland and the Northern Territory, there are at least 6 generations per year. There are fewer generations in most of coastal New South Wales.

▶ Habits

Adult Queensland fruit flies are strong fliers capable of travelling many kilometres from forested areas to commercial crops.

Flies usually over-winter as inactive adults, and while they may be observed in the orchard, there is often little egg laying until August–September. During spring, fruits other than citrus, e.g. peaches, nectarines, loquats, mulberries and native fruits, are important hosts. Fruit flies may breed on these hosts, and then large numbers may migrate onto nearby citrus in spring–summer.

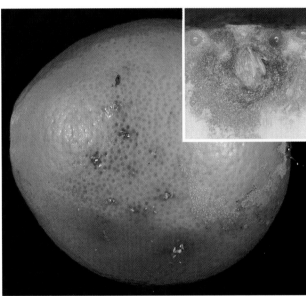

Plate 16.3 Queensland fruit fly 'stings' on green mature Meyer lemon. A clear gum has exuded from the stings, and the rind around them has yellowed. Inset, a cross-section through the rind at a sting site shows the eggs deposited beneath the rind.

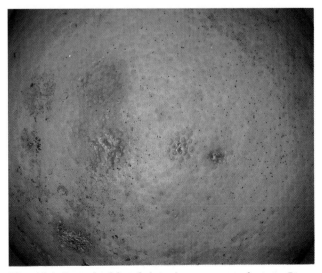

Plate 16.4 Queensland fruit fly 'stings' on mature navel orange. Rots are developing around the egg-laying sites.

Fly populations are highest in citrus during summer–autumn, when many varieties are maturing. However, numbers can build up rapidly on susceptible fruit, e.g. on Meyer lemons, in spring–summer.

▶ Hosts

An extremely wide range of tropical, subtropical and temperate fruits are attacked.

▶ Origin and distribution

Queensland fruit fly is native to the east coast of Australia.

Damage

Fruit
Destroyed by larvae feeding inside the fruit, and by rots.

Citrus fruit are not usually 'stung' until green mature, and showing some colour on one side of the fruit. The area around the 'sting' yellows, and often clear or brown gum exudes from the sting.

Stung mature fruit become infested with maggots and rot, and may drop from the tree. Maggots also develop successfully in ripe and fallen citrus fruit. However, compared to fruits like guavas and peaches, citrus fruits are not good hosts for fruit flies, as the citric acid in the fruit pulp and the oil in the rind kill quite a few of the eggs and larvae.

▶ Varieties attacked

All citrus varieties are attacked, with Meyer lemon and grapefruit being most susceptible, and Lisbon lemon least susceptible. The periods when fruit are most vulnerable to attack are listed below.

Table 16.1 Periods when Queensland fruit fly attacks fruit in Queensland and New South Wales

Fruit	Period when attacked
Queensland and northern NSW	
Meyer lemons (green mature)	early November – June
kumquats	early November–June
grapefruit	early January – June
navel oranges	early February – June
early mandarins, e.g. Imperials	early February – June
Siletta and Joppa oranges	early April – June
Hickson and Ellendale mandarins	early April – June
Murcott mandarins	May–June, early August (if still on the tree)
late Valencia and Seville oranges	early August (if still on the tree)
NSW	
grapefruit	spring (if still on tree), autumn
navel oranges	spring (if still on tree), autumn
Lisbon lemons	from October (if winter crop fruit still on tree)

Natural enemies

▶ Parasites

The main natural enemies of Queensland fruit fly are the braconid wasps *Opius perkinsi*, *Fopius deeralensis*, *Fopius arisanus* and *Diachasmimorpha tryoni*. These wasps are about 8 mm long. Females use their long ovipositor to parasitise fruit fly larvae inside fruit. At times these wasps are quite common, but they do not significantly reduce fruit fly numbers.

▶ Predators

The assassin bug (*Pristhesancus plagipennis*), praying mantises, spiders and birds prey on adult Queensland fruit flies.

Management

▶ Monitoring

Male Queensland fruit flies are attracted by the synthetic pheromone sold as 'Cue-lure', which resembles the substance females produce to attract males. It can be used in small traps to detect increases in fly activity. (These traps do not contribute significantly to controlling the fruit flies, as only males are caught.)

The Cue-lure trap will attract males within a radius of 400 metres or more. A trap placed near the edge of an orchard can give a false estimate of fly numbers in the orchard itself.

- Place one Cue-lure trap per 5–10 hectares, in the centre of the block (in a shady spot inside one of the trees) to determine if fly numbers remain low or are increasing.
- Empty traps weekly (or fortnightly) and record the numbers of flies caught.
- The synthetic pheromone, usually in a cotton wick, will last for 6 months. The insecticide also in the wick (to kill flies entering the trap), needs renewing once or twice a season.
- In addition, in small blocks of highly susceptible cultivars, especially in coastal areas, check for adults in trees or on fruit while monitoring for other pests or while doing orchard maintenance.

▶ Action level

The relationship between the numbers of male Queensland fruit flies caught in traps, and the numbers of females in an orchard is not clear. However, assume that less than 10 male

Plate 16.5 Cue-lure trap used to catch male Queensland fruit flies.

flies per trap per week usually indicates low activity, while a surge to 50 flies per week would require action to protect susceptible fruit.

Take action if even one female fly is seen stinging fruit within the orchard.

▸ Appropriate action

Victoria, South Australia, inland New South Wales

The inland citrus-producing areas of New South Wales, Victoria, and South Australia are maintained as fruit-fly-free areas. Departments of agriculture in these states operate continual fruit-fly detection programs. As soon as a certain number of flies are detected, an eradication program begins.

Queensland, coastal New South Wales

Fruit fly can be controlled by baiting, or by using cover sprays. Baiting is preferred because it is as effective as cover spraying, less costly and, importantly, much less disruptive to natural enemies of other pests.

The bait consists of yeast autolysate mixed with an appropriate insecticide. It is applied in a coarse, low-pressure spray low on the tree skirt at a rate of about 100 mL per tree or 20–30 L/ha every 7 days during periods when fruit are susceptible to attack.

Absence of fly activity 7 days after baiting indicates that baiting can be done at 7-day intervals. Activity after 5 days indicates that the baits should be applied twice weekly until activity decreases.

Baiting must start at least a month before any fruit reach the stage where they are susceptible to attack, and be continued regularly until harvest. For most citrus varieties, baiting needs to start in early January and continue until harvest.

▸ Additional management notes

Small areas of trees (under 0.25 ha) are difficult to protect from fruit fly by using baits.

Small blocks of Meyer lemons (0.25–1.0 ha) will need to be bait-sprayed twice weekly during periods of heavy fly activity from December to February.

Continual heavy rain can reduce the efficacy of bait sprays or cover sprays, and spraying should be carried out twice weekly if rain persists. If possible apply sprays during a break in the weather.

Orchard hygiene is important in controlling fruit flies. Remove fallen fruit and soak it for at least 3 days in water topped with kerosene. Alternatively, slash between rows to destroy it.

Mediterranean fruit fly (medfly)

Ceratitis capitata (Wiedemann), Diptera: Tephritidae

Figure 16.3 Importance and distribution of Mediterranean fruit fly (medfly).

Plate 16.6 Adult Mediterranean fruit fly (medfly).

Description

▸ General appearance

Mediterranean fruit fly, or medfly, adults are 4–5 mm long, and slightly smaller than Queensland fruit fly. Medfly has a yellow body marked with white, brown, blue and black, mottled wings, and pale-green eyes.

Larvae, or maggots, are cream-coloured with a pointed head and squarish rear end. Mature larvae are 8 mm long.

▸ Distinguishing features

Medfly is the only fruit fly attacking citrus in Western Australia.

In the field it is impossible to distinguish between eggs, maggots and pupae of medfly and Queensland fruit fly, but adults can easily be told apart. The Queensland fruit fly has been eradicated from Western Australia, and the medfly occurs only in Western Australia.

Smaller ferment flies (*Drosophila* spp.), whose larvae are found only in fruit that are already damaged or rotting, can often be seen flying around rotting fruit, and should not be confused with primary fruit flies (i.e. major pest species that cause initial damage to fruit).

▶ Life cycle

Female medflies can lay up to 1000 eggs during their life span. These are deposited (in batches of 2–30) beneath the skin in the albedo of the ripening fruit. Eggs are 1 mm long. Larvae hatch from the eggs and tunnel into the fruit pulp to complete their development.

There is heavy mortality of eggs and young larvae, particularly in immature fruit, caused by oil released from oil cells in the rind ruptured during egg laying. In thicker skinned varieties, larval death follows the formation of gum in and around the egg-laying site.

When fully grown, larvae drop from the fruit onto the ground. Pupation takes place in the top 50 mm of soil. The adult then emerges from the brown, barrel-like pupa and forces its way to the surface.

Adult female flies must feed on protein, e.g. from bacteria growing on fruit and plant surfaces, and on sugars, e.g. in honeydew and nectar, for several days before they can mature and lay their eggs. They can live for up to 6 months.

The whole life cycle takes 4–17 weeks, depending on temperature.

▶ Seasonal history

There are 4–5 generations per year.

▶ Habits

Medfly adults over-winter in citrus trees. Numbers fall in winter, and start increasing in spring. Populations are highest in late summer and early autumn. The flies cause most damage in citrus at this time, especially to early maturing varieties. This coincides with the end of the season for deciduous fruits. Mature deciduous fruit are a good breeding place for fruit flies, which, at the end of the season when there are no more fruit, then migrate onto ripening citrus fruit.

Flies may move away from an orchard if there are no more ripe fruit, but medflies are not as mobile as Queensland fruit flies. Flies are often seen basking on the sunny side of trees.

▶ Hosts

Medfly attacks a very wide range of deciduous and subtropical fruits, with over 200 hosts recorded. In Western Australia, stonefruit and loquats are particularly susceptible.

▶ Origin and distribution

Medfly is thought to be native to north Africa. It is a pest throughout subtropical regions and regions with a Mediterranean climate, with the exception of those in South-East Asia and North America. In Australia, it is found only in Western Australia between Esperance and Carnarvon.

Damage

Fruit
Destroyed by larvae feeding inside the fruit, and by rots.

Medfly attacks fruit that are beginning to colour. The damage results from puncturing of the rind during egg laying, and larvae feeding on the fruit pulp. In addition, organisms such as green mould enter the fruit through the punctures, and rots develop.

▶ Varieties attacked

All varieties of citrus can be attacked. Thin-skinned mandarins and oranges are most susceptible, while lemons are rarely damaged.

Natural enemies

Although numerous attempts have been made to establish wasp parasites of medfly in Western Australia, none have been successful.

No other natural enemies have been identified.

Management

▶ Monitoring

Some idea of the numbers of medflies in an orchard can be obtained by trapping flies.

Traps baited with Trimedlure or Capilure (either sticky traps, or dry traps where flies are killed with a pest strip) can be used to capture male medflies.

Wet traps baited with protein, ammonia or sugar can be used to capture female flies, but these traps are much less effective than the traps for males.

▶ Action levels

No action levels have been developed for medfly in citrus in Western Australia. Growers begin regular baiting either when the fruit is half-grown, or 6 weeks before the fruit begins to colour.

▶ Appropriate action

Good control of medfly can be achieved in most years by baiting with yeast autolysate or protein hydrolysate in combination with an insecticide. Baiting is preferred to cover spraying with insecticides, as cover spraying kills important beneficial organisms that control other pests, such as mites, scales and mealybugs.

Growers generally apply approximately 100 mL of bait mixture in a strip low on the tree, once a week. At the peak of the season, it may be necessary to apply bait twice a week. If the bait is applied in the early morning at the correct rate, leaf burn will be minimised.

▶ Additional management notes

Orchard hygiene is important in controlling fruit flies. Remove fallen fruit, and soak it for at least three days in water topped with kerosene. Alternatively, slash between rows to destroy it.

Medfly can be difficult to control when numbers are very high, or in orchards close to towns or neglected orchards. In these situations, a community baiting scheme is necessary to ensure effective control.

Neglected orchards should be eradicated, and backyard and hobby growers should control fruit flies by applying bait sprays.

Papaya fruit fly

Bactrocera papayae Drew and Hancock, Diptera: Tephritidae

Figure 16.4 Importance and distribution of papaya fruit fly.

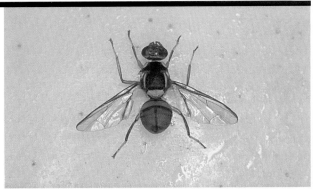

Plate 16.7 Adult papaya fruit fly.

Description

▸ General appearance
The adult papaya fruit fly is similar in appearance to Queensland fruit fly, and related species. The eggs, larvae and pupae are virtually indistinguishable from those of related species.

▸ Distinguishing features
The papaya fruit fly has a dark T-shaped mark on the top of the abdomen, with the stem of the T running from the fifth to third segments. Microscopic examination is necessary to distinguish papaya fruit fly from solanum fruit fly (*Bactrocera cacuminatus*) and the rainforest species *Bactrocera endriandrae*. (These two species are not normally found on citrus, but may be attracted from nearby hosts into traps placed in citrus.)

Smaller ferment flies (*Drosophila* spp.), whose larvae are found only in fruit that are already damaged or rotting, can often be seen flying around rotting fruit, and should not be confused with primary fruit flies (i.e. important pest species that cause initial damage to fruit).

▸ Life cycle
The life cycle of papaya fruit fly is similar to that of Queensland fruit fly.

The adult female fly feeds for up to a week on protein, e.g. on bacteria growing on plant surfaces, and on sugars, e.g. in honeydew and nectar, before laying eggs.

Eggs are white, banana-shaped and 1 mm long. They are deposited in batches of 10–12 into the spongy albedo of the fruit. The rind puncture is not visible at first, but later a yellow area develops around the 'sting' site.

Eggs hatch in 2–3 days and the larvae burrow into the fruit pulp. The number of larvae per fruit varies from one to 12 or more. In summer, larvae complete their development in about 10 days. Pupation occurs in the soil, and during summer can take as little as 9 days. Pupae are brown, barrel-shaped and about 4–5 mm long.

▸ Seasonal history
Papaya fruit fly has only recently been found in far north Queensland, and little is known of its seasonal history.

▸ Habits
The habits of papaya fruit fly are similar to those of Queensland fruit fly.

Adults are strong fliers capable of travelling many kilometres. Flies usually over-winter as inactive adults, and while they may be observed in the orchard, there is often little egg laying until August–September. During spring, fruits other than citrus, e.g. peaches, nectarines, loquats, guavas, mulberries and native fruits, are important hosts. Fruit flies may breed on these hosts, and then large numbers may migrate onto nearby citrus in spring–summer.

Fly populations are highest in citrus during summer–autumn, when many varieties are maturing. However, numbers can build up rapidly on susceptible fruit in spring–summer.

Papaya fruit fly infests a similar range of fruits to Queensland fruit fly, and a much wider range of vegetable crops. A key difference is the ability of papaya fruit fly to infest greener fruit than Queensland fruit fly can.

▸ Hosts
Papaya fruit fly attacks most fruits and many vegetables (Solanaceae and Cucurbitaceae).

▸ Origin and distribution
Papaya fruit fly is native to South-East Asia, and is found in south Thailand, Malaysia, Singapore, Borneo, Indonesia, and Papua New Guinea. In Australia, it is found in only a small area of far north Queensland, and attempts are being made to eradicate it.

Damage

Fruit
Destroyed by larvae feeding inside and by rots.

▸ Varieties attacked
All citrus varieties are susceptible, and fruit are attacked at an early stage of growth.

Natural enemies

Parasites
In South-East Asia, the main natural enemies of papaya fruit fly are braconid wasps, *Fopius* spp. and *Diachasmimorpha* spp. These wasps are about 8 mm long. Females use their long ovipositor to parasitise fruit fly larvae inside fruit.

Management

Monitoring
Male papaya fruit flies are monitored by using the synthetic pheromone methyl eugenol (which is like the substance females produce to attract males). This chemical can attract flies from up to a kilometre away. It must be replaced about once every 8 weeks.

Methyl eugenol can be used in small traps to detect increases in fly activity, in the same manner that Cue-lure is used to monitor Queensland fruit fly. (These traps do not contribute significantly to controlling the fruit flies, as only males are caught.)

Action level
Any fly activity is undesirable.

Appropriate action
It is hoped that killing male flies, and baiting susceptible crops will eradicate this fly from Australia. Male flies are killed when they are attracted to blocks of material impregnated with synthetic pheromone and pesticide. Several of these blocks are placed in every square kilometre of trees.

There are also plans for ongoing mass release of sterilised male flies. Females mate once only, and when they mate with a sterile male, produce infertile eggs.

Yeast autolysate bait sprays, as used for Queensland fruit fly, give effective control in citrus orchards. The bait consists of yeast autolysate mixed with an appropriate insecticide. It is applied in a coarse, low-pressure spray low on the tree skirt at a rate of about 100 mL per tree or 20–30 L/ha every 7 days during periods when fruit are susceptible to attack.

Absence of fly activity 7 days after baiting indicates that baiting can be done at 7-day intervals. Activity after 5 days indicates that the baits should be applied twice weekly until fly activity decreases.

Baiting must start at least 6 weeks before any fruit reach the stage where they are susceptible to attack (this is 2 weeks earlier than for Queensland fruit fly). Baiting should be continued regularly until harvest or until fly activity ceases.

Citrus blossom midge

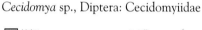
Cecidomya sp., Diptera: Cecidomyiidae

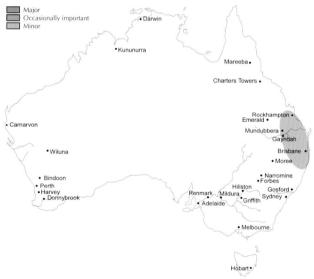

Figure 16.5 Importance and distribution of citrus blossom midge.

Plate 16.8 Swollen flower bud dissected to reveal numerous translucent larvae of citrus blossom midge.

Description

General appearance
The adult citrus blossom midge is a small, fragile fly about 2 mm long. The larva is cream-coloured, 2–3 mm long and without legs. The head of the larva is not readily visible, and the body tapers toward the front end.

Distinguishing features
This is the only midge found in citrus flowers.

Life cycle
The short-lived female citrus blossom midge lays up to 50 eggs inside the flower bud. Eggs hatch within 1–2 days, and larvae feed on the developing flower for about a week before pupating. The whole life cycle takes about 14 days.

Seasonal history
There are at least 12 generations per year.

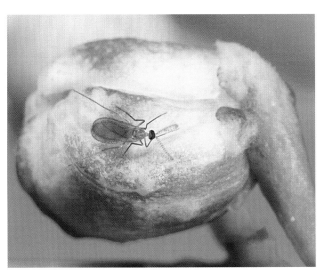

Plate 16.9 Adult male citrus blossom midge. This midge is resting on a swollen flower bud containing midge larvae.

Plate 16.10 Swollen flower bud (right) infested with citrus blossom midge larvae. There are three normal buds on the left.

▶ Habits
The large number of larvae (up to 50) inside the flower bud causes it to swell to two to three times normal size. When the bud is broken open, the larvae 'skip' or flick away (moving in a way similar to fruit fly larvae).

▶ Hosts
Citrus is the only known host.

▶ Origin and distribution
Citrus blossom midge is native to eastern Australia.

Damage

 Flowers
Floral parts destroyed, and the flowers drop.

Over 50% of flowers can be attacked on some trees. However citrus blossom midge is not a serious problem, because of the large numbers of flowers usually set on individual trees. Infestations also tend to be sporadic.

▶ Varieties attacked
All citrus varieties are attacked, but particularly lemons.

Natural enemies
No natural enemies have been identified.

Management

▶ Monitoring
- Citrus blossom midge occurs on all citrus varieties, but particularly lemons.
- The pest should be monitored every 14 days during flowering in September.
- Using a ×10 hand lens, check for larvae in flowers on 5 randomly selected flower racemes per tree.
- The number of trees from which samples should be taken depends on block size (see chapter 25).
- Additional care should be taken when monitoring blocks that have a history of infestation.

▶ Action level
Action is required when 50% or more of flowers are infested on all trees monitored.

▶ Appropriate action
Apply a single spray of a selective insecticide.

17 Wasps (Hymenoptera)

Wasps are insects with a narrow waist and two pairs of membranous wings, the smaller hindwings being interlocked with the forewings. The mouthparts are adapted for chewing, lapping or sucking. The adult females have an ovipositor modified for sawing, piercing or stinging, as well as for egg laying.

Larvae have a distinct head, and chewing mouthparts. The larvae of some species have thoracic legs, others have abdominal legs, others have both, and some have no legs.

Most wasps active in Australian citrus are beneficial parasites of citrus pests, but the citrus gall wasp damages citrus trees. The parasitic wasps and the citrus gall wasp are minute to small in size (0.5 mm long to 7 mm long).

Citrus gall wasp

Bruchophagus fellis (Girault), Hymenoptera: Eurytomidae

Figure 17.1 Importance and distribution of citrus gall wasp.

Plate 17.1 Female citrus gall wasp laying eggs into a young twig.

Description

▶ General appearance
The adult citrus gall wasp is small, shiny-black and 2.5 mm long. Larvae are 2 mm long, thickset, white and have no legs.

▶ Distinguishing features
This is the only wasp that directly attacks citrus. The woody galls it forms are quite distinctive (see plate 17.2).

▶ Life cycle
Adults emerge in most areas from mid-September until early November. They mate and immediately begin egg laying, as they live only about a week. The female wasp lays eggs between the bark and wood of young spring-flush twigs. Twigs in which eggs have been deposited have scar-like flecks on the bark. Egg laying continues throughout October and finishes by about mid-November.

Eggs hatch in 2–4 weeks, and are all hatched by early December. The young larvae burrow into the bark, and a flask-like sheath of soft host tissue develops around each larva. By late December, woody tissue begins to form around the sheath of soft tissue, the twig swells and begins to develop the characteristic gall.

Larval development continues, and pupation takes place within the galls in late winter. The pupal stage lasts 2–3 weeks, after which adults emerge.

▶ Seasonal history
There is one generation per year.

▶ Habits
Over the last century, since the introduction and widespread planting of cultivated citrus varieties, this pest has spread from the border ranges between Queensland and New South Wales to most citrus areas of Queensland and northern New South Wales. The larval development of citrus gall wasp in the young twig can produce galls up to 250 mm long and 25 mm thick which can contain hundreds of larvae.

▶ Hosts
The native host is the finger lime (*Microcitrus australasica*).

▶ Origin and distribution
Citrus gall wasp is native to Australia.

Damage

 Twigs
Galls up to 250 mm long and 25 mm thick; dieback if infestation is heavy.

Heavily infested trees can be covered with galls, resulting in very little leaf or fruit production, and severe dieback.

Plate 17.2 Galls formed on grapefruit twigs as a result of attack by citrus gall wasp. Note the small holes through which adults have emerged.

▶ Varieties attacked

All citrus varieties can be attacked, but there are marked differences in susceptibility. Rough lemon and Troyer citrange rootstocks are very susceptible, while grapefruit is the most susceptible cultivated variety.

Lemons and oranges can be seriously affected. Mandarins are the least susceptible.

Natural enemies

▶ Parasites

Two small wasps (*Megastigmus brevivalvus* and *Megastigmus trisulcus*), about 2.5 mm long, are very important natural enemies of citrus gall wasp (*M. brevivalvus* in particular).

Females are honey-coloured with red eyes. *M. brevivalvus* has a plumper abdomen and a short black ovipositor about a third of the length of the body, while *M. trisulcus* has a more slender abdomen and an ovipositor about the same length as its body. Males are black on top and brown underneath.

M. brevivalvus begins to emerge from parasitised gall wasp larvae about a fortnight after the first gall wasps emerge, with most emerging in October. *M. trisulcus* emerges from late October to early November. After emerging, the parasites locate gall wasp eggs in the young twigs with their ovipositors, and lay their eggs within the gall wasp eggs. The parasite egg hatches after the gall wasp larva has hatched, and the parasite larva develops slowly in the host larva for several months. Its development accelerates in the spring, and it finally destroys its host. Over 90% of gall wasp larvae can be parasitised.

When the parasites are active, the gall wasp population is much reduced, and galls tend to be small (20–30 mm long), and fewer in number.

Management

▶ Monitoring

- Citrus gall wasp is more common in grapefruit, oranges and lemons.
- Monitor 3 young branches (300 mm long) per tree in mid-winter for the presence or absence of galls.

Plate 17.3 Female wasp Megastigmus brevivalvus *parasitising gall wasp eggs in a twig. Note the thin ovipositor inserted into the twig.*

Plate 17.4 Female wasp Megastigmus trisulcus, *a parasite of citrus gall wasp. Note the long ovipositor at the rear.*

- To check levels of parasitism, collect about 20 galls (2 or 3 from each of 10 trees per block) in late August, and keep them in a plastic container with a fine mesh lid. Gall wasps will emerge first for about 10 days, then *Megastigmus brevivalvus* (if present) will emerge for 2–4 weeks. If present, *Megastigmus trisulcus* will emerge last, towards the end of October.
- If no parasites emerge by mid-October from the galls being monitored, *Megastigmus* wasps should be released immediately. Both species are available commercially for purchase.
- The number of trees from which samples should be taken depends on block size (see chapter 25).
- Additional care should be taken when monitoring blocks that have a history of infestation.

▶ Action level

The action level is 33% or more of branches infested with galls exceeding 50 mm in length, and parasites present in low numbers.

▶ Appropriate action

If action is necessary, there are two choices. Release parasites in October if none are present, and forego spraying. Alternatively, apply a recommended pesticide between the last week in November and the first week of December, especially when there is a serious infestation. The timing of pesticide application is critical: spray when citrus gall wasp eggs have hatched, and before woody tissue has started to form around larvae.

▶ Additional management notes

In spring, heavily flecked young twigs indicate that citrus gall wasps are laying large numbers of eggs in the current spring growth (see plate 17.5).

Heavily galled trees will benefit from a heavy pruning during winter.

Avoid planting large areas of grapefruit in coastal areas of Queensland and New South Wales.

Do not allow shoots to develop on rough lemon or Troyer citrange rootstocks in the orchard, as these can become heavily infested with citrus gall wasp.

Plate 17.5 Gall wasp parasite Megastigmus brevivalvus *laying eggs into the eggs of citrus gall wasp in a twig. The extensive brown stippling on the twig is caused by large numbers of citrus gall wasps laying eggs.*

Paper wasps

Common paper wasp *Polistes humilis synoecus* Saussure, yellow paper wasp *Polistes dominulus* (Christ), *Polistes tepidus* (Fabricius), Hymenoptera: Vespidae

Figure 17.2 Importance and distribution of paper wasps.

Plate 17.6 Nest of the common paper wasp.

Plate 17.7 Nest of an unidentified species of paper wasp in a Queensland citrus tree.

Description

▶ General appearance

The common paper wasp and the yellow paper wasp are slightly longer than a honeybee, and *Polistes tepidus* is markedly longer. These wasps have more slender bodies than a honeybee, and a distinct narrow waist. The tip of the abdomen is more pointed than that of the honeybee. The wings are longitudinally folded when the insect is at rest.

▶ Distinguishing features

The body of the common paper wasp is light brown, with bands of black and yellow on the abdomen.

The yellow paper wasp is yellow and black. The antennae of this wasp are black at the base and orange-brown for the rest of their length.

Polistes tepidus is a large wasp, about 25 mm in length.

▶ Life cycle

The paper wasp colony starts in spring or summer with a fertile female, who constructs the first few cells of the nest from chewed wood fibres. An egg is laid in each cell and hatches after a few days. The young larvae are fed on nectar and chewed insect prey. When the larvae are fully grown, they spin a cocoon within the cell. This gives the exposed ends of the cells a whitish cap.

After pupation, adult wasps emerge. The new wasps are worker females, who take over nest construction, food gathering and feeding the developing brood.

Later in the season, fertile females and drones are produced. After mating, these reproductive wasps leave the nest, which is abandoned by all the wasps at the beginning of winter. The reproductive females then hibernate over the winter, before starting new colonies in the spring.

Nests are about 125 mm in diameter, and attached to a twig. Paper wasps build their nests in trees, such as citrus. In orchards, they attack pickers and pruners when disturbed.

▶ **Seasonal history**
There are several generations per year.

▶ **Habits**
Adult paper wasps feed mainly on nectar from flowers, while larvae are fed caterpillars and other insect prey. These wasps also need plenty of moisture, and can often be seen around ponds and pools of water.

▶ **Hosts**
Paper wasps nest in a wide range of shrubs and trees.

▶ **Origin and distribution**
The common paper wasp was introduced into south-west Western Australia from the eastern states in about 1950.

The only state where yellow paper wasp is found is Western Australia. The wasp was introduced into the south-west of the state from overseas in 1977. It is more numerous than the common paper wasp.

Polistes tepidus is common in the Bundaberg area of Queensland.

Damage

Paper wasps do not cause any damage to citrus plants. They are regarded as pests in the orchard because they sting orchardists and their staff.

Natural enemies

Nothing is known about the natural enemies of these paper wasps.

Management

▶ **Monitoring**
Check regularly for the presence of active nests.

▶ **Action level**
Destroy nests as they are found.

▶ **Appropriate action**
Wear protective clothing when destroying wasp nests. The most suitable time to destroy a nest is in the late afternoon, near dark. Give the nest an extended burst of spray from a can of commercially available household insecticide. Be prepared to leave the area in a hurry. Remove and destroy the nest the next day.

18 Thrips (Thysanoptera)

Thrips are small, slender insects with short antennae. Some grow to a length of about 12 mm, but most species grow to a length of 3–4 mm. Some species have long, narrow wings with a fringe of long hairs on the hind edge, and others are wingless. Thrips have mouthparts adapted for sucking and rasping.

The life cycle of thrips is unusual. The juvenile stages, or instars, look similar to the adults, and have similar feeding habits. The first two instars have no wing pads. These suddenly appear in the third instar, and enlarge greatly in the fourth. The fourth instar is quiescent and does not feed. In some species which have a fifth instar, the fourth instar enters the soil and forms a cocoon in which the quiescent fifth instar is passed. The quiescent stage can be referred to as a pupa, the first two instars as larvae, and the third instar as a propupa.

Many species of thrips are economically significant pests of a wide range of crops. Flowers are often favoured feeding sites, but thrips also feed on developing fruit. Some species of thrips are predatory, feeding on small insects and mites.

Monitoring
The following points apply to the monitoring of thrips and their natural enemies:

- The number of trees from which samples are taken depends on block size (see chapter 25 for details).
- Additional care should be taken when monitoring blocks that have a history of economic damage.

Scirtothrips

Scirtothrips albomaculatus Bianchi, *Scirtothrips dorsalis* Hood, Thysanoptera: Thripidae

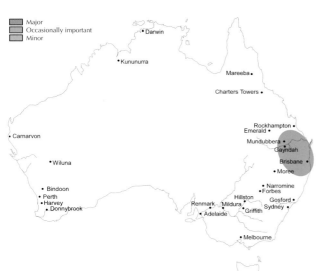

Figure 18.1 Importance and distribution of scirtothrips.

Plate 18.1 Adult scirtothrips. Note the feathery wings folded along the back.

Description

▶ General appearance
Scirtothrips adults are small, slender, orange to yellow insects about 1.5 mm long, with two pairs of dark wings fringed with long hairs.

▶ Distinguishing features
Scirtothrips is similar in appearance to rust thrips, but rust thrips has two dark marks at the base of the wings.

▶ Life cycle
The eggs (0.3 mm long) are usually laid into soft tissue on very young fruit near the calyx, or in young leaves (near the midrib or main veins). Larvae develop through two stages to become a propupa. They then pupate, and emerge as adults. The complete life cycle takes about 5 weeks.

▶ Seasonal history
There are at least 6 generations per year.

▶ Habits
Scirtothrips feed on young plant tissue—leaves, shoots, blossoms and small fruit. On fruit, they prefer to feed in protected areas such as under the calyx, or between touching fruit.

▶ Hosts
These two species of scirtothrips have a wide host range, including mango, strawberry and mangosteen.

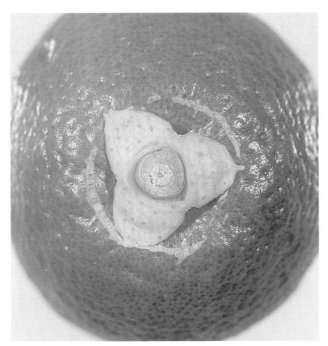

Plate 18.2 Ring scarring around the stem end of a navel orange caused by Scirtothrips dorsalis.

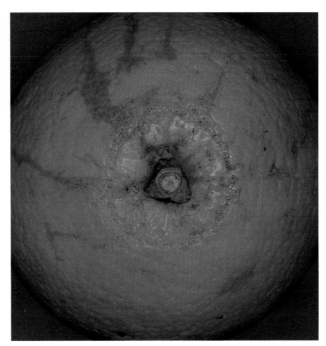

Plate 18.3 Scirtothrips damage to a mature navel orange. There is ring scarring around the stem end, and scarring extending down the sides.

▶ Origin and distribution
Scirtothrips spp. are serious pests of citrus in South-East Asia, South Africa and California, but in Australia so far have been a problem only in Queensland and northern New South Wales.

Damage

 Fruit
Disfiguring of young fruit (up to 3 months old), especially by ring scarring (halo damage) and scurfing.

 Leaves
Young foliage scarred and twisted.

Ring scarring at the stem end of fruit is characteristic of scirtothrips damage. If infestations are heavy, scarring can extend down the sides of the fruit. Fruit can also become scarred in protected areas between fruit clusters where scirtothrips have been feeding.

Megalurothrips causes similar scarring. The damage caused by scirtothrips and megalurothrips tends to remain grey to white in colour, while that caused by citrus rust thrips is a rusty brown.

Heavy infestations of scirtothrips cause twisting and grey scarring of young foliage similar to the damage caused by broad mite (see pages 26–27).

Scirtothrips damage can easily be confused with wind damage (see page 13). However, wind damage does not produce the characteristic ring scarring at the stem end of the fruit.

▶ Varieties attacked
In Queensland, navel orange has been most affected. Occasional outbreaks occur around flowering, but damage can also occur between touching fruit later in the season. Other varieties attacked include lemons, limes and Eleanor mandarin.

Natural enemies

▶ Parasites
Parasitic wasps are known to attack thrips, but their importance is unknown.

▶ Predators
Predatory thrips such as *Haplothrips* spp., predatory phytoseiid mites, predatory bugs, and spiders have been found in association with thrips populations, but no direct evidence of their effect on the thrips has been collected.

▶ Pathogens
Thrips are susceptible to fungal diseases caused by *Beauveria* spp., *Metarhizium* spp. *Entomophthora* spp., and *Paecilomyces fumosa rosea*. These diseases are more likely to be effective control agents in humid regions.

Management

▶ Monitoring
- Scirtothrips attack lemons, limes and mandarins, but are more common on navel oranges.
- In areas where thrips have previously been a problem, sample 5 young fruit per tree every week from petal fall to calyx closure.

▶ Action level
Action is required when 5% or more of fruit are infested with scirtothrips.

▶ Appropriate action
Bait sprays may be the key to thrips management and their use is being evaluated. The use of organophosphate pesticides should be minimised because they are disruptive to biological control systems, and thrips may quickly become resistant to them.

Megalurothrips

Megalurothrips kellyanus (Bagnall), Thysanoptera: Thripidae

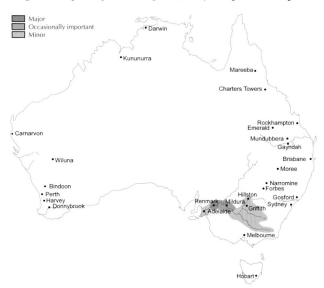

Figure 18.2 Importance and distribution of megalurothrips.

Description

▶ **General appearance**
Adult megalurothrips are black, 2–3 mm long, and have two pairs of wings, fringed with long hairs. The larvae resemble the adults, but are smaller, ranging in size up to 2 mm. They are lemon-yellow and have no wings.

▶ **Distinguishing features**
Adult megalurothrips are black, with black legs and dark wings, and this colouring distinguishes them from other thrips found on citrus.

▶ **Life cycle**
The life cycle of megalurothrips is similar to that of scirtothrips, and comprises the following stages: egg; two larval stages; propupa; pupa; and adult.

▶ **Seasonal history**
There are at least 6 generations per year.

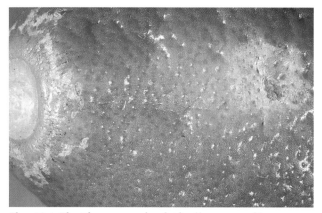

Plate 18.4 Blemish on stem end and side of lemon caused by megalurothrips.

Plate 18.5 Adult megalurothrips and rind scarring on lemon.

▶ **Habits**
Megalurothrips is active all year round in the Sunraysia and Riverland districts of southern Australia, and causes problems only in these areas. Most damage occurs in late spring but can occur in summer and autumn also.

Megalurothrips starts feeding on fruitlets shortly after flowering has finished. The insects congregate in sheltered sites, such as between touching fruit and under the calyx.

▶ **Hosts**
Citrus is the only known host.

▶ **Origin and distribution**
Megalurothrips is native to Australia.

Damage

 Fruit
Disfiguring of young fruit (up to 3 months old), especially by stem-end scarring, ring scarring (halo damage) and scurfing; rind discolouration on mature fruit.

Ring scarring is mainly caused by young megalurothrips from November to April. Rind discolouration on mature fruit is caused by young megalurothrips and adults from April to harvest. This damage is caused when the thrips feed in protected areas between touching fruit.

Damage caused by megalurothrips is similar to that caused by scirtothrips. The damage caused by scirtothrips and megalurothrips tends to remain grey to white in colour, while that caused by citrus rust thrips is a rusty brown.

Damage caused by megalurothrips is also similar to wind damage, but wind damage does not result in ring scarring around the calyx, or in the whitish to yellow discolouration between touching fruit.

▶ **Varieties attacked**
All citrus varieties can be attacked, but lemons are particularly susceptible. Navel oranges may be more seriously affected

where lemons are grown close by. Serious damage can occur on navels in late spring to early summer. Damage is less common on mandarins and other varieties with small calyxes.

Natural enemies

Parasites
Parasitic wasps are known to attack megalurothrips, but their importance is unknown.

Predators
Predatory thrips such as *Haplothrips* spp., predatory phytoseiid mites, predatory bugs, and spiders have been found in association with thrips populations, but no direct evidence of their effect on the thrips has been collected.

Pathogens
Thrips are susceptible to fungal diseases caused by *Beauveria* spp., *Metarhizium* spp. *Entomophthora* spp., and *Paecilomyces fumosa rosea*. These diseases are likely to be more effective control agents in humid areas.

Management

Monitoring
- All citrus varieties are attacked, with young lemons being most susceptible. Of the mature fruit, Leng navel oranges are most susceptible to damage.
- Sample 5 young fruit per tree every week from petal fall to calyx closure. Check between touching fruit, particularly late in the season.
- A minimum of 50 fruit should be sampled.

Action level
Action is required when 5% or more of fruit are infested before April, and 10% from April onwards.

Appropriate action
Bait sprays may be the key to thrips management and their use is being evaluated. Minimise the use of organophosphate pesticides, because they are disruptive to biological control systems, and thrips may quickly become resistant to them.

Citrus rust thrips (orchid thrips)

Chaetanaphothrips orchidii (Moulton), Thysanoptera: Thripidae

Figure 18.3 Importance and distribution of citrus rust thrips.

Plate 18.6 Adult citrus rust thrips. Note the feathery wings with dark spots at the base, folded along the back.

Plate 18.7 Larvae of citrus rust thrips on navel orange.

Description

General appearance
Adult thrips are orange to yellow, cigar-shaped insects, about 1.5 mm long. They have narrow, fringed wings with black markings.

Distinguishing features
Rust thrips is similar in appearance to scirtothrips, but rust thrips has two dark marks at the base of the wings.

Life cycle
Eggs are laid in the fruit or leaf tissue. Larvae develop through two stages before becoming a propupa, then a pupa. Pupation takes place in the soil, and then the adult emerges. The complete life cycle takes 3–5 weeks in summer.

Seasonal history
There are several generations per year.

Habits
Citrus rust thrips prefer protected sites between touching fruit, or where leaves touch fruit surfaces.

Hosts
Citrus rust thrips also attacks a number of glasshouse plants.

Origin and distribution
This insect is also found in Europe, Florida and Israel.

Plate 18.8 Characteristic brown marks on Siletta oranges made by citrus rust thrips. Damage is usually between touching fruit (top), but occasionally at the stem end of fruit (bottom).

Damage

Fruit
Circular rust marks between touching fruit or where leaves touch.

The presence of citrus rust thrips is characterised by the brown rusty marks they make on the rind of fruit. This damage is similar to that caused by citrus rust mite, but limited to the areas where fruit touch or where leaves touch fruit.

The damage caused by scirtothrips and megalurothrips tends to remain grey to white in colour.

▸ Varieties attacked
All citrus varieties are attacked, but particularly navel orange, grapefruit and Siletta orange.

Natural enemies

▸ Parasites
Parasitic wasps are known to attack thrips, but their importance is unknown.

▸ Predators
Predatory thrips such as *Haplothrips* spp., predatory phytoseiid mites, predatory bugs, and spiders have been found in association with thrips populations, but no direct evidence of their effect on the thrips has been collected.

▸ Pathogens
Thrips are susceptible to fungal diseases caused by *Beauveria* spp., *Metarhizium* spp., *Entomophthora* spp., and *Paecilomyces fumosa rosea*. These diseases are more likely to be effective control agents in humid regions.

Management

▸ Monitoring
- Citrus rust thrips attack all citrus varieties, but particularly navel orange, grapefruit and Siletta orange. It mostly damages semi-mature and mature fruit which are touching other fruit or leaves.
- Where thrips have previously been a problem, sample 5 touching fruit per tree weekly from January until harvest.

▸ Action level
Action is required if 10% or more of fruit are infested.

▸ Appropriate action
Apply a single spray of a selective pesticide. Petroleum spray oil can be used (see chapter 27). Thorough coverage is very important. Spraying should be carried out in midsummer, before young green fruit come into contact with each other, and repeated if required.

Early harvest of Valencia oranges (before Christmas in New South Wales) will help reduce damage.

Greenhouse thrips

Heliothrips haemorrhoidalis (Bouché), Thysanoptera: Thripidae

Figure 18.4 Importance and distribution of greenhouse thrips.

Plate 18.9 Greenhouse thrips: left, adult; right, pupa.

Description

▸ General appearance
Adult greenhouse thrips are black, about 1.5 mm long, with yellowish legs and two pairs of narrow wings fringed with hairs. Larvae are white or pale yellow. Newly emerged adults are black and brown. Males have been found only in Peru.

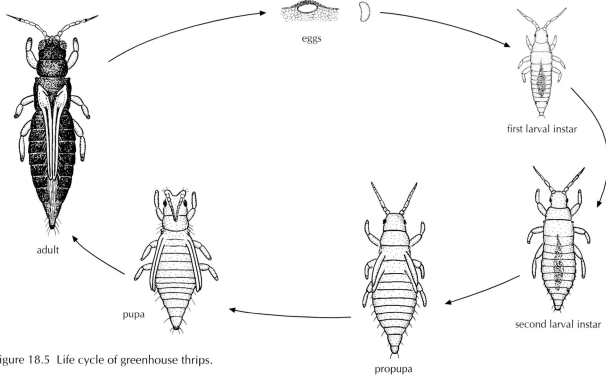

Figure 18.5 Life cycle of greenhouse thrips.

▶ Distinguishing features
Adult megalurothrips are also black, but have black legs.

There are some predatory thrips which have black adult stages. These should not be confused with greenhouse thrips.

The feeding sites of greenhouse thrips are often littered with dark faeces.

▶ Life cycle
The small eggs (0.3 mm long) are laid singly in leaf or fruit tissue. They hatch into larvae which are almost transparent. There are two larval stages followed by a propupa, pupa and the adult. The adults lay 2–3 eggs per day, and live for up to 3 months. The complete life cycle takes around 5 weeks.

▶ Seasonal history
There are at least 6 generations per year.

▶ Habits
Greenhouse thrips prefer to feed between touching surfaces of leaves or fruit, or under spider webbing. When infestations are heavy, thrips feed over the entire surface of fruit and leaves.

▶ Hosts
Other hosts include mango, avocado and a range of subtropical fruits. Greenhouse thrips are also very common on ornamentals and in glasshouses.

▶ Origin and distribution
Greenhouse thrips are found throughout the world.

Damage

Fruit
Grey scars, generally between touching fruit, sprinkled with black spots of excreta.

Leaves
Grey scarring and yellowing.

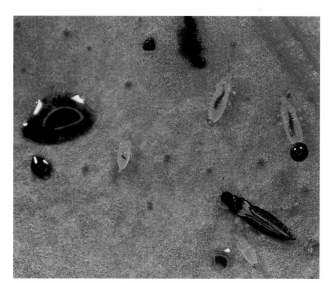

Plate 18.10 Adult and larval greenhouse thrips with inky excrement held at the tip of the abdomen. There is also excrement, some coalesced into large drops, on the leaf surface.

Plate 18.11 Damage to two mature Valencia oranges caused by greenhouse thrips. The raised scarring is on the areas of rind that were touching on the tree.

Plate 18.12 Grey scarring of leaves caused by greenhouse thrips. Note the numerous black spots of thrips excreta.

Plate 18.13 The wasp Thripobius semiluteus, *a parasite of greenhouse thrips.*

The rind of damaged fruit becomes soft and flaccid after harvest. Young and mature fruit are equally susceptible to attack.

▶ Varieties attacked
All citrus varieties are attacked, but late Valencia oranges are worst affected.

Natural enemies

▶ Parasites
The major parasites of greenhouse thrips are the wasps *Thripobius semiluteus* and *Megaphragma mymaripenne*. Another parasite, *Goetheana shakespeari* (= *parvipennis*) may have been displaced by *T. semiluteus*. Parasitism is highest in autumn, and appears to peak too late in the season to prevent significant damage. Greenhouse thrips are also parasitised by the wasp *Ceranisus* sp.

▶ Predators
Predatory phytoseiid mites, e.g. *Euseius elinae*, attack thrips eggs and small thrips larvae. Predatory thrips may also be important.

▶ Pathogens
Unidentified fungal diseases occasionally reduce populations of greenhouse thrips.

Management

▶ Monitoring
- Greenhouse thrips attack all citrus varieties, but are more common on Valencia oranges.
- Where greenhouse thrips have previously been a problem, sample 5 pairs of touching fruit per tree, from October to early March.

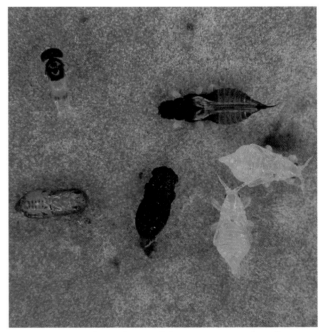

Plate 18.14 Greenhouse thrips and the wasp parasite Thripobius semiluteus. *Top right, adult thrips; bottom right, two thrips propupae; bottom left, newly formed wasp pupa; bottom centre, wasp pupa; top left, adult wasp.*

▶ Action level
The action level is 10% or more of fruit pairs infested.

▶ Appropriate action
Apply a single spray of a selective pesticide. Petroleum spray oil can be used (see chapter 27). Thorough coverage is very important.

Spraying should be carried out in midsummer, after the harvest of mature fruit and before young green fruit come into contact with each other, and may be repeated if required.

Early harvest of Valencia oranges (before Christmas in New South Wales) will help reduce damage.

Plague thrips

Thrips imaginis Bagnall, Thysanoptera: Thripidae

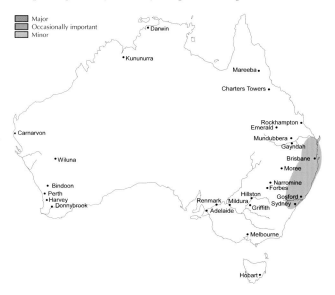

Figure 18.6 Importance and distribution of plague thrips.

Plate 18.15 Plague thrips feeding in a citrus flower.

Description

▶ **General appearance**
Adult female plague thrips are narrow-bodied, light brown or grey, and about 1.2 mm long. Males are smaller and yellow. Both sexes have two pairs of narrow, delicate wings fringed with long hairs.

▶ **Distinguishing features**
Plague thrips are commonly found in and near blossom in spring, often in huge numbers.

▶ **Life cycle**
The eggs are laid in soft tissues of blossoms and young leaves. The larva develops through two stages before becoming a propupa, then a pupa. Pupation takes place in the soil, and then the adult emerges. The complete life cycle takes about 10 days in summer.

▶ **Seasonal history**
There are at least 12 generations per year.

▶ **Habits**
Plague thrips is present throughout the whole year, but frequently increases to plague numbers in spring. Plagues generally occur in dry sunny springs after mild winters preceded by autumns with above-average rainfall.

Infestations on citrus are usually the result of adult thrips migrating, abruptly and in very large numbers, from a wide range of weeds, grasses and other flowering plants.

▶ **Hosts**
Plague thrips infest a wide range of host plants, including other fruit trees, ornamentals, grasses and weeds.

▶ **Origin and distribution**
Plague thrips is native to Australia. It is also found in New Zealand, Fiji, New Caledonia and New Guinea.

Damage

 Flowers
Possible petal browning caused by feeding; small, brown, scab-like blisters on pistils, calyxes and peduncles (flower stalks) caused by egg laying.

 Leaves
Chlorosis (yellow spotting) and scar-like blistering on young leaves.

No damage is caused to fruit, and fruit set is not affected. Damage to leaves is barely detectable when the leaves mature.

Plague thrips does not cause economic damage to citrus. However, it is important to recognise this pest and the damage it causes, so that unnecessary measures are not taken to control it.

Plate 18.16 Scar-like blistering on mandarin calyxes and flower stalks caused by egg laying of plague thrips.

▶ Varieties attacked
All citrus varieties can be infested.

Natural enemies

Natural enemies are unlikely to have significant effects on infestations of plague thrips on citrus. However, heavy rain can reduce infestations spectacularly.

▶ Predators
A number of predators have been found in association with plague thrips, but no direct evidence of their effects has been documented.

▶ Pathogens
Thrips are susceptible to fungal diseases caused by *Beauveria* spp., *Metarhizium* spp., *Entomophthora* spp., and *Paecilomyces fumosa rosea*.

Management

Plague thrips are found in and around the flowers of all citrus varieties, most commonly in spring. However, as these thrips do not cause economic damage, monitoring and control are not required.

Plate 18.17 Chlorotic spots caused by plague thrips on young mandarin leaves.

Termites (Isoptera) 19

Termites are social insects which live in colonies in nests. The colonies are made up of insects of different castes: reproductives, soldiers and workers. The soldiers and workers are wingless. Workers, often numbering in the millions in one colony, forage for food, travelling along tunnels or galleries that may extend over a hundred metres from the central nest.

Giant northern termite

Mastotermes darwiniensis Froggatt, Isoptera: Mastotermitidae

Figure 19.1 Importance and distribution of giant northern termite.

Plate 19.1 Giant northern termites tunnelling in the trunk of a citrus tree. These termites belong to the worker caste.

Description

▶ General appearance

The giant northern termite is the largest termite in Australia, with workers 15 mm long, and queens (reproductive females) 20 mm long. The workers are white, and sometimes appear translucent. They are wingless, with round heads, medium-length antennae, and small eyes. The soldiers' bodies are similar to those of workers, but they have enlarged heads and massive mandibles, or jaws.

The queen is similar to the workers, but larger, with well-developed eyes. Initially, she has two pairs of transparent wings, but after she mates, her wings fall off. She then settles down to become an egg-producing factory, with a greatly distended abdomen.

▶ Distinguishing features

No other termites commonly attack citrus trees.

▶ Life cycle

Queens lay eggs in pods of about 20. Nymphs hatch from the eggs. Only workers or soldiers are produced early in the life of the colony. Winged reproductive termites do not develop until the colony is several years old.

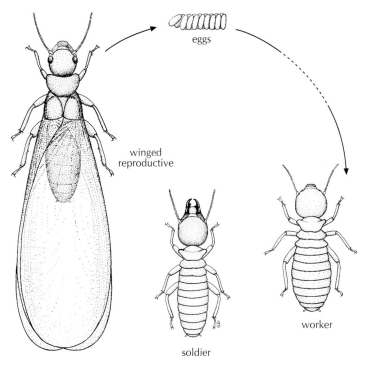

Figure 19.2 Life cycle of giant northern termite.

Colonies can become very large, comprising millions of individuals. In a big colony, there are numerous egg-producing secondary queens. New colonies readily branch off by establishing their own queen.

Plate 19.2 *Citrus tree affected by giant northern termite. Note the gallery on the trunk.*

▶ Seasonal history
Queen termites continually produce eggs. Winged reproductives are produced each year in mature colonies.

▶ Habits
Giant northern termites are active all year round in northern Australia. Their nests are found at or below ground level, usually in a log or a stump. The giant termite does not occur in rainforest.

This termite invades and quickly destroys citrus trees, moving in from the soil and up through the trunk and limbs, and down into the roots.

The range of operation of a colony can be over 100 metres. The galleries are mostly within 250 mm of the surface. Colonies are often difficult to find.

▶ Hosts
The giant northern termite attacks a wide range of fruit and ornamental trees.

▶ Origin and distribution
The giant northern termite is native to Australia, but also occurs in New Guinea.

Damage

Branches
Main limbs destroyed.

Trunk
Centre eaten out and tree destroyed.

Roots
Major roots destroyed.

A tree infested with termites is killed when it collapses, or when the termites ringbark it from the inside. Evidence of termite activity can usually be seen on the outside of the tree in the form of surface galleries.

▶ Varieties attacked
All citrus varieties are susceptible.

Natural enemies

The echidna probably preys on giant northern termites. Little is known about other natural enemies.

Management

▶ Monitoring
In the tropics it is important to inspect the trunks of trees during routine operations, or when monitoring for other pests. Check also for signs of wilting that may be caused by termite activity.

▶ Action level
Once an infestation is detected, infested trees must be given prompt attention.

▶ Appropriate action
Apply the appropriate pesticide to the trunk and base of the tree.

▶ Additional management notes
Pieces of timber such as logs or stumps are potential nest sites for colonies of termites. If possible, remove any such timber from within 100 metres of the orchard.

Keep the ground under the trees weed-free.

Other insects and mites

Insects

The following insect pests have been recorded on citrus in Australia. However, their relatively minor importance does not warrant detailed information about their biology and control.

Table 20.1 Other insect pests of citrus in Australia

COMMON NAME	SCIENTIFIC NAME	NOTES ON DISTRIBUTION AND DAMAGE
Armoured scales		
Oriental scale	*Aonidiella orientalis* (Newstead), Hemiptera: Diaspididae	Queensland and Northern Territory
Soft scales		
Seychelles scale	*Icerya seychellarum* (Westwood), Hemiptera: Margarodidae	Northern Territory, most of Asia, Pacific islands, and southern Africa; may be native to Australia
—	*Icerya aegyptiaca* (Douglas), Hemiptera: Margarodidae	Northern Territory; possibly native to India or east Asia
Mealybugs		
Golden mealybug	*Nipaecoccus aurilanatus* (Maskell), Hemiptera: Pseudococcidae	Northern Territory
True bugs		
Fruitspotting bug	*Amblypelta lutescens lutescens* (Distant), Hemiptera: Coreidae	On Clementine mandarin in the Northern Territory, mandarins in north Queensland, and kumquats throughout Queensland
Passionvine bug	*Fabrictilis gonagra* (Fabricius), Hemiptera: Coreidae	Queensland, New South Wales; feeds on fruit and shoots
Rutherglen bug	*Nysius vinitor* Bergroth, Hemiptera: Lygaeidae	New South Wales, Queensland; swarms in citrus, originating from drying weeds in spring, can cause rapid severe wilting, leaf drop, and twig dieback
Grey cluster bug	*Nysius clevelandensis* Evans, Hemiptera: Lygaeidae	Northern Territory, Queensland, New South Wales
Moths		
Leaf case moth	*Hyalarcta huebneri* (Westwood), Lepidoptera: Psychidae	Coastal New South Wales and Queensland; feeds on leaves, causing serious damage to individual trees
Cluster caterpillar	*Spodoptera litura* (Fabricius), Lepidoptera: Noctuidae	Northern Territory
Fruit-tree borer	*Maroga* spp., Lepidoptera: Oecophoridae	Northern Territory; wood borer
Northern citrus butterfly	*Princeps fuscus canopus* (Westwood), Lepidoptera: Papilionidae	Northern Territory
Beetles		
Pitted apple beetle	*Geloptera porosa* Lea, Coleoptera: Chrysomelidae	Coastal New South Wales; chews holes in leaves and green bark in early summer
Mottled flower scarab beetle	*Protaetia fusca* (Herbst), Coleoptera: Scarabaeidae	Northern Territory, Queensland, New South Wales
Dried-fruit beetle	*Carpophilus* spp., Coleoptera: Nitidulidae	All states; usually associated with damaged or rotting fruit
Large auger beetle	*Bostrychopsis jesuita* (Fabricius), Coleoptera: Bostrychidae	Northern Territory, Queensland; destructive borer; one generation per year
—	*Myllocerus* sp., Coleoptera: Curculionidae	Northern Territory; leaf-feeding weevil
Whitestriped weevil	*Perperus lateralis* Boisduval, Coleoptera: Curculionidae	Coastal New South Wales; foliage feeder

(continued)

Table 20.1 Other insect pests of citrus in Australia (continued)

COMMON NAME	SCIENTIFIC NAME	NOTES ON DISTRIBUTION AND DAMAGE
Thrips		
Onion thrips	*Thrips tabaci* Lindeman, Thysanoptera: Thripidae	Central coast of New South Wales; rare; reported to cause skin blemish shortly after fruit set
Tomato thrips	*Frankliniella schultzei* (Trybom), Thysanoptera: Thripidae	Northern Territory
Fruit flies		
Northern Territory fruit fly	*Bactrocera aquilonis* (May), Diptera: Tephritidae	Northern Territory; increasingly important
Jarvis' fruit fly	*Bactrocera jarvisi* (Tryon), Diptera: Tephritidae	Queensland and Northern Territory
Island fruit fly	*Dirioxa pornia* (Walker), Diptera: Tephritidae	Coastal Queensland and New South Wales; common; not a true pest; infests damaged fruit
Grasshoppers		
Spur-throated locust	*Austacris* (= *Nomadacris*) *guttulosa* (Walker), Orthoptera: Acrididae	Queensland, New South Wales and Northern Territory; foliage feeder

Mites

The vegetable spider mite (*Tetranychus neocaledonicus* (André), Acarina: Tetranychidae) is a minor pest that has been found on citrus in the Northern Territory. As for the insects in the table above, its relatively minor importance does not warrant detailed information about its biology and control.

Many mites found on citrus, however, are NOT pest species and cause no problems to the citrus grower. They feed on fungi, or decaying plant and animal material on the bark, leaves and fruit, but do not damage any part of the citrus plant. For example, the mites listed in the table below have been found under the calyx of oranges in inland citrus districts of Australia. None of them are pests, and the predatory species are all beneficial, even though they may not all be major predators of pest species. (For more information on some of the fungi-feeding and scavenging mites, see pages 35 and 36.)

Table 20.2 Mite species found under the calyx of oranges in Australian inland citrus districts

FAMILY	SPECIES	RATE OF OCCURRENCE*
Fungi-feeding and scavenging mites		
Acaridae	various *Acarus* species	common
Glycyphagidae	various *Glycyphagus* species	rare
Oripodidae	various species	common
Tarsonemidae	*Tarsonemus waitei*	rare
	other *Tarsonemus* species	uncommon
Tydeidae	*Tydeus californicus*	common
	other *Tydeus* species	common
	Pronematus species	common
Predatory mites		
Bdellidae	*Bdella* species	uncommon
Cunaxidae	*Cunaxa* species	rare
Eupalopsellidae	*Eupalopsis jamesi*	uncommon
	other *Eupalopsis* species	uncommon
Eupodidae	*Eupodes* species	rare
Hemisarcoptidae	*Hemisarcoptes* species	rare
Parasitidae	*Parasitus* species	rare
Phytoseiidae	*Amblyseius* species	uncommon
Stigmaeidae	*Agistemus* species	common

* Rate of occurrence

very common	1 in 5 fruit
common	1 in 20 fruit
uncommon	1 in 100 fruit
rare	1 in 1000 fruit

Spiders

Spiders are arachnids, a class of arthropods that also includes scorpions, ticks and mites. Their bodies are in two parts: a cephalothorax, and an abdomen. They have four pairs of legs. Most spiders have eight eyes arranged in two rows. Spiders prey on insects and other small animals.

Spiders (Araneida) — 21

Brown house spider (webbing spider)

Badumna longinqua L. Koch, Araneida: Desidae

Figure 21.1 Importance and distribution of brown house spider.

Plate 21.1 Webs of the brown house spider in a citrus tree.

Description

▶ **General appearance**
Brown house spiders are small, grey-brown and gather together in groups.

▶ **Distinguishing features**
Citrus foliage is webbed together by these spiders.

▶ **Life cycle**
These spiders lay eggs and pass through spiderling stages before becoming adults.

▶ **Seasonal history**
One generation takes 1–2 years.

▶ **Habits**
Brown house spiders spin webs to catch insects for food. In citrus orchards, there are usually more of these spiders and webs in boundary rows next to scrub. They are more common in drier inland areas.

▶ **Hosts**
Brown house spiders also infest other fruit trees.

▶ **Origin and distribution**
The brown house spider is native to Australia.

Damage

Besides catching pest insects, the webs of the brown house spider snare numerous natural enemies of pests. Fruit inside the webs are often infested with scales and mealybugs.
There are numerous spiders in the webs, and they are a nuisance to fruit pickers.

▶ **Varieties attacked**
This spider occurs on all citrus varieties.

Natural enemies

Nothing is known about the natural enemies of the brown house spider.

Management

▶ **Monitoring**
Watch for excessive webbing in trees.

▶ **Action level**
Action is required if fruit picking is interfered with.

▶ **Appropriate action**
Apply an appropriate spray with coarse jetting to break up the webs. Spot-spraying by hand can also be successful. Thick webs may also need to be broken up with a stick.

Nematodes

Nematodes are one of the most numerous kinds of animals in the world. They are small, mostly microscopic, worms. Some live in animal tissues, some in water, some in the soil (where they feed mainly on detritus or decaying matter), and some are parasites of plants, including citrus.

The most important and widespread nematode which parasitises citrus in Australia is the citrus nematode (*Tylenchulus semipenetrans*). Others which sometimes cause problems, are the root lesion nematode (*Pratylenchus* spp.) and the stubby root nematode (*Paratrichodorus* sp.). Several other species of plant parasitic nematodes have been found in association with citrus trees in Australia and other countries, but they have not been found to be economic pests. These include *Paratylenchulus* sp., *Xiphinema* sp. and *Hemicriconemoides* sp.

22 Nematodes (Nematoda)

Citrus nematode

Tylenchulus semipenetrans Cobb., Nematoda: Tylenchida

Description

▶ General appearance

Citrus nematodes exist in the soil as infective larvae, which are visible only under a microscope. As they mature, they swell into a shape resembling a lemon.

▶ Distinguishing features

Unlike root-knot nematodes which become completely buried in root tissue, citrus nematodes remain with their hindquarters in the soil.

▶ Life cycle

Mature female citrus nematodes produce up to 100 eggs in a sticky gelatinous egg mass which is extruded into the surrounding soil. The eggs hatch within 12 days in the presence of moisture, and the infective larvae enter the soil and travel up to 3 metres to a new root to feed.

The mature larvae develop into females, partially embedded in the root, which are capable of producing offspring in the absence of males (i.e. parthenogenetically). In times of stress when water or food is scarce, infective larvae develop into male nematodes which do not feed on the roots, but mate with the females to produce offspring with different gene combinations.

The complete life cycle takes 6–8 weeks when soil temperatures are between 25° and 30°C.

▶ Seasonal history

There are 5–7 generations per year.

▶ Habits

The larva of citrus nematode pierces a citrus root, usually just behind the growing tip, and then buries its head in the root. Saliva released from the mouth of the nematode causes the root to form a feeding site made up of three or so giant nurse cells around the head of the nematode. The nematode

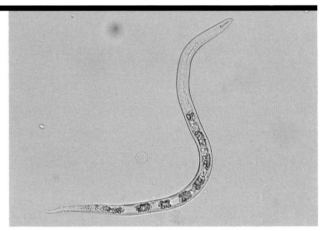

Plate 22.1 Citrus nematode larva. The fine stylet with stylar knob at the anterior end (top of photograph) is used to pierce plant roots. The body tapers towards the rear end.

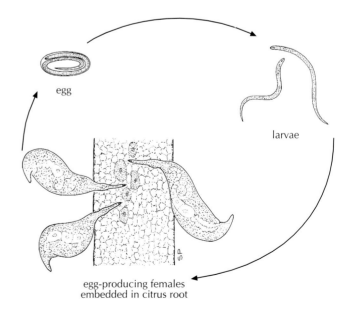

Figure 22.1 Life cycle of citrus nematode.

feeds on these cells by piercing them with its needle-like mouthparts and sucking out some of the cell contents, but not enough to kill the cells.

Citrus nematodes take many years to kill a citrus tree, but yields of trees on susceptible rootstocks can be noticeably reduced early in the life of the trees.

Citrus nematodes are most abundant in the warmer months of September through to April, though they can still be found in low numbers in winter.

▶ Hosts

There are several different strains (biotypes) of citrus nematode, each with a different host range. Hosts other than citrus include grape, olive and persimmon.

▶ Origin and distribution

Citrus nematodes are found in all citrus growing areas of the world.

Damage

  Roots
Capacity to carry water and nutrients reduced.

▶ Varieties attacked

All varieties of citrus are attacked. However, some rootstocks are highly resistant to citrus nematode attack (e.g. trifoliate orange), some are moderately tolerant (e.g. Troyer and Carrizo citrange), and some are highly susceptible (e.g. sweet orange).

Natural enemies

Nematodes are attacked by a range of pathogens and predators in the soil, including nematode-trapping fungi, spore-forming bacteria, and tardigrades (plump, mite-like animals less than 0.5 mm long, which live in water films).

Management

▶ Monitoring

It is difficult to diagnose attack by citrus nematodes, as the nematodes cannot be seen with the naked eye. However, roots which have been attacked usually have a knobbly, gritty, dirty appearance when the soil is shaken from them. This is because soil adheres to the sticky egg masses extruded by the females.

Larvae may be detected in the soil by extracting the nematodes from the soil with special procedures, and identifying them under a microscope. Nematode numbers are often expressed as a number per 500 g of soil.

If you suspect nematode problems in your citrus trees or in replant soils, a soil and root sample should be submitted to a reputable laboratory for diagnosis. Most government departments of agriculture, and some consultants, provide such services.

The soil sample should be taken from the root zone, between the trunk of the tree and the drip zone, in the top 300 mm of soil. It should contain some roots and soil. Most diagnostic laboratories require between 500 g and 1000 g of soil.

Take smaller samples from at least 10 sites in each area where a nematode count is required and mix these into one sample for each area. Separate samples should be taken from areas where soil, drainage or past crops differ.

Store the samples in plastic bags in the shade, or in the refrigerator if they must be stored for some time before delivering to the laboratory. A separate sample, of roots only, may be required if numbers of root lesion nematodes are also to be estimated.

▶ Action level

Citrus nematodes are considered to be at damaging levels if there are more than 10 000 larvae per 500 g of soil. Up to 50 000 larvae per 500 g of soil have been found in some citrus orchards.

Another way of assessing nematode numbers is to stain the nematodes on the roots and count them. A count of 60–70 adult citrus nematodes per gram of root is considered to be damaging.

▶ Appropriate action

Most recent citrus plantings in Australia are on rootstocks tolerant of or resistant to nematodes. Tolerant rootstocks allow nematodes to exist on their roots but are able to perform well in spite of the nematodes. Rootstocks such as Troyer and Carrizo citrange fall into this category. Resistant rootstocks, e.g. trifoliate orange, are poor hosts for the nematode, and populations are low on these rootstocks.

The nematode tolerance and resistance of common citrus rootstocks used in Australia is given in table 1.1 (page 4). If nematode numbers are high, then replanting on tolerant or resistant rootstocks is recommended as the best way to control nematodes in citrus.

In new planting sites care must be taken to avoid the introduction of nematodes on the planting material. Replanting with susceptible rootstocks is not recommended in sites with high nematode numbers but if it is necessary, preplant fumigation or growth of a non-host crop for 1–3 years will help to reduce nematode numbers in the soil.

Cultural practices which optimise tree health will reduce the effects of nematodes. Nematodes do not usually kill mature trees but interfere with water and nutrient uptake, so relatively more nutrients may be required when large numbers of nematodes are present. Increasing the organic matter in the soil by using cover crops and organic fertilisers may enhance the performance of biological control organisms.

Pesticides are available for use against nematodes in established orchards but they are expensive and can reduce the levels of beneficial organisms in the soil.

Plate 22.1 Swollen rear end of a female citrus nematode protruding from a root.

Root lesion nematodes

Pratylenchus spp., Nematoda: Tylenchida

Description

▶ General appearance
Root lesion nematodes are small worms, visible only under a microscope.

▶ Distinguishing features
Blackening and sloughing of the root sheath are evidence of attack by root lesion nematode. However, these symptoms are similar to those caused by phytophthora root rot.

▶ Life cycle
Root lesion nematodes can go through all stages of their life cycle completely within the tree root but, when food supplies become scarce, infective larvae and adults will emerge into the soil to travel to new roots.

The species of *Pratylenchus* most commonly found in Australia are *P. brachyurus* and *P. vulnus*. Both species usually reproduce without mating. Males are rare.

Eggs are laid singly in roots and/or soil. Larvae pass though four stages before becoming adults. The life cycle takes 30–80 days, depending on soil temperature.

▶ Seasonal history
There are usually 2–3 generations per year.

▶ Habits
Root lesion nematodes are amongst the most primitive of the nematode species attacking citrus. They are migratory endoparasites, and do not set up a feeding site in one spot on the citrus root, but travel through the root, feeding as they go and severely damaging root tissue.

They are most abundant in the warmer months of September through to April, though they can still be found in low numbers in winter.

▶ Hosts
Some species of root lesion nematode attack cover crops and weeds in the citrus orchard.

▶ Origin and distribution
Throughout the world nine species of root lesion nematode have been found to attack citrus, and several of these species are present in the soils of Australian citrus groves.

Damage

Roots
Damaged root tissue dies and forms lesions; roots darken and have large areas of brown discolouration.

Root lesion nematodes have not been directly associated with poor performance and decline of citrus trees in Australia, but they have in other countries. However, other more damaging organisms, such as the root-rotting fungi *Phytophthora* spp. and *Fusarium* spp., may enter roots through the injuries caused by root lesion nematodes.

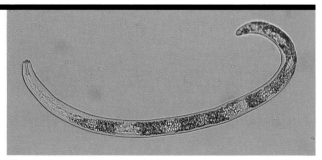

Plate 22.3 Root lesion nematode. Note the strong stylet, with stylar knob, at the anterior end (left).

▶ Varieties attacked
Most species of citrus are susceptible to root lesion nematodes and the resistance and tolerance of commonly used rootstocks is not known.

Natural enemies

Nematodes are attacked by a range of pathogens and predators in the soil, including nematode-trapping fungi, spore-forming bacteria, and tardigrades (plump, mite-like animals less than 0.5 mm long, which live in water films).

Management

▶ Monitoring
The presence of root lesion nematodes may be indicated by dark lesions on the roots, but the best way to determine if there are large numbers present is to incubate root samples and extract the nematodes. Some government departments of agriculture and some commercial laboratories offer this service.

The presence of root lesion nematodes can also be verified by extracting them from soil samples, using methods similar to those for citrus nematodes. However, soil extraction cannot be used to estimate numbers of these nematodes; the process of root extraction must be used instead.

▶ Action level
Action levels have not been accurately determined for root lesion nematodes in citrus in Australia, but they are thought to be around 500 nematodes per 500 g soil.

▶ Appropriate action
Little is known about the resistance or tolerance to root lesion nematodes of the commonly used citrus rootstocks. However, it is probable that rootstocks which are tolerant of citrus nematodes will also be tolerant of root lesion nematodes.

Cultural practices which optimise tree health will reduce the effects of nematodes. Nematodes do not usually kill mature trees, but interfere with water and nutrient uptake, so relatively more nutrients may be required when large numbers of nematodes are present. Also, damage caused by root lesion nematode increases the tree's susceptibility to root diseases. Increasing soil organic matter by using cover crops and organic fertilisers may enhance the performance of biological control organisms.

Pesticides are available for use against nematodes in established orchards, but they are expensive and can reduce the levels of beneficial organisms in the soil.

Stubby root nematodes

Trichodorus spp., *Paratrichodorus* spp., Nematoda: Tylenchida

Description

▶ **General appearance**
Stubby root nematodes are small worms, visible only under a microscope.

▶ **Life cycle**
Stubby root nematodes are ectoparasitic, spending their entire life in the soil. Little is known about the life cycle. However, eggs are laid singly in the soil. The complete life cycle takes 16–22 days between the temperatures of 22°C and 30°C.

▶ **Seasonal history**
The seasonal history of stubby root nematodes has not been studied in Australia.

▶ **Habits**
Stubby root nematodes are migratory, travelling through the soil and browsing on the roots with their needle-like mouth parts. They usually attack growing roots just behind the root tips and cause severe stunting of root systems, because damaged roots are unable to grow.

▶ **Hosts**
Little is known about the hosts of stubby root nematode.

▶ **Origin and distribution**
Stubby root nematodes are found throughout the world and have been associated with severe damage to citrus in some countries, including the USA.

Damage

 Roots
Possible severe root stunting if large numbers of nematodes present.

Stubby root nematodes occasionally cause problems in citrus nurseries in Australia, but do not usually cause problems in mature citrus trees.

▶ **Varieties attacked**
Most varieties of citrus are susceptible to stubby root nematodes and the resistance and tolerance of commonly used rootstocks is not known.

Natural enemies

Nematodes are attacked by a range of pathogens and predators in the soil, including nematode-trapping fungi, spore-forming bacteria, and tardigrades (plump, mite-like animals less than 0.5 mm long, which live in water films).

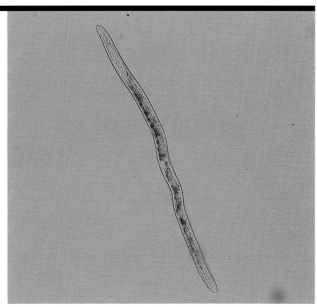

Plate 22.4 Stubby root nematode. The stylet at the anterior end (top of photograph) is slightly curved and has no knob. The body is cigar-shaped.

Management

▶ **Monitoring**
The presence of stubby root nematodes is indicated when a tree has short, stubby roots. The nematodes themselves are detected using soil extraction methods. Some government departments of agriculture and some commercial laboratories offer this service. If the presence of stubby root nematodes is suspected, take care to handle the soil sample gently, as nematodes in the sample can be killed by being knocked or jolted.

▶ **Action level**
The action level for stubby root nematodes in citrus in Australia has not been determined.

▶ **Appropriate action**
Little is known about the resistance or tolerance of the commonly used citrus rootstocks to stubby root nematodes.

Cultural practices which optimise tree health will reduce the effects of nematodes. Nematodes do not usually kill mature trees but interfere with water and nutrient uptake, so relatively more nutrients may be required when large numbers of nematodes are present. Increasing the organic matter in the soil by using cover crops and organic fertilisers may enhance the performance of biological control organisms.

Pesticides are available for use against nematodes in established orchards but they are expensive and can reduce the levels of beneficial organisms in the soil.

Snails

Terrestrial snails are molluscs belonging to the class Gastropoda, which also includes marine and freshwater snails, and slugs. Snails have a spiral shell and a large muscular foot. They have a mouth with a feeding organ called a 'radula' made up of rows of tiny teeth, for rasping. Land snails feed on plants and may do severe damage, even to trees.

23 Snails (Stylommatophora)

Common garden snail

Helix aspersa (Muller), Stylommatophora: Helicidae

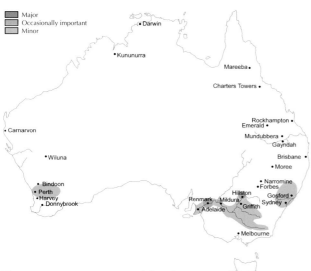

Figure 23.1 Importance and distribution of common garden snail.

Plate 23.1 *Common garden snail and damage to fruit.*

Description

▶ General appearance

The common garden snail has a brown shell with a distinctive, bold, striped pattern on the shell. It grows to a diameter of about 25 mm.

▶ Distinguishing features

This snail is rarely confused with native snails, because it is usually the only species present in orchards in significant numbers.

▶ Life cycle

Snails are hermaphrodites, and all those of reproductive age can lay several batches of 50–100 eggs a year, during prolonged periods when conditions are favourable, usually in spring and autumn.

The eggs are white, spherical, about 3 mm in diameter, and have a soft shell. They hatch into small snails in about 2 weeks. Development to maturity generally takes about a year, but the rate of development varies considerably depending on the conditions.

▶ Seasonal history

There is normally one generation per year.

▶ Habits

The common garden snail feeds on a wide variety of cultivated plants, and thrives in cultivated areas such as gardens and orchards. Sprinkler irrigation and no-till weed control with herbicides create ideal conditions for their development.

Snails feed at night, in the early morning, and during cool, damp overcast conditions at all times, including during irrigation. At other times they hide in sheltered places such as on tree trunks or in the tree skirt area under leaf litter. Damage is worst on the tree skirt, but where snail numbers are high, damage will extend into the upper part of the tree.

Snails are most active in the spring and autumn. During hot, dry weather, they seal the opening of their shell with a parchment-like membrane, and aestivate (enter a dormant or resting phase) in sheltered places. Cavities under stones and timber or branches lying on the ground make ideal resting sites. Aestivating snails are often found in crotches of trees during summer months.

▶ Hosts
Common garden snails attack many different kinds of plants. In the citrus orchard, many broadleaf weeds are snail hosts.

▶ Origin and distribution
The common garden snail originated in Europe, but has spread throughout the world.

Damage

 Fruit
Scarring of the rind and holes chewed through to the flesh.

 Leaves
Scarred and skeletonised.

 Twigs
Scarred and skeletonised.

The common garden snail can cause extensive damage to citrus fruit by scarring the rind and chewing holes through to the flesh. The damage is worst on navel oranges, particularly Thompson navel. Most of the damage is caused from March to June. In some seasons, if snails are not controlled, they can ruin the whole fruit crop.

Although leaves and twigs can be damaged as described above, this damage is not economically significant. Slight scarring can also occur on branches, but this is also insignificant.

▶ Varieties attacked
Damage is most severe on navel oranges, but all citrus varieties can be affected.

Natural enemies

▶ Parasites
Larvae of sciomyzid flies may parasitise garden snails.

▶ Predators
Some species of birds, e.g. thrushes, eat snails, as does the blue-tongue lizard.

Management

▶ Monitoring
Wooden boards about 30 cm × 25 cm with 30 mm cleats at either end can be used to monitor snail numbers. Place one board under the skirt of each of several trees throughout the orchard. The board provides shelter and moisture for the snails and they cluster on the underside.

Inspect the boards and count the snails on them once a month. The snails on the boards can then be destroyed. (However, this is too labour-intensive to be an effective control method.)

▶ Action level
No action levels have been established. Once fruit damage is observed, it is generally too late to achieve effective control. Generally, if economic damage has occurred in the previous

Plate 23.2 Common garden snail and damage to leaves.

season, it can be expected again and control measures are warranted.

▶ Appropriate action
Cultural control
The number of snails climbing trees can be reduced by skirt pruning and by applying a band of Bordeaux slurry or dry lime around the trunks of the trees once a year. Alternatively, copper bands can be fixed around the tree trunks. Once in place, copper bands can keep snails out of trees for 5 years or more, provided the tree skirt is kept clear of the ground.

Snails are rarely a problem in orchards that are regularly cultivated, such as furrow-irrigated orchards. Snail numbers are highest when overhead sprinklers and chemical weed control are used. With low-throw sprinklers, damage is often confined to the skirt of the tree.

Chemical control
Baiting and spraying are effective. Problem areas usually need treatment once a year. Treatment is most effective if carried out in March, before the snail's main activity period in autumn. Baiting and spraying are also effective during cool, damp weather when snails are active, especially at the end of a long period of hot, dry weather.

Distribute the bait in open areas around the tree skirts. After feeding, the snails are too weak to move back into the shade and are killed by the sun. If cool, moist weather persists, however, many of the affected snails may recover, especially after feeding on metaldehyde baits. (This is because the metaldehyde causes the snails to dehydrate, but in continuing moist conditions, they may be able to rehydrate.)

Spraying trees with copper-based fungicides helps to prevent snails from climbing the trees. Some carbamate insecticides are also effective, but their use should be minimised, as they disrupt IPM programs.

24 Natural enemies of citrus pests

Parasites

The most significant and best known parasites of citrus pests are wasps (insects of the order Hymenoptera) or flies (insects of the order Diptera). These parasites develop as larvae inside the body tissues of their host (in which case they are called 'endoparasites') or feed on the body tissues of their host from the outside (in which case they are called 'ectoparasites'). Both kinds of parasite ultimately kill the host.

The adult parasites are free-living, and fly in search of food and of hosts in or on which to deposit their eggs. Most adults feed on honeydew or nectar. Eggs are deposited in, on or near the host with a sword-like or tube-like ovipositor. The ovipositor in some small wasps is multi-functional, being used to first bore a hole into the host, then to paralyse the host, then to deposit the egg.

Some of these wasps also use their ovipositor to wound their hosts and then feed on the oozing body fluids. This is called 'host feeding' and it can cause up to 50% mortality in some scales, e.g. when aphytis wasps feed on red scale.

After hatching from the eggs, parasite larvae feed in or on the host, and pass through several developmental stages. They then pupate, either in or near the dying host. The adults emerge from the pupae, and the cycle recommences.

Most parasites attack specific hosts. For example, most encyrtid wasps parasitise scales and mealybugs only; some wasps, e.g. *Aphytis lingnanensis*, a scale parasite, attack three or four species; and others, e.g. *Anicetus beneficus*, attack only one host. The success of all parasites depends on being able to find a host, parasitise it, and reproduce quickly.

Insect parasites and the common names of their main hosts in Australian citrus orchards are listed in table 24.1. This list has been compiled from a range of published references (see reference list), and field collections.

Table 24.1 Important parasites attacking insect pests of citrus in Australia

ORDER AND FAMILY	GENUS	MAIN SPECIES	INSECT HOSTS
Hymenoptera (wasps)			
Aphelinidae	Aphelinus	sp.	aphids
	Aphytis	spp.	chaff scale, Glover's scale
	Aphytis	chrysomphali	red scale, yellow scale
	Aphytis	columbi	circular black scale
	Aphytis	holoxanthus	circular black scale
	Aphytis	lepidosaphes	mussel scale
	Aphytis	lingnanensis	red scale, citrus snow scale
	Aphytis	melinus	red scale
	Cales	noacki	Australian citrus whitefly
	Centrodora	darwini	spined citrus bug
	Centrodora	scolypopae	passionvine hopper
	Coccophagus	ceroplastae	pink wax scale, long soft scale
	Coccophagus	gurneyi	citrophilous mealybug
	Coccophagus	lycimnia	soft brown scale, citricola scale
	Coccophagus	near rusti	green coffee scale
	Coccophagus	semicircularis	soft brown scale, citricola scale
	Encarsia	citrina	red scale, yellow scale, citrus snow scale, circular black scale, oriental scale, mussel scale
	Encarsia	perniciosi	red scale
	Encarsia	spp.	Australian citrus whitefly, green coffee scale, pink wax scale
	Eretmocerus	spp.	whiteflies
	Euryischomyia	flavithorax	soft brown scale

ORDER AND FAMILY	GENUS	MAIN SPECIES	INSECT HOSTS
Encyrtidae	*Achalcerinys*	sp.	citrus planthopper
	Ageniaspis	*citricola*	citrus leafminer
	Anagyrus	*agraensis*	spherical mealybug
	Anagyrus	*fusciventris*	longtailed mealybug, citrophilous mealybug
	Anagyrus	*pseudococci*	citrus mealybug
	Anagyrus	sp.	citrus mealybug
	Anicetus	*beneficus*	pink wax scale
	Anicetus	*communis*	white wax scale
	Coccidoxenoides	*peregrinus*	citrus mealybug
	Comperiella	*bifasciata*	red scale, yellow scale
	Comperiella	*lemniscata*	oriental scale
	Diversinervus	*elegans*	soft brown scale, nigra scale, Florida wax scale
	Encyrtus	*infelix*	hemispherical scale
	Encyrtus	*lecaniorum*	soft brown scale
	Leptomastidea	*abnormis*	citrus mealybug
	Leptomastix	*dactylopii*	citrus mealybug
	Metaphycus	*bartletti*	black scale
	Metaphycus	*helvolus*	soft brown scale, black scale
	Metaphycus	*lounsburyi*	black scale
	Metaphycus	near *varius*	pink wax scale
	Microterys	*flavus*	soft brown scale, Florida wax scale
	Microterys	*triguttatus*	soft brown scale
	Ooencyrtus	spp.	fruitpiercing moths, fruitspotting bugs, planthoppers
	Paraceraptrocerus	*nyasicus*	white wax scale
	Tetracnemoidea	*brevicornis*	longtailed mealybug, citrophilous mealybug
	Tetracnemoidea	*peregrina*	longtailed mealybug
	Tetracnemoidea	*sydneyensis*	longtailed mealybug
Pteromalidae	*Acroclisoides*	*tectacorisi*	spined citrus bug
	Moranila	*californica*	wax scales
	Moranila	sp.	Florida wax scale
	Ophelosia	spp.	citrophilous mealybug, cottony cushion scale
	Pachyneuron	*kingsleyi*	citrus butterflies
	Pteromalus	*puparum*	citrus butterflies
	Scutellista	*caerulea*	black scale, hemispherical scale, nigra scale, pink wax scale, Florida wax scale, white wax scale, hard wax scale
Scelionidae	*Gryon*	*meridionis*	fruitspotting bugs
	Scelio	*flavicornis*	giant grasshopper
	Telenomus	spp.	fruitpiercing moths, corn earworm, native budworm
	Trissolcus	*basalis*	green vegetable bug
	Trissolcus	*oenone*	spined citrus bug
	Trissolcus	*ogyges*	spined citrus bug
Eulophidae	*Citrostichus*	*phyllocnistoides*	citrus leafminer
	Cirrospilus	near *ingenuus*	citrus leafminer
	Cirrospilus	*quadristriatus*	citrus leafminer
	Euplectrus	*kurandaensis*	banana fruit caterpillar
	Quadrastichus	sp.	citrus leafminer
	Semielacher	*petiolatus*	citrus leafminer
	Sympiesis	sp.	citrus leafminer

(continued)

Table 24.1 Important parasites attacking insect pests of citrus in Australia *(continued)*

ORDER AND FAMILY	GENUS	MAIN SPECIES	INSECT HOSTS
Eulophidae	*Tetrastichus*	*ceroplastae*	white wax scale, Florida wax scale
	Zaommomentedon	*brevipetiolatus*	citrus leafminer
Eupelmidae	*Anastatus*	spp.	bronze orange bug, fruitspotting bugs
	Anastatus	*biproruli*	spined citrus bug, bronze orange bug
	Goetheana	*shakespeari*	greenhouse thrips
	Thripobius	*semiluteus*	greenhouse thrips
Torymidae	*Megastigmus*	*brevivalvus*	citrus gall wasp
	Megastigmus	*trisulcus*	citrus gall wasp
Braconidae	*Apanteles*	spp.	lightbrown apple moth
	Aphidius	spp.	aphids
	Diachasmimorpha	*tryoni*	Queensland fruit fly
	Dolichogenidea	*arisanus*	lightbrown apple moth
	Fopius	*arisanus*	Queensland fruit fly
	Fopius	*deeralensis*	Queensland fruit fly
	Microplitis	*demolitor*	corn earworm, native budworm
	Opius	*perkinsi*	Queensland fruit fly
Ichneumonidae	*Lissopimpla*	*excelsa*	corn earworm
	Lissopimpla	*semipunctata*	banana fruit caterpillar
	Paniscus	*testaceous*	banana fruit caterpillar
	Phytodietus	sp.	orange fruitborer
	Xanthopimpla	sp.	lightbrown apple moth
Trichogrammatidae	*Megaphragma*	*mymaripenne*	greenhouse thrips
	Trichogramma	near *brassicae*	corn earworm, native budworm
	Trichogramma	*chilonis*	corn earworm, native budworm
	Trichogramma	*funiculatum*	lightbrown apple moth
	Trichogramma	spp.	citrus butterflies
	Trichogrammatoidea	spp.	fruitpiercing moths
	Trichogrammatoidea	spp.	corn earworm, native budworm
Platygasteridae	*Allotropa*	sp.	citrophilous mealybug
	Aphanomerus	spp.	citrus planthopper
	Fidiobia	*citri*	Fuller's rose weevil
Diptera (flies)			
Tachinidae	*Argyrophylax*	*proclinata*	yellow peach moth
	Blepharipa	*fulviventris*	small citrus butterfly
	Blepharipa	sp.	large citrus butterfly
	Chaetophthalmus	*biseriatus*	corn earworm and native budworm
	Exoristae	*sorbillans*	fruitpiercing moths
	Goniozus	sp.	lightbrown apple moth
	Monoleptophaga	*caldwelli*	monolepta beetle
	Palexoristus	*solemis*	banana fruit caterpillar
	Sturmia	sp.	banana fruit caterpillar
	Voriella	sp.	lightbrown apple moth
	Zosteromyia	sp.	lightbrown apple moth
Cryptochetidae	*Cryptochetum*	*iceryae*	cottony cushion scale
	Cryptochetum	spp.	cottony cushion scale

Predators

Insects from a wide range of orders, many mites, and all spiders are predators. In contrast to an individual parasite, which usually feeds off a single host, an individual predator usually feeds on several hosts. Predators tend to be scarce or absent when prey numbers are low, and more numerous when the numbers of prey increase.

Predators may actively search for and capture their prey, e.g. ladybirds or predatory mites, or they may lie in wait and seize them when they come within range or become trapped, e.g. praying mantises, assassin bugs and some spiders. Some predators, e.g. beetles, have mouthparts for biting and chewing; others, e.g. bugs, mites and spiders, have mouthparts for piercing and sucking. Lacewing larvae have large, sickle-shaped mandibles with a channel along the inner side. The prey is impaled on the mandibles and its body fluids drawn up through the channel.

Some predators have a wider range of prey than the range of hosts that parasites attack. Some lacewings, for example, will feed on mealybugs, juvenile scales, mites, aphids and moth and butterfly eggs. Spiders, some predatory mites and assassin bugs also feed on a wide range of prey species, and can be called generalist feeders.

Many predators, however, feed on a limited number of prey species. The mealybug ladybird (*Cryptolaemus montrouzieri*) feeds on several species of mealybug and several species of soft scales. Most ladybirds of the genus *Stethorus* feed only on tetranychid mites, while the vedalia ladybird (*Rodolia cardinalis*) usually preys only on cottony cushion scale.

Predators and their main prey in Australian citrus orchards are listed in table 24.2. This list has been compiled from a range of published references (see reference list) and field collections.

Table 24.2 Important predators attacking citrus pests in Australia

ORDER AND FAMILY	GENUS	MAIN SPECIES	HOSTS
Coleoptera (beetles)			
Coccinellidae (ladybirds)	*Chilocorus*	*baileyi*	oriental scale
	Chilocorus	*circumdatus*	citrus snow scale, red scale, oriental scale
	Coccinella	*transversalis* (=*repanda*)	black citrus aphid, young citricola scale, black scale, soft brown scale, moth eggs, thrips
	Coelophora	*inaequalis*	black citrus aphid, leafhopper and planthopper nymphs
	Cryptolaemus	*montrouzieri*	citrus mealybug, longtailed mealybug, citrophilous mealybug, spherical mealybug, cottony cushion scale, cottony citrus scale, green coffee scale, soft brown scale, black scale, citricola scale
	Diomus	*notescens*	mealybugs, aphids, moth eggs, citricola scale
	Halmus	*chalybeus*	aphids, corn earworm eggs, mealybugs, citrus red mite, rust mites, red scale, wax scales
	Harmonia	*conformis*	black citrus aphid, some scales and mealybugs
	Harmonia	*testudinaria*	black citrus aphid
	Micraspis	*frenata*	hard wax scale, aphids
	Parapriasus	*australasiae*	young black scale, citricola scale, soft brown scale
	Rhyzobius	*ruficollis*	longtailed mealybug, citrophilous mealybug
	Rhyzobius	*lophanthae*	black scale, citricola scale, soft brown scale, red scale, yellow scale
	Rhyzobius	near *lophanthae*	red scale, soft brown scale, citricola scale, black scale
	Rhyzobius	*ventralis*	black scale, citricola scale, soft brown scale, pink wax scale
	Rodolia	*cardinalis*	cottony cushion scale
	Rodolia	*koebelei*	*Icerya* sp.
	Rodolia	spp.	*Icerya* spp.
	Scymnodes	*lividigaster*	cottony cushion scale, citrus aphids

(continued)

Table 24.2 Important predators attacking citrus pests in Australia *(continued)*

ORDER AND FAMILY	GENUS	MAIN SPECIES	HOSTS
Coccinellidae (ladybirds)	Scymnus	sp.	black citrus aphid, broad mite, citrophilous mealybug, longtailed mealybug
	Serangium	bicolor	citrus red mite, rust mites
	Stethorus	fenestralis	oriental spider mite
	Stethorus	histrio	oriental spider mite
	Stethorus	nigripes	citrus red mite
	Stethorus	vagans	oriental spider mite
	Telsimia	sp.	two-spotted mite, oriental spider mite, citrus snow scale, red scale, oriental scale
Nitidulidae	Cybocephalus	aleurodophagus	Australian citrus whitefly
	Cybocephalus	sp.	citrus snow scale
Cantharidae (soldier beetles)	Chauliognathus	pulchellus	eggs of corn earworm and native budworm
Cleridae (predatory beetles)	several species		wide range of small prey species
Carabidae (ground beetles)	several species		soil insects, armyworms, and cutworms
Cicindelidae (tiger beetles)	several species		soil insects, armyworms and cutworms
Melyridae (soft-winged flower beetles)	several species		moth eggs
Staphylinidae (rove beetles)	several species		scales
Hemiptera (bugs)			
Lygaeidae (big-eyed bugs)	Geocoris	lubra	mites
Reduviidae (assassin bugs)	Pristhesancus	plagipennis	spined citrus bug, bronze orange bug, citrus butterflies, mealybugs, aphids, Fuller's rose weevil, katydids, Queensland fruit fly
Pentatomidae (shield bugs)	Amyotea	hamatus	bronze orange bug
	Cermatulus	nasalis	citrus butterflies, corn earworm, native budworm, spined citrus bug
	Oechalia	schellembergii	corn earworm, native budworm, lightbrown apple moth, citrus butterflies, spined citrus bug
Miridae (brown smudge bugs)	Deraeocoris	signatus	corn earworm, native budworm
Nabidae (damsel bugs)	Nabis	kinbergii	corn earworm, native budworm, aphids, leafhoppers, spider mites
Anthocoridae (flower bugs)	Orius	armatus	corn earworm, native budworm
	Orius	tantillus	two-spotted mite, aphids, thrips
Diptera (flies)			
Asilidae (robber flies)	Asilus	ferrugineiventris	flies and beetles
Cecidomyiidae (midges)	Diadiplosis	koebelei	citrophilous mealybug
	Arthrocnodax	sp.	brown citrus rust mite, citrus rust mite, scales
Cryptochetidae (cryptochetid flies)	Cryptochetum	iceryae	cottony cushion scale
Syrphidae (hoverflies)	Simosyrphus	grandicornis	black citrus aphid
	Melangyna	viridiceps	aphids, leafhoppers, mealybugs, soft scales

ORDER AND FAMILY	GENUS	MAIN SPECIES	HOSTS
Dolichopodidae (dolichopodid flies)	*Psilopus*	sp.	moth and butterfly eggs
Thysanoptera (thrips)			
Phlaeothripidae	*Aleurodothrips*	*fasciapennis*	red scale
	Karnyothrips	*flavipes*	oriental scale, red scale
	Haplothrips	sp.	greenhouse thrips
Thripidae	*Scolothrips*	*sexmaculatus*	two-spotted mite
Dermaptera (earwigs)			
Labiduridae	*Labidura*	*riparia truncata*	moth and butterfly larvae and pupae
Hymenoptera (wasps)			
Vespidae	*Polistes*	spp.	soft scales (especially pink wax and white wax scales), moth and butterfly larvae
Neuroptera (lacewings)			
Chrysopidae	*Mallada*	spp.	aphids, scales, whiteflies, citrophilous mealybug, longtailed mealybug, moth eggs
	Micromus	*tasmaniae*	black citrus aphid, black scale, citricola scale, soft brown scale
	Oligochrysa	*lutea*	mealybugs, aphids, mites
	Plesiochrysa	*ramburi*	black citrus aphid, black scale, citricola scale, soft brown scale, mealybugs
Lepidoptera (moths and butterflies)			
Cosmopterigidae	*Batrachedra*	*arenosella*	citrus snow scale, red scale, oriental scale
Noctuidae	*Catoblemma*	*dubia*	black scale, nigra scale, soft brown scale, citricola scale, wax scales
Mantodea (praying mantises)	several species		general predators of bugs, moths, caterpillars and beetles
Acarina (mites)			
Phytoseiidae	*Euseius*	*elinae*	brown citrus rust mite, citrus rust mite, citrus bud mite, broad mite, oriental spider mite, two-spotted mite, Australian citrus whitefly, scale eggs and crawlers, greenhouse thrips
	Euseius	*victoriensis*	as for *E. elinae*, minus greenhouse thrips
	Amblyseius	*herbicolus*	two-spotted mite, brown citrus rust mite, citrus rust mite, citrus red mite, broad mite
	Amblyseius	*lentiginosus*	as for *A. herbicolus*
	Phytoseiulus	*persimilis*	two-spotted mite
	Typhlodromus	*occidentalis*	two-spotted mite
Stigmaeidae	*Agistemus*	sp.	brown citrus rust mite, citrus rust mite, flat mites
Anystidae	*Anystis*	sp.	mites, aphids, thrips, leafhopper nymphs
Cheyletidae	*Hemicheyletia*	sp.	mites, scale eggs
Hemisarcoptidae	*Hemisarcoptes*	*mali*	citrus snow scale, oriental scale
Bdellidae	*Bdella*	sp.	mites
Eupalopsellidae	*Eupalopsis*	*jamesi*	red scale
Araneida (spiders)			general predators, e.g. of corn earworm, native budworm, spined citrus bug, and any other prey small enough to handle
Thomisidae (flower spiders)	*Diaea*	spp.	as above
Lycosidae (wolf spiders)	*Lycosa*	spp.	as above

(continued)

Table 24.2 Important predators attacking citrus pests in Australia (continued)

ORDER AND FAMILY	GENUS	MAIN SPECIES	HOSTS
Oxyopidae (lynx spiders)	*Oxyopes*	spp.	as above
Heteropodidae (huntsman spiders)	*Holconia*	spp.	as above
Salticidae (jumping spiders)	*Opisthoncus*	spp.	as above
Passeriformes (birds—only a few representative species are listed)			
Meliphagidae eastern spinebill	*Acanthorhynchus*	*tenuirostris*	white wax scale
Zosteropidae silvereye	*Zosterops*	*lateralis*	white wax scale
Acanthizidae yellow-rumped thornbill	*Acanthiza*	*chrysorrhoa*	white wax scale

Plate 24.1 Lynx spider Oxyopes sp.

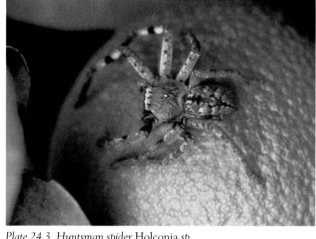

Plate 24.3 Huntsman spider Holconia sp.

Plate 24.2 Flower spider Diaea sp.

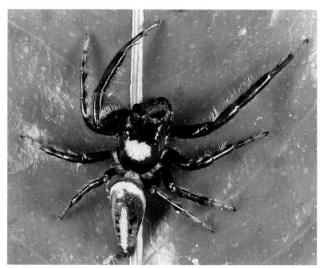

Plate 24.4 Jumping spider Opisthoncus sp.

Pathogens

Pathogens, or microbial control agents, may be fungi, bacteria, viruses or microsporidia (protozoans). Parasitic nematodes can also be included. Some pathogens that attack citrus pests in Australia are listed in table 24.3.

Table 24.3 Important pathogens attacking Australian citrus pests

TYPE OF PATHOGEN	SPECIES	HOST
Fungi		
	Verticillium lecanii	hard scales, soft scales
	Fusarium coccophilum (*Nectria flammea*)	red scale
	Fusarium henningsii	aphids
	Fusarium lavarum (*Nectria aurantiicola*)	red scale
	Fusarium moniliforme var. *subglutinans*	hard wax scale
	Fusarium stilboides	white wax scale
	Order Entomophthorales	corn earworm, native budworm
	Nomuraea rileyi	corn earworm, native budworm
	Beauveria bassiana	beetles, bugs, planthoppers
	Hirsutella kirchneri	mites
	Hirsutella thompsonii	mites
	Metarhizium spp.	beetles
	Paecilomyces fumosa rosea	citrus rust thrips
Bacteria		
	Bacillus thuringiensis	moths and butterflies
Viruses		
	nuclear polyhedrosis	corn earworm, native budworm
	granulosis	moths and butterflies
Protozoans		
	Nosema locustae	grasshoppers
Parasitic nematodes		
	Steinernema carpocapsae	trunk borers, soil pests
	Heterorhabditis spp.	trunk borers, soil pests, corn earworm, native budworm

Integrated pest management (IPM) in citrus

IPM is a strategy which encourages the reduction of pesticide use by employing a variety of pest control options in harmonious combination to contain or manage pests below economic injury levels. The aim is to produce high-quality marketable produce at minimal cost by intelligently managing pests using the various control options.

These options include:

- *biological control*
- *cultural control*
- *varietal selection*
- *chemical control.*

IPM aims to maximise the use of biological and cultural controls. Other measures must play a supportive rather than disruptive role. This especially applies to pesticides, which should not be used on a 'calendar' basis, but strictly when needed as defined by systematic pest monitoring. Selective chemicals are preferred over broad-spectrum chemicals.

For more information on the concepts of IPM, see chapter 2, pages 14–15.

25 Practical integrated pest management (IPM)

A template for IPM in citrus

Several important components make up a practical IPM program. These include:
- identification of pests and their natural enemies
- monitoring of pests and their natural enemies
- data recording and reporting
- decision making
- taking appropriate action to manage pests
- reappraisal and research.

Identification

Correct identification of both the pests and their natural enemies is essential. There are many situations where incorrect identification can lead to a totally inappropriate response.

For example, brown citrus rust mite and citrus rust mite both cause damage to citrus fruit in Australia. These rust mites look very similar, but are controlled to a different degree by both natural enemies and miticides. It generally takes much experience and training before a person is sufficiently familiar with both species to make quick and accurate assessments in the field.

When starting to implement an IPM program and to identify pests and natural enemies, it is important to collect specimens for their identification to be confirmed by an experienced pest scout or entomologist. When in the orchard, carry small tubes and plastic bags for specimens.

A ×10 hand lens is an essential tool for monitoring. This can be worn on a cord around the neck to leave the hands free for examining leaves, fruit and twigs, and for recording the results of sampling.

Choose a hand lens with good-quality double lenses that can be separated for cleaning. It is best to have a large field of view (about 180 mm^2).

Access to a binocular microscope is also desirable, and some growers have chosen to purchase one.

Monitoring

The most important element of IPM is monitoring. Assessment of pest and natural enemy populations must be both quick and accurate.

The following information outlines an existing commercial monitoring program used in Australian citrus. Because pests vary between regions, this program should be used as a guide only.

There are many citrus pests, and they in turn have many natural enemies, so it is advisable to employ a qualified IPM scout or consultant to monitor your orchard.

If a citrus grower or a staff member is going to monitor an orchard, this person must be trained and committed to routine monitoring. As a guide to the time necessary, an average 2

Plate 25.1 Monitoring in Meyer lemon orchard, Glasshouse Mountains, Queensland.

hectare block will take at least 35 minutes to be properly sampled by an inexperienced person (see table 25.1), compared to 25 minutes by an experienced person. A new scout will need several months of full-time monitoring before being able to sample confidently and accurately.

When the sampling interval for a suitable IPM program is selected (e.g. every 14 days), time must be set aside for the monitoring work. Monitoring should not be deferred due to other commitments.

▶ Important monitoring issues

Monitor pests and natural enemies

Both pests *and* their natural enemies need to be monitored and their levels assessed in order to understand the relationship between them and to make informed decisions about pest control. If assessments are based on pest levels only, decisions tend to be pesticide-oriented.

Monitor quickly and efficiently

Monitoring of commercial citrus orchards must be quick and efficient. For this reason, checking for the presence or absence of pests is usually preferable to doing precise counts.

Obtain accurate and repeatable results

If possible, the results of monitoring should be accurate and repeatable. It is important to remember that each sample is only one of many for that block for the season. Trends should be considered before taking each new sample.

If the reliability of an assessment is in doubt, another sample should be taken shortly afterwards. Regular, frequent monitoring is generally preferable to infrequent monitoring, even if the latter is more intensive.

Use common sense

While it is necessary to have a monitoring system, this is only one of many tools available to the grower and the pest scout to help them make good pest management decisions. Good decisions are the result of useful monitoring systems tempered with experience, keen observation and common sense. Knowledge of the crop, the particular orchard, and the grower's preferences are other important factors.

▸ Number of trees in sample
When monitoring, check a random sample of trees. The number of trees in the sample depends on block size (see table 25.1).

A block of trees is considered as an area that is treated (and sprayed) as a unit. It usually comprises trees of the one variety planted in the same year and with uniform soil type, rootstock, and irrigation and nutritional practices.

On large uniform blocks, samples can consist of a lower proportion of the total number of trees, and still give reliable estimates of pest and natural enemy populations.

▸ Selecting trees for the monitoring sample
Sample trees should normally be selected at random from within the block, but the whole block should be covered. However, pest biology and a previous history of infestation may require sampling to be concentrated in certain areas.

Consecutive samples should be taken from the four quadrants of each tree (north, south, east and west). Avoid non-representative trees (e.g. unhealthy or smaller trees).

▸ Parts of the tree to be sampled
The parts of trees that are sampled are called 'units'. A unit may be a fruit, a group of leaves, a shoot etc. In a block where the sample size is 15 trees, 75 individual units of each type may be assessed on each monitoring date, i.e. 5 sample units are usually taken from each tree.

In large trees, one in five of each unit should be taken by climbing into the top centre of the tree where possible. This is to check for inadequate spray coverage, and for hard-to-find pest species, such as citrus rust mite.

It is best to check the type of unit where pests or natural enemies are likely to be most abundant or most damaging. For example, pink wax scale generally occurs on the upper surface of mature leaves, and a unit with these leaves (e.g. a shoot) should be sampled.

Rust mites feed on leaves and fruit, but since economic damage is confined to fruit, the sampling unit is the fruit. However, early in the season before fruit are large enough to sample, leaves may be sampled instead.

Within a sampling unit different areas or surfaces may need to be checked. For example, brown citrus rust mite is most likely to be found on the exposed surfaces of fruit and the upper surfaces of leaves. Conversely, citrus rust mite prefers the protection offered by the shaded sides of fruit and the undersurfaces of leaves. In addition, the predatory mite *Euseius victoriensis*, which feeds on rust mites, rests during the day on the undersurfaces of medium-age to mature leaves within the canopy. These preferences must be considered.

Some species are difficult to monitor in the field. For example, to assess the proportion of red scale parasitised by *Aphytis* spp., a microscope must be used to examine red scale on fruit that have been removed from trees. See figure 25.1 for an example of a scale parasitism assessment sheet suitable for commercial monitoring. Such an assessment will give information on the structure and viability of the scale population, as well on the activity of key parasites.

▸ Time needed for monitoring
It should be possible to check a sample of 10 trees in 20–30 minutes (see table 25.1). The actual time taken will depend on the time of year and the organisms being sampled.

▸ Frequency of monitoring
The frequency of monitoring will vary depending upon the time of year, the location and the levels of pest activity. It may be as short as 3–6 days during high-risk periods, but may stretch to 3–6 weeks or more in winter.

▸ Monitoring calendars
A seasonal monitoring calendar is used to show when each species should be monitored and what the sampling frequency should be. (For example, see the monitoring charts on pages 216–225.) Pest activity generally relates to season and stage of development of the crop.

▸ Monitoring tools
Many tools are available to assist with monitoring, e.g. pheromone traps and coloured sticky cards can be used to collect samples of pests or natural enemies, which can then be used to estimate the field populations.

Data recording and reporting

▸ Sample cards
Sample cards are used for recording the results of monitoring. They should be large enough to write on legibly, and allow sufficient space for additional notes. They should be small enough to be carried easily in the field and fit into a pocket or under one's arm while using a hand lens.

A card measuring 99 mm × 210 mm (1/3 A4 page) is a convenient size for field work. A firm backing board is an added benefit.

An example of a sample card is shown in figure 25.2. Each column is reserved for data on one species of pest or natural enemy that may be sampled. The species sampled will vary depending upon the variety of citrus and the time of year. The names of species sampled routinely may be preprinted onto the card, but blank columns will allow inclusion of other species when necessary. If you wish to sample the incidence of diseases as well, columns could also be allowed for particular diseases.

Table 25.1 Number of trees in samples for monitoring, and approximate monitoring times, for blocks of different sizes

Number of trees in block	Area (hectares)	Number of trees sampled	Time taken (minutes) to check sample
0–500	<2	10	20–30
501–750	2–3	12	25–35
751–1000	3–4	15	30–45
1001–2000	4–8	20	40–60
2001–4000	8–16	25	50–75
>4000	>16	30	60–90

The shorter times are those required by a professional pest scout, while the longer times are an estimate of those required by a less experienced person. Highly variable blocks warrant sampling of additional trees.

Each cell on the sample card is filled in with the number of fruit, leaves, shoots etc. infested with a pest on one tree, which is usually in the range 0–5. The column for each species is totalled and the percentage of the sample units infested is entered in the row at the bottom of the card.

Space is available below the totals row for any comments the scout wishes to make. These may include observations about a species status in the block, signs of damage, hot spots, or anything else in the block that may be worth reporting to the grower, e.g. faulty irrigation equipment.

▶ Orchard report forms

The results of sampling, as calculated at the bottom of the sample card, are entered onto an orchard report form (figure 25.3). Comments or recommendations upon which the grower should act are entered in the column provided.

The columns in the report match those of the sample card exactly to minimise risk of error when transferring the data. (This also applies to computer spreadsheets.) The report is completed and presented to the grower as soon as possible after the orchard has been monitored.

▶ Recording pesticide applications and release of natural enemies

Growers must keep accurate records of pesticide application and natural enemy insect releases. This information is essential if scouts and researchers are to improve orchard management. An example of a suitable spray record sheet is shown in figure 25.4.

The information from the spray record sheet is entered onto a wall chart in abbreviated form at the end of each week (a section of a sample wall chart is given in figure 25.5). The wall chart provides a summary or overview of the treatments for each block of citrus at any point in the season and is a useful decision-making tool for the grower and the pest scout.

The need for accuracy in keeping these records cannot be over-emphasised. The charts are not only useful for day-to-day decision making, they can also be valuable for research, and provide reliable information on pesticide application, showing how growers are using pesticides responsibly, and also minimising their use.

▶ Computer spreadsheets and population graphs

At a convenient time, data are transferred to a computer spreadsheet. The software used should have a good graphing facility. Initially, data are entered by date so that all the entries for one visit are located together (see figure 25.6). Then the data for the entire season can be sorted by block (see figure 25.7). Data for individual blocks can then be presented as graphs.

Data for application of sprays and release of natural enemies from the wall chart can be added to graphs at this stage (see figure 25.8, where such information is shown by arrows at the appropriate times).

Graphs produced in colour give a good overview of the movement of pest and natural enemy species throughout a season.

The graphs also help to show the effects of various treatments on both pests and natural enemies during a season. For example, certain fungicides are known to harm predatory mites. Graphs can show the extent and duration of suppression of predatory mites by fungicides.

When graphs are collated for individual blocks over many seasons, trends can be observed, e.g. a tendency in some species to resistance to or tolerance of certain pesticides. The graphs also give an insight into the way in which pests and natural enemies interact, and the timeframes within which control of certain pests can be expected.

Decision making

Every time monitoring is carried out, the pest scout must use the information gained to make appropriate decisions and recommendations, which are discussed with the grower. This is the most difficult part of the job.

Action levels are one decision-making tool available to the pest scout. Other information used by good pest scouts includes their previous experience with pests, knowledge of the block history, estimates of crop volume and market value, as well as knowledge of the grower's preferences. All of these factors need to be weighed up before the scouts discuss recommendations with their clients.

▶ Action levels

Action levels are helpful in the decision-making process. An action level is the point at which action should be taken to avoid unacceptable damage (i.e. damage that causes economic loss). The action levels given in this book, which are included in the information on each pest, and also in the monitoring charts on pages 216–225, have been determined by research (see method 1 below) and practical experience (see method 2 below) in the citrus growing areas of Australia.

Method 1

Determine the action level by research, evaluating such parameters as population densities of pests and natural enemies, damage levels, and the economics of crop production, including the costs of spraying and the expected market returns for the crop.

Method 2

It is possible to implement worthwhile IPM programs even if action levels have not been determined for all pests by means of detailed research.

Start with a 'best estimate' of the action level. For example, assign an interim action level of 10% of fruit infested with the pest. Pest scouts then use their experience to continually refine the accuracy of the action level.

Interim levels can also take numbers of natural enemies into account. For example, 10% of fruit infested with a pest could be considered the action level if predators are absent. However, if a major predator is active, the action level can be raised according to the counts of predators obtained when monitoring. For example, it could be decided that, if predators are present on more than 40 out of 100 fruit examined, the action level for the pest could be raised to 20% of fruit infested.

Appropriate action

In most cases, the results of monitoring indicate that no action is required.

When the pest scout decides that some action must be taken, several options may be considered. These could include the release of commercially reared natural enemies, some form of cultural control, e.g. sod culture, or the application of a chemical. If the decision is made to use a chemical, choose one that is least likely to harm natural enemies.

Plate 25.2 *Rearing of oleander scale on butternut pumpkins for the commercial mass production of aphytis wasps, parasites of red scale. The oleander scales act as hosts of the aphytis wasps.*

Plate 25.3 *Releasing commercially reared aphytis wasps into a Queensland mandarin orchard.*

Reappraisal and research

Post-harvest assessments

At the end of the growing season, the quality of harvested fruit should be assessed to provide feedback on the success of the IPM program. Fruit that are downgraded to second grade or juice can be sampled to determine the primary cause of rejection, e.g. wind blemish, sooty mould, red scale, mite damage, and so on. (See figure 25.9 for a sample post-harvest assessment sheet.)

It is also useful if the grower gives feedback based on subjective judgements of the fruit from each block as it passes through the packing shed. This will help the scout get a clearer idea of the grower's perceptions and expectations. Preprinted forms, ready to fill in, can be supplied to growers at the start of the harvest season (see figure 25.10).

The information from these assessments can be included with the graph for each block (see the sample graph in figure 25.8) to enable comparison between the results of monitoring and the final quality assessment. A study of this information will help fine-tune action levels.

Research

Ongoing research is essential to the success of practical IPM, as pests and natural enemies are part of a complex and dynamic biological system. New pests appear from time to time (e.g. citrus leafminer recently appeared in Western Australia), and the status of existing pests may alter with changes in grower practices, varietal selection and so on. Ultimately, the success of practical IPM depends upon a cooperative relationship between growers, pest scouts, researchers and others in the citrus industry.

Figure 25.1 Red scale parasitism assessment sheet.

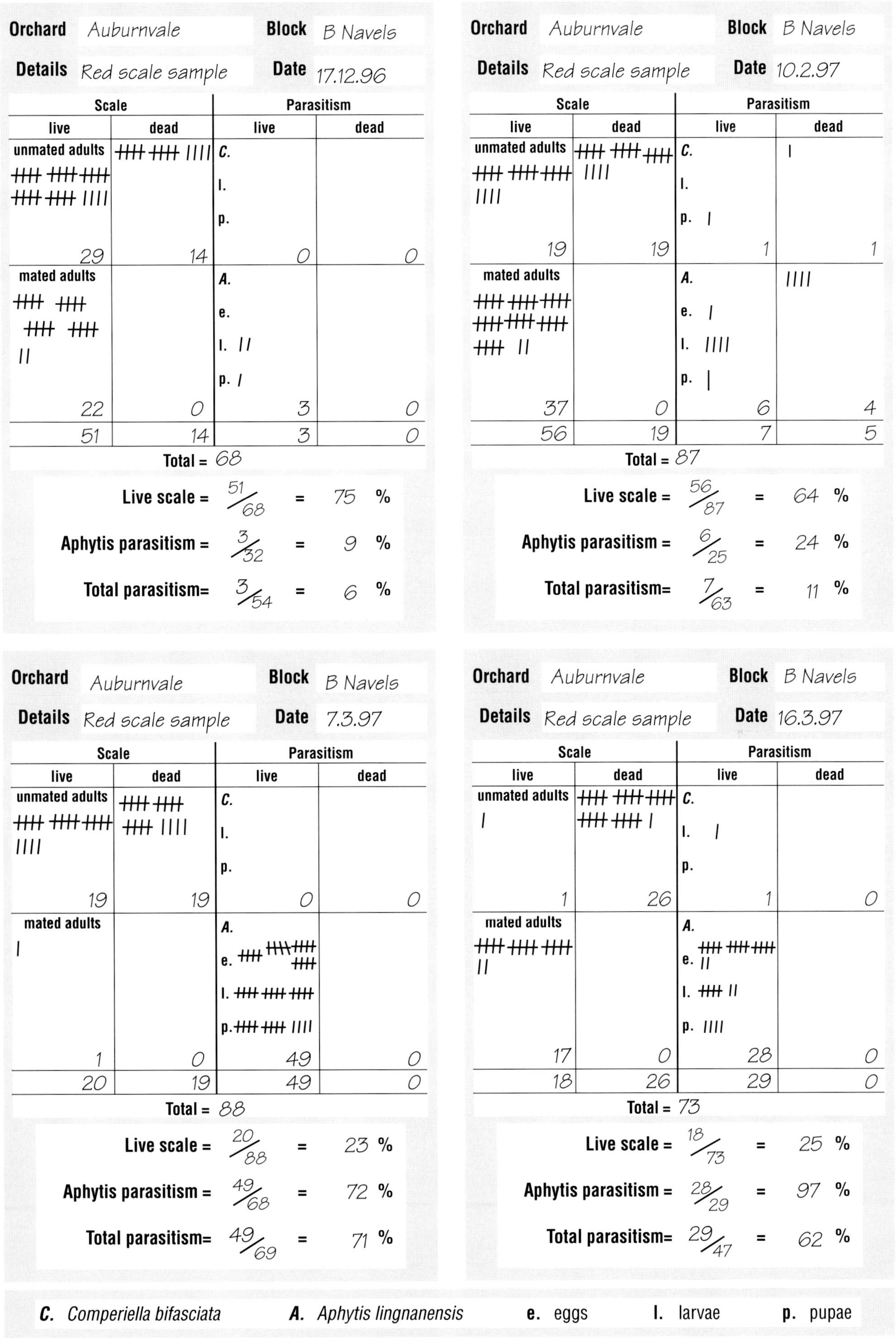

Orchard	Auburnvale		Block	B Navels
Details	Red scale sample		Date	17.12.96

Scale		Parasitism	
live	dead	live	dead
unmated adults 卌 卌 卌 卌 卌 卌 卌 卌 IIII	卌 卌 IIII	C. l. p.	
29	14	0	0
mated adults 卌 卌 卌 卌 II		A. e. l. II p. I	
22	0	3	0
51	14	3	0

Total = 68

Live scale = 51/68 = 75 %

Aphytis parasitism = 3/32 = 9 %

Total parasitism = 3/54 = 6 %

Orchard	Auburnvale		Block	B Navels
Details	Red scale sample		Date	10.2.97

Scale		Parasitism	
live	dead	live	dead
unmated adults 卌 卌 卌 卌 卌 卌 IIII	卌 卌 卌 IIII	C. l. p. I	I
19	19	1	1
mated adults 卌 卌 卌 卌 卌 卌 卌 II		A. e. I l. IIII p. I	IIII
37	0	6	4
56	19	7	5

Total = 87

Live scale = 56/87 = 64 %

Aphytis parasitism = 6/25 = 24 %

Total parasitism = 7/63 = 11 %

Orchard	Auburnvale		Block	B Navels
Details	Red scale sample		Date	7.3.97

Scale		Parasitism	
live	dead	live	dead
unmated adults 卌 卌 卌 IIII	卌 卌 卌 IIII	C. l. p.	
19	19	0	0
mated adults I		A. e. 卌 卌 卌 卌 l. 卌 卌 卌 p. 卌 卌 IIII	
1	0	49	0
20	19	49	0

Total = 88

Live scale = 20/88 = 23 %

Aphytis parasitism = 49/68 = 72 %

Total parasitism = 49/69 = 71 %

Orchard	Auburnvale		Block	B Navels
Details	Red scale sample		Date	16.3.97

Scale		Parasitism	
live	dead	live	dead
unmated adults I	卌 卌 卌 卌 卌 I	C. l. I p.	
1	26	1	0
mated adults 卌 卌 卌 II		A. e. 卌 卌 卌 II l. 卌 II p. IIII	
17	0	28	0
18	26	29	0

Total = 73

Live scale = 18/73 = 25 %

Aphytis parasitism = 28/29 = 97 %

Total parasitism = 29/47 = 62 %

C. *Comperiella bifasciata*   **A.** *Aphytis lingnanensis*   **e.** eggs   **l.** larvae   **p.** pupae

Figure 25.2 Sample card for recording the results of monitoring one block of trees.

Orchard Esmeralda **Date** 24.2.97
Block 2 Gf **Sampler** Malcolm

	pm	crm	bcrm	rs	cmb	lx	pl		
1	4	0	0	1	2	0	0		
2	5	1	0	0	0	0	2		
3	3	0	0	1	2	0	0		
4	6	0	0	2	1	1	0		
5	2	2	0	0	1	0	0		
6	8	0	0	1	0	0	0		
7	10	0	0	0	0	0	0		
8	7	0	0	0	1	1	0		
9	11	0	0	0	0	0	0		
10	3	0	0	0	0	0	0		
11									
12									
13									
14									
15									
16									
17									
18									
19									
20									
TOTALS	118*	3/50 = 6%	0%	5/50 = 10%	7/50 = 14%	28%**	2/50 = 4%		

Key
- **pm** predatory mite (*Euseius*)
- **bcrm** brown citrus rust mite
- **crm** citrus rust mite
- **rs** red scale
- **cmb** citrus mealybug
- **pl** planthoppers
- **lx** *Leptomastix dactylopii* (wasp parasite)

Notes

* 59 pm on 50 leaves = 118 pm on 100 leaves

** 2 lx on 7 fruit infested with mealybug = 2/7 = 28%

Figure 25.3 Orchard report form, used for compiling the results of monitoring different blocks from sample cards.

Orchard report

Orchard: Esmeralda   Date: 24.1.97

Block	Trees no. sampled/ no. climbed	pm (no/100 leaves)	crm %	bcrm %	rs %	cmb %	lx %	pl %	scb %	cm %	css %	Notes
1E	10/2	30	0	0	0	4		2				
2E	10/2	116	0	0	4	2		0			4	
3E	12/3	218	0	0	5	2		20			5	spray endosulfan @ 30 mL/100 L
4E	10/2	22	0	2	0	0		2			2	
5E	10/2	0	0	4	4	0		0				
6E	10/2	68	2	2	0	2		0				booster release of aphytis required
7E	10/2	48	0	2	32	0		2			2	
8E	10/2	112	2	0	6	0		0				
9E	5	32	0	0	0	0		0				
10E	10	114	0	0	16	8		0				
11E	7	15	30	6	15	21		3				spray mancozeb @ 200 g/100 L for crm
1Gf	10/2	0	6	0	2	22		0			4	
2Gf	10/2	0	0	6	2	2		0			2	
3Gf	10/2	12	2	18	0	18		0			2	predatory mites should control; leave bcrm till next visit
1GI	10	18	0	0	12	0		0				bcrm very low numbers; leave till next visit
2GI	10/2	0	0	14	4	0		0				

Key

ab	alternaria brown spot	**cm**	*Cryptolaemus montrouzieri* (mealybug ladybird)	**frw**	Fuller's rose weevil
aph	aphytis	**cmb**	citrus mealybug	**hel**	heliothis
bcrm	brown citrus rust mite	**crm**	citrus rust mite	**kd**	katydid
bm	broad mite	**crt**	citrus rust thrips	**ltm**	longtailed mealybug
bs	black spot	**css**	citrus snow scale	**lx**	*Leptomastix dactylopii* (wasp parasite)
cb	cell breakdown	**fm**	flat mite	**mel**	melanose
clm	citrus leafminer			**pl**	planthoppers

pm	predatory mite (*Euseius*)
pt	predatory thrips
pwx	pink wax scale
rs	red scale
scb	spined citrus bug
ss	soft scales

Figure 25.4 Spray record sheet for one block of trees.

Spray record sheet

Block name/number	Block size (ha)	Tree height (m)	Variety
Savages	1.3	4	Imperial

Date	Operator	Spray cart	Capacity
28.9.97	Craig	red cart	3000 L

Chemicals	Amount/rate
1 Kocide (100 g/100 L)	3 kg
2 hydrated lime	3 kg
3 wetter	400 mL
4	
5	

Vat count

⟋⟋⟋⟋⟋

Total vats: 5

Tractor gear	RPM	Pressure
Iseki/3rd low	1200	3500 kPa (500 psi)

Weather/comments

Wind 5 km/h from south-east.

Figure 25.5 Wall chart, used for compiling records of spraying and natural enemy release for different blocks.

Benyenda spray record, natural enemy releases, 96–97 season

Block	Variety	12 Sep	19 Sep	26 Sep	3 Oct	10 Oct	17 Oct	24 Oct	31 Oct
1.2.PP	Navel			Cu, Zn, oil				aphytis	
Diesel	Navel		Cu, Zn, oil					aphytis	aphytis
Dump	Navel			Cu, Zn, W				aphytis	
Corner	Navel			Cu, Zn, oil			aphytis		
Shed	Navel		Cu, Zn, oil					aphytis	
Middle	Navel		Cu, Zn, oil					aphytis	
Ranch	Navel			Cu, Zn, oil					
Front	Navel			Cu, Zn, W					aphytis
River	Navel			Cu, Zn, W				aphytis	
Mid River	Navel		Cu, Zn, W						
River	Imperial			Cu, Zn, W				aphytis	
Cleo	Imperial			Cu, Zn, W				aphytis	
River	Ellendale				Cu, Zn, W				
Front	Ellendale				Cu, Zn, W				
Troyer	Ellendale				Cu, Zn, W				
Middle	Ellendale					Cu, Zn, oil			
Pump	Ellendale				Cu, Zn, W	½ Cu, Zn, W			
1+2	Valencia		Cu, Zn, oil			½ Cu, Zn, W			
Front	Valencia		Cu, Zn, oil		Cu, Zn, W				
Reserve	Valencia			Cu, Zn, oil					
	Joppa								

Key

Fungicides
An Antracol
B benomyl
Cu copper
Mz mancozeb
Pa phosphonate

Insecticides
En Endosulfan
L Lorsban
oil (1%)
Pir Pirimor
Sc Supracide

Miticides
K Kelthane
S sulphur
T Torque

Nutritional
KNit pot. nitrate
Mn manganese
Zn zinc

Growth regs etc
2,4-D (50% amine)
Eth Ethrel
GA gibberellic acid

Adjuvants
Coda Codacide oil
M Menthene
Nf Nufilm
oil (0.6%)
W wetter

Figure 25.6 Printout from spreadsheet program, showing results of monitoring different blocks on one date (i.e. data sorted by date).

Esmeralda Block	Var	Date	pm (no mites/100 leaves)	crm %	bcrm %	rs %	cmb %	lx %	pl %	scb %	bm %	css %	aph %	clm %
1	E	1–Dec	54	0	0	0	0		2			4		
2	E	1–Dec	2	0	0	0	0		2			2		
3	E	1–Dec	8	0	0	8	0		0			4		
4	E	1–Dec	32	4	30	6	2		0					
5	E	1–Dec	0	10	12	6	0		0					
6	E	1–Dec	4	0	2	0	0		0					
7	E	1–Dec	16	0	6	2	0		2					
8	E	1–Dec	0	6	10	0	0		2					
9	E	1–Dec	48	0	0	0	0		0					
10	E	1–Dec	126	0	0	4	8		0					
11	E	1–Dec	38	0	0	6	9		0					
1	GF	1–Dec	0	0	0	0	16		0					
2	GF	1–Dec	0	2	0	0	12		2			18		
3	GF	1–Dec	0	0	0	0	6		0					
1	I	1–Dec	4	0	0	10	0		0					
2	I	1–Dec	104	0	0	4	0		6					
3	I	1–Dec	21	0	12	15	0		9					
4	I	1–Dec	20	0	0	2	2		0					
1	MC	1–Dec	40	0	4	12	12		3					
2	MC	1–Dec	21	0	0	9	3		0					
1	N	1–Dec	0	0	0	2	2		0			16		
2	N	1–Dec	0	0	0	0	0		0			22		
3	N	1–Dec	71	0	0	2	2		0			2		
4	N	1–Dec	23	0	0	5	0		0			7		
5	N	1–Dec	84	0	0	0	25		2					
6	N	1–Dec	22	0	6	2	2		0					
7	N	1–Dec	100	0	0	0	24		0					
8	N	1–Dec	0	0	0	6	16		0					
9	N	1–Dec	30	0	0	4	12		0	12		2		
1	V	1–Dec	0	0	0	2	14		0			14		
2	V	1–Dec	56	0	0	0	8		0					
3	V	1–Dec	66	0	0	2	16		2					

Key

ab	alternaria brown spot	**css**	citrus snow scale	**pwx**	pink wax scale
aph	aphytis	**fm**	flat mite	**rs**	red scale
bcrm	brown citrus rust mite	**frw**	Fuller's rose weevil	**scb**	spined citrus bug
bm	broad mite	**hel**	heliothis	**ss**	soft scales
bs	black spot	**kd**	katydid		
cb	cell breakdown	**ltm**	longtailed mealybug	***Variety***	
clm	citrus leafminer	**lx**	*Leptomastix dactylopii*	**E**	Ellendale
cm	*Cryptolaemus montrouzieri* (mealybug ladybird)		(wasp parasite)	**GF**	Grapefruit
		mel	melanose	**I**	Imperial
cmb	citrus mealybug	**pl**	planthoppers	**MC**	Murcott
crm	citrus rust mite	**pm**	predatory mite (*Euseius*)	**N**	Navel
crt	citrus rust thrips	**pt**	predatory thrips	**V**	Valencia

Figure 25.7 Printout from spreadsheet program, showing monitoring results for one block for an entire season (i.e. data sorted by block).

Esmeralda														
Block	Var	Date	pm (no mites/100 leaves)	crm %	bcrm %	rs %	cmb %	lx %	pl %	scb %	bm %	css %	aph %	clm %
5	E	7-Oct												
5	E	22-Oct												
5	E	4-Nov	0											
5	E	19-Nov	0	2	8	2	0		0					
5	E	1-Dec	0	10	12	6	0		0					
5	E	14-Dec	0	0	0	4	0		0					
5	E	23-Dec	0	0	0	4	0		0					
5	E	11-Jan	0	0	0	10	0		2			2		
5	E	24-Jan	0	0	4	4	0		0					
5	E	8-Feb	0	0	18	10	0		0			2		
5	E	23-Feb	0	0	0	16	0		0			6		
5	E	16-Mar	4	0	12	10	0		2			8		
5	E	29-Mar	8	0	22	16	4		0					
5	E	14-Apr	32	2	0	8	2		0			4		
5	E	29-Apr	12	0	18	0	0		0					
5	E	16-May	40	0	8	6	2		0					
5	E	26-May	22	0	0	0	0		0			8		
6	E	7-Oct												
6	E	22-Oct												
6	E	4-Nov	0											
6	E	19-Nov	0	0	0	2	0		4					
6	E	1-Dec	4	0	2	0	0		0					

Key

ab	alternaria brown spot	css	citrus snow scale	pwx	pink wax scale
aph	aphytis	fm	flat mite	rs	red scale
bcrm	brown citrus rust mite	frw	Fuller's rose weevil	scb	spined citrus bug
bm	broad mite	hel	heliothis	ss	soft scales
bs	black spot	kd	katydid		
cb	cell breakdown	ltm	longtailed mealybug	***Variety***	
clm	citrus leafminer	lx	*Leptomastix dactylopii* (wasp parasite)	E	Ellendale
cm	*Cryptolaemus montrouzieri* (mealybug ladybird)			GF	Grapefruit
		mel	melanose	I	Imperial
cmb	citrus mealybug	pl	planthoppers	MC	Murcott
crm	citrus rust mite	pm	predatory mite (*Euseius*)	N	Navel
crt	citrus rust thrips	pt	predatory thrips	V	Valencia

Figure 25.8 Graph from spreadsheet program, showing monitoring results, when sprays were applied and when natural enemies were released over a whole season on one block.

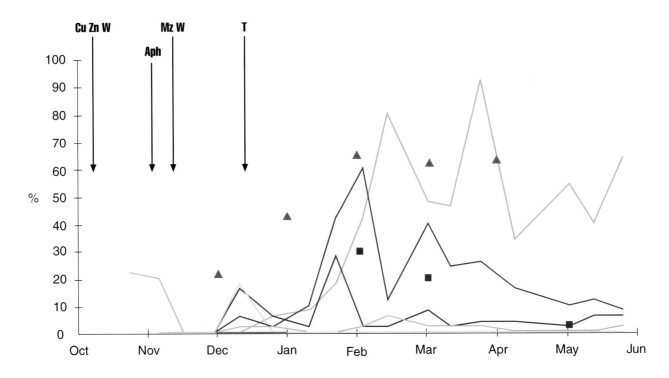

Key

■	Leptomastix
▲	Aphytis
● (light)	Broad mite
●	Citrus snow scale
▲ (light)	Spined citrus bug
——	Flatids
——	Mealybug
——	Red scale
——	Brown citrus rust mite
——	Citrus rust mite
——	Predatory mites

Aph	Aphytis
B	Benlate
Cu	Copper
E	Endosulfan
K	Kelthane
L	Lorsban
Lx	Leptomastix
Mz	Mancozeb
S	Sulphur
Sc	Supracide
T	Torque
W	Wetting agent

Figure 25.9 Post-harvest assessment sheet.

Post-harvest assessment

Orchard
Block
Date

No. 1st grade cartons
No. 2nd grade cartons
No. juice bins

2nd grade

Rejection cause	No. of fruit/box	% downgrading
red scale		
mite damage		
thrips		
sooty mould		
black spot		
melanose		
rots		
torn buttons		
mechanical		
sunburn		
poor colour		
ethylene burn		
bruising		
other defects		
Total		

Juice grade

Rejection cause	No. of fruit/box	% downgrading
red scale		
mite damage		
thrips		
sooty mould		
black spot		
melanose		
rots		
torn buttons		
mechanical		
sunburn		
poor colour		
ethylene burn		
bruising		
other defects		
Total		

Figure 25.10 Post-harvest evaluation form for grower.

Orchard block evaluation

Orchard: Mesner's
Season:

Block	Var.	Wind rub (l m h)	Red scale (l m h)	Mealybug/sooty mould (l m h)	Mite damage (l m h)	General comments/other pests
1	N					
2	N					
1	E					
1	I					
1	H					

l low m medium h high
N navel I Imperial H Hickson E Ellendale

Monitoring guides

Monitoring is a vital component of IPM. It requires a systematic, speedy and accurate assessment of both pest and natural enemy populations.

Important components of monitoring are:
- recognising the pest and its damage
- knowing when to take action to prevent economic loss
- recognising the pest's natural enemies, and understanding their potential to assist in control.

The charts on the next 10 pages are a guide to monitoring in three regions:

- Queensland and coastal New South Wales
- inland New South Wales, Sunraysia and the Riverland
- Western Australia.

Major and minor pests are included, as well as information on damage caused by diseases and disorders that may be confused with damage caused by pests. Each chart is for early season, mid-season or late season. As shown in the following sample from one of the charts, the charts provide photographs to help identify the pest and/or the damage, and other monitoring information.

HOW TO USE THE MONITORING GUIDES

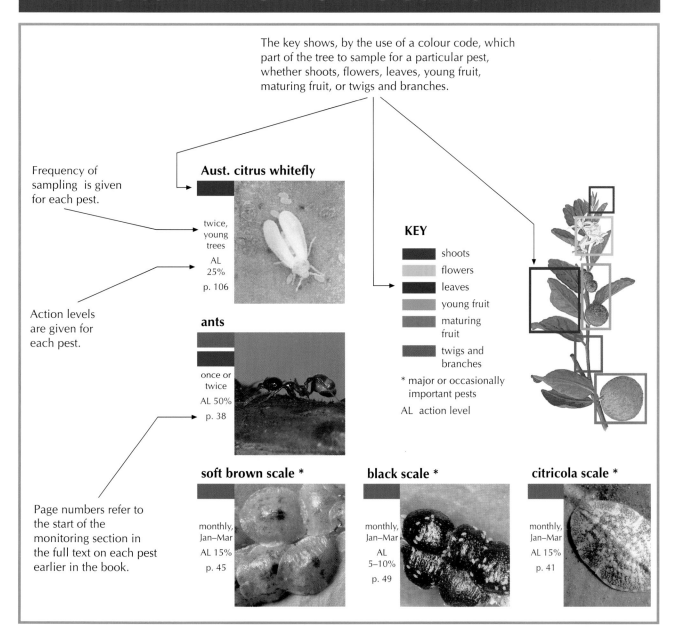

216   25 Practical IPM: monitoring guides

MONITORING GUIDE: QUEENSLAND AND COASTAL NEW SOUTH WALES

EARLY SEASON September–October Sample all varieties fortnightly for these pests. See also page 220 for fungal diseases and fruit disorders which may be confused with pest damage.

bud and flower moths

AL 50%
p. 133

citrus blossom midge

AL 50%
p. 167

corn earworm *

AL 25%
p. 136

spiraea aphid
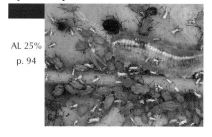
AL 25%
p. 94

black citrus aphid *

AL 25%
p. 91

monolepta beetle
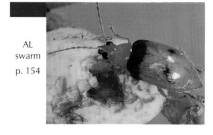
AL swarm
p. 154

broad mite *
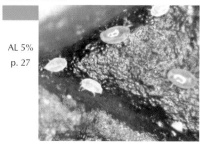
AL 5%
p. 27

citrus bud mite *

AL 10%
p. 25

KEY
- shoots
- flowers
- leaves
- young fruit
- maturing fruit
- twigs and branches

* major or occasionally important pests

AL action level

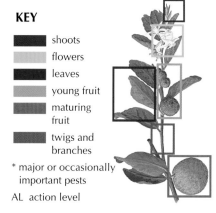

rhyparida beetle

AL swarm
p. 154

bronze orange bug

AL 25% of trees
p. 113

katydids *

AL 5%
p. 157

scirtothrips *
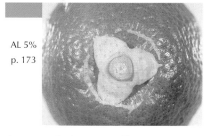
AL 5%
p. 173

banana fruit caterpillar

AL 25%
p. 137

lightbrown apple moth
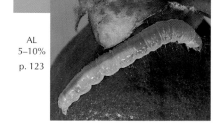
AL 5–10%
p. 123

citrus leafeating weevil

AL 5 beetles per branch
p. 143

citrus red mite (NSW only) *

AL 50%
p. 33

MID-SEASON October–December Sample all varieties fortnightly for these pests. See also page 220 for fungal diseases and fruit disorders which may be confused with pest damage.

MONITORING GUIDE: QUEENSLAND AND COASTAL NEW SOUTH WALES

pink wax scale *
AL 5%
p. 55
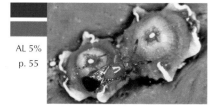

Florida wax scale (Q only) *
AL 5%
p. 57

cottony citrus scale (Q only) *
AL 5%
p. 52

soft brown scale
AL 15%
p. 45
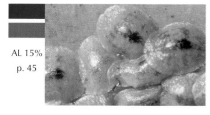

green coffee scale *
AL 5%
p. 43

aleurocanthus whitefly
AL 25%
p. 107
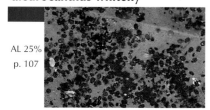

Australian citrus whitefly
AL 25%
p. 106
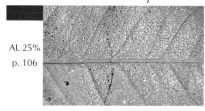

white wax scale
AL 5%
p. 59

planthoppers
AL 20%
p. 98

KEY
- shoots
- flowers
- leaves
- young fruit
- maturing fruit
- twigs and branches

* major or occasionally important pests
AL action level

citrus mealybug *
AL 10–20%
p. 85
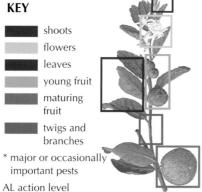

spherical mealybug (Q only) *
AL 5%
p. 87

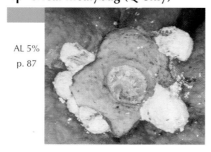

black scale
AL 5–10%
p. 48

hemispherical scale
AL 5%
p. 50

nigra scale
AL 5%
p. 51

long soft scale (Q only) *
AL 5%
p. 47
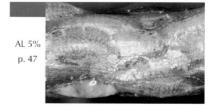

cottony cushion scale
AL 5%
p. 62

rastrococcus mealybug
AL 25%
p. 88

longtailed mealybug
AL 10–20%
p. 81

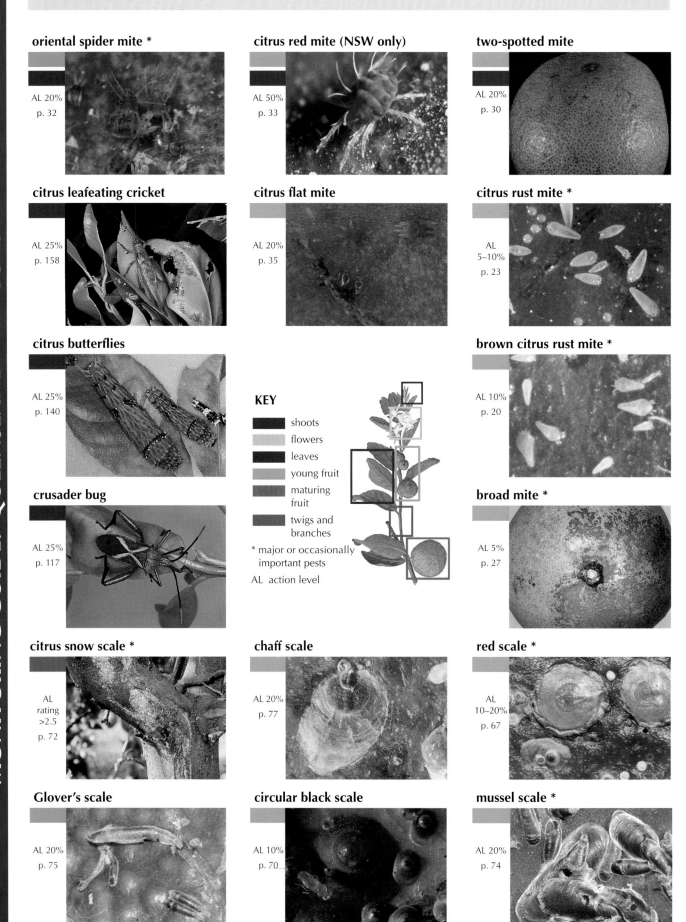

MONITORING GUIDE: QUEENSLAND AND COASTAL NEW SOUTH WALES

LATE SEASON January–May Sample all varieties for these pests once. See also page 220 for fungal diseases and fruit disorders which may be confused with pest damage.

Check for these pests again (see page 218)

citrus rust mite *	red scale *
brown citrus rust mite *	citrus snow scale *
oriental spider mite *	mussel scale *
two-spotted mite	chaff scale
citrus flat mite	circular black scale
broad mite	Glover's scale
citrus red mite *	

Check for these pests again (see page 217)

pink wax scale *	cottony cushion scale
white wax scale	long soft scale *
Florida wax scale *	citrus mealybug *
cottony citrus scale	spherical mealybug *
green coffee scale *	longtailed mealybug
soft brown scale	rastrococcus mealybug
black scale	citrus planthopper *
nigra scale	green planthopper *
hemispherical scale	

giant northern termite
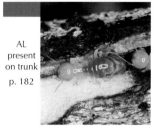
AL present on trunk
p. 182

citrus gall wasp *

AL 33%
p. 169

speckled longicorn
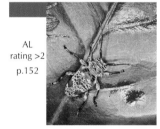
AL rating >2
p.152

citrus branchborer

AL branch wilting
p. 149

hard wax scale *
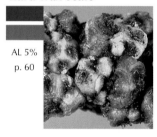
AL 5%
p. 60

KEY
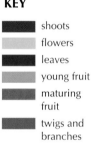
- shoots
- flowers
- leaves
- young fruit
- maturing fruit
- twigs and branches

* major or occasionally important pests
AL   action level

fig longicorn
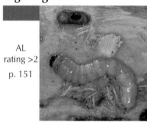
AL rating >2
p. 151

citrus leafminer *
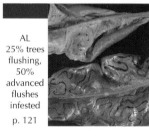
AL 25% trees flushing, 50% advanced flushes infested
p. 121

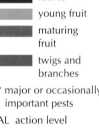

Queensland fruitfly *
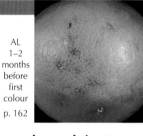
AL 1–2 months before first colour
p. 162

spined citrus bug *

AL 10% of trees
p. 111

citrus leafhopper *
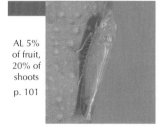
AL 5% of fruit, 20% of shoots
p. 101

citrus rust thrips *
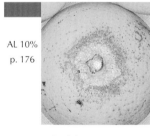
AL 10%
p. 176

greenhouse thrips *
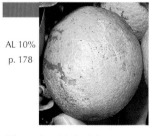
AL 10%
p. 178

citrus rindborer *
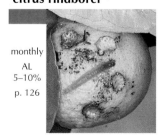
monthly
AL 5–10%
p. 126

fruitpiercing moths *

AL 5%
p. 131

orange fruitborer
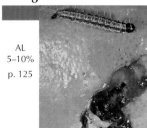
AL 5–10%
p. 125

blastobasid fruitborers

AL 5–10%
p. 129

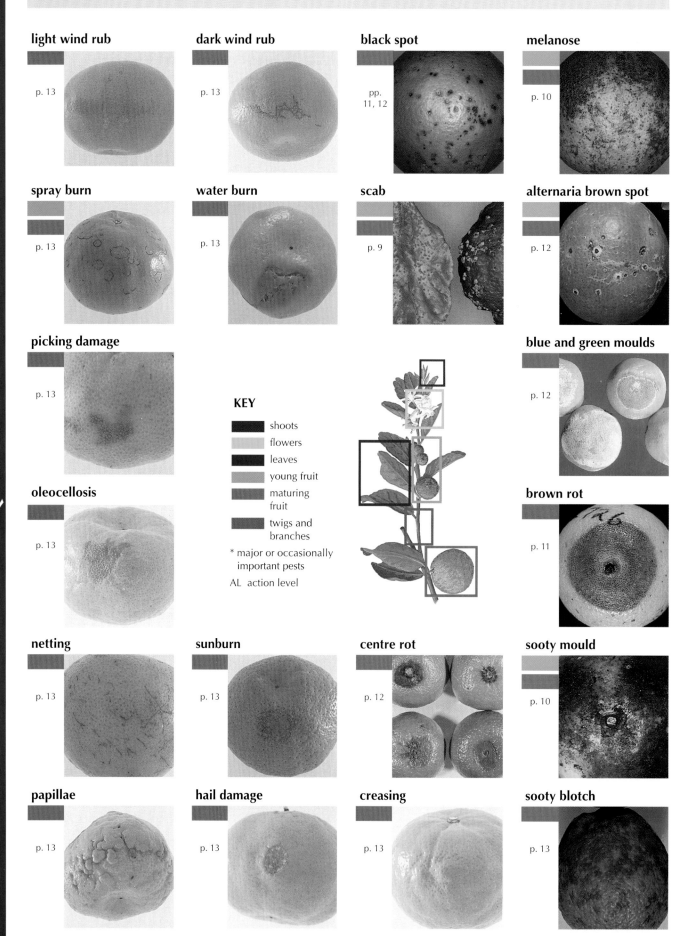

25 Practical IPM: monitoring guides

EARLY SEASON September–December Sample all varieties for these pests at the time intervals given below. See also page 223 for fungal diseases and fruit disorders which may be confused with pest damage.

MONITORING GUIDE: INLAND NEW SOUTH WALES, SUNRAYSIA AND RIVERLAND

Fuller's rose weevil
once or twice
AL present
p. 142

katydids *
fortnightly
AL 5%
p. 157

longtailed mealybug *
monthly
AL 10–20%
p. 81

citrophilous mealybug *
fortnightly
AL 10–20%
p. 80

lightbrown apple moth *
fortnightly
AL 5–10%
p. 123

spined citrus bug *
monthly
AL 10%
p. 111

red scale *
monthly
AL 10–20%
p. 67
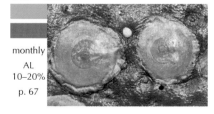

soft brown scale *
once, Dec
AL 15%
p. 45
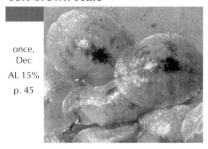

citricola scale *
once or twice, Feb–Mar
AL 15%
p. 41

KEY
- shoots
- flowers
- leaves
- young fruit
- maturing fruit
- twigs and branches

* major or occasionally important pests
AL action level

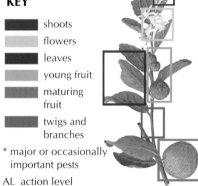

megalurothrips *
fortnightly
AL 5%
p. 175

common garden snail
monthly
AL history of infestation
p. 191
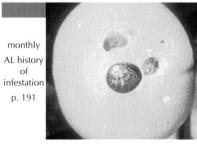

black scale *
once, Dec
AL 5–10%
p. 49

white wax scale
once, Dec
AL 5%
p. 59

ants *
once, Dec
AL 50%
p. 38
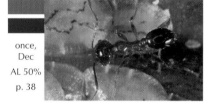

crusader bug
twice
AL 25% young trees
p. 117

black citrus aphid
fortnightly
AL 25% young trees
p. 91

citrus butterflies
twice
AL 25% young trees
p. 140

two-spotted mite
twice
AL 25% young trees
p. 30
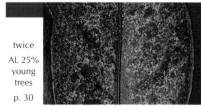

MONITORING GUIDE: INLAND NEW SOUTH WALES, SUNRAYSIA AND RIVERLAND

MID-SEASON – LATE SEASON January–May Sample all varieties for these pests at the time intervals given below. See also page 223 for fungal diseases and fruit disorders which may be confused with pest damage.

longtailed mealybug *

monthly
AL 10–20%
p. 81

red scale *

monthly
AL 10–20%
p. 67

citrophilous mealybug *

monthly
AL 10–20%
p. 80

spined citrus bug *
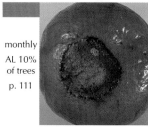
monthly
AL 10% of trees
p. 111

brown citrus rust mite

fortnightly or monthly, Jan–Mar
AL 10%
p. 20

lightbrown apple moth
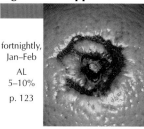
fortnightly, Jan–Feb
AL 5–10%
p. 123

blastobasid fruitborers

fortnightly, Feb to harvest
AL 5–10%
p. 129

Fuller's rose weevil
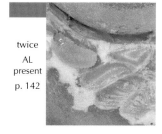
twice
AL present
p. 142

Aust. citrus whitefly

twice, young trees
AL 25%
p. 106

megalurothrips
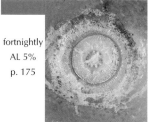
fortnightly
AL 5%
p. 175

KEY
- shoots
- flowers
- leaves
- young fruit
- maturing fruit
- twigs and branches

* major or occasionally important pests
AL action level

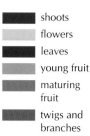

ants

once or twice
AL 50%
p. 38

black citrus aphid

twice, young trees
AL 25%
p. 91

soft brown scale *

monthly, Jan–Mar
AL 15%
p. 45

black scale *

monthly, Jan–Mar
AL 5–10%
p. 49

citricola scale *
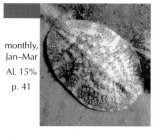
monthly, Jan–Mar
AL 15%
p. 41

citrus butterflies

twice, young trees
AL 25%
p. 140

hard wax scale

once or twice, Jan–Feb
AL 5%
p. 60

cottony cushion scale

twice, young trees
AL 5%
p. 62

planthoppers
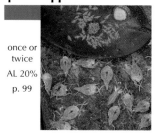
once or twice
AL 20%
p. 99

citrus leafminer *
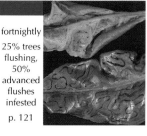
fortnightly
25% trees flushing, 50% advanced flushes infested
p. 121

DISEASES AND DISORDERS
Symptoms of these diseases and disorders are sometimes confused with pest damage.

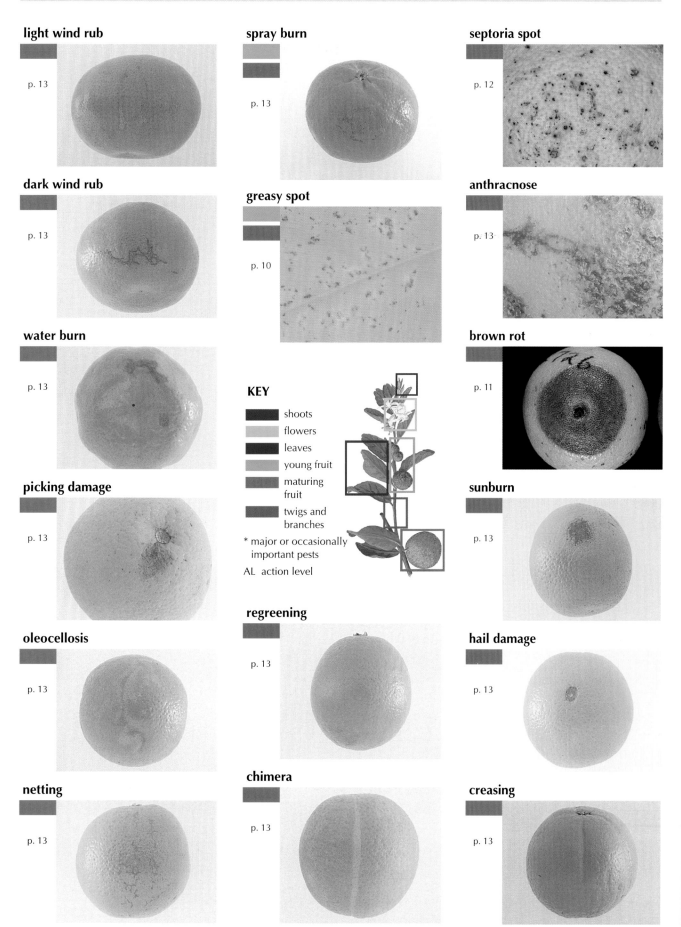

MONITORING GUIDE: WESTERN AUSTRALIA

EARLY SEASON September–December Sample all varieties for these pests at the time intervals given below. See also page 223 for fungal diseases and fruit disorders which may be confused with pest damage.

Fuller's rose weevil
once or twice, Nov–Dec
AL present
p. 142

red scale *
fortnightly
AL 10–20%
p. 67

soft brown scale
once, Nov–Dec
AL 15%
p. 45

broad mite
fortnightly
AL 5%
p. 27

brown citrus rust mite
fortnightly
AL 10%
p. 20

black scale
once, Nov–Dec
AL 5–10%
p. 49

citrus bud mite
once or twice
AL 10%
p. 25

KEY
- shoots
- flowers
- leaves
- young fruit
- maturing fruit
- twigs and branches

* major or occasionally important pests
AL action level

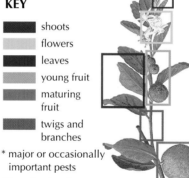

citricola scale
once, Nov–Dec
AL 15%
p. 41

ants
Nov–Dec
AL 50%
p. 38

Australian citrus whitefly
once or twice
AL 25%
p. 106

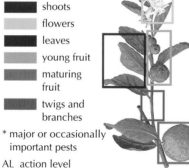

greenhouse thrips
once or twice
AL 10%
p. 178

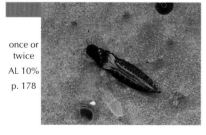

crusader bug
Sep–Oct
AL 25% young trees
p. 117

pink wax scale
once, Dec
AL 5%
p. 55

Mediterranean fruit fly *
traps weekly
AL 1–2 months before first colour
p. 164

black citrus aphid
fortnightly
AL 25%
p. 91

white wax scale
once, Dec
AL 5%
p. 59

25 Practical IPM: monitoring guides

MID – LATE SEASON January–May Sample all varieties for these pests at the time intervals given below. See also page 223 for fungal diseases and fruit disorders which may be confused with pest damage.

MONITORING GUIDE: WESTERN AUSTRALIA

Fuller's rose weevil *

once or twice, Jan–Feb
AL present
p. 142

longtailed mealybug

fortnightly
AL 10–20%
p. 81

brown citrus rust mite *

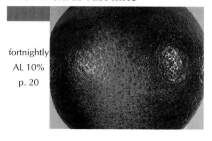

fortnightly
AL 10%
p. 20

common garden snail

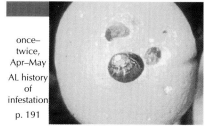

once–twice, Apr–May
AL history of infestation
p. 191

ants

Jan–Feb
AL 50%
p. 38

citrus leafminer *

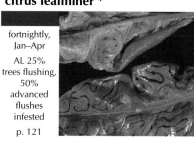

fortnightly, Jan–Apr
AL 25% trees flushing, 50% advanced flushes infested
p. 121

red scale *

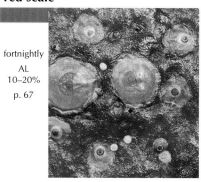

fortnightly
AL 10–20%
p. 67

KEY
- shoots
- flowers
- leaves
- young fruit
- maturing fruit
- twigs and branches

* major or occasionally important pests

AL action level

crusader bug

once–twice weekly, Apr–May
AL 25% young trees
p. 117

black citrus aphid

fortnightly, Apr–May
AL 25%
p. 91

Mediterranean fruit fly *

traps weekly
AL 1–2 months before first colour
p. 164

greenhouse thrips

once, Jan
AL 10%
p. 178

Australian citrus whitefly

once, Feb
AL 25%
p. 106

white wax scale

once, Feb
AL 5%
p. 59

pink wax scale

once, Feb
AL 5%
p. 55

hard wax scale

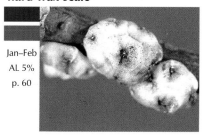

Jan–Feb
AL 5%
p. 60

26 Pesticide application

When IPM is being implemented by citrus growers, it is important that pests be effectively controlled when action levels are reached. If the appropriate action is spraying trees with a chemical, it is of the utmost importance to use spraying equipment which gives excellent coverage of the trunk, limbs, twigs, leaves and fruit. Good timing and good coverage will maximise pest kills.

Pesticide application is usually required only when the pest populations are out of balance with their natural enemies. Pesticides kill pests, but also further reduce the numbers of natural enemies. (This problem can be made worse if spray coverage is poor, as pests that are sedentary or in protected positions will be less likely to come into contact with the spray and will be poorly controlled, while natural enemies, that are more mobile, and more likely to come into contact with the spray, will be killed.) However, populations of natural enemies have a greater chance of recovering when only occasional sprays are required, and there is the longest possible interval between sprays.

Citrus is one of the most difficult tree crops to spray efficiently because of the tree shape and the dense foliage. Poor spray coverage can be due to a number of reasons relating to tree type, problems with equipment and its operation, and environmental conditions. They include:

- inappropriate sprayer
- poor machine design
- incorrect calibration
- poor maintenance of nozzles, pumps and hoses
- excessive spraying speed
- insufficient spray volume
- poor spray penetration in large, dense trees
- poor spray coverage of the inside top of large trees
- poor spraying conditions (windy and/or wet).

It is imperative that attention be given to correctly selecting, setting up and operating pesticide application equipment. Regular tree pruning and skirting are also important to facilitate spray penetration.

Information on spraying equipment and its operation can be obtained from pesticide application and safety courses, IPM scouts, pesticide application specialists and sprayer manufacturers. It is recommended that all spray operators attend pesticide application and safety courses.

Because many citrus pests are sedentary (e.g. scale insects) or slow-moving (e.g. mealybugs and mites), a high level of pesticide coverage (90%) is necessary, otherwise the pests will not come into contact with the pesticide. (Scales and mealybugs are also protected by a water-repelling, waxy, leathery or mealy covering.)

To achieve high mortality, the pests must be thoroughly wetted with pesticides. On mature trees up to 5 m high, this requires high-volume spraying (10 000 or more L/ha). Low-volume sprayers usually give unsatisfactory control of these sedentary or slow-moving pests, unless adapted to apply high volumes.

High-volume spraying is not so important for more mobile pests, such as true bugs, beetles and caterpillars, and for sprays to combat diseases or to supply nutrients. In these cases, either high-volume or low-volume sprays are acceptable.

Sprayer types

Oscillating booms

For the spraying of scales, mealybugs and mites on citrus, the oscillating boom has built a good reputation over the last 50 years. When efficiently operated, the oscillating boom gives at least 90% coverage of the whole tree and, with the normal spraying arm on the top, is the most effective sprayer for large trees up to 6 m high. Oscillating booms normally apply sprays at rates of 7000–15 000 L/ha, but can apply as little as 3000 L/ha with appropriate nozzles.

Oscillating booms have become the yardstick by which other types of sprayers are measured for efficient control of scales, mites and mealybugs in citrus. Oscillating booms are highly recommended by commercial pest scouts. Most growers use one-sided units.

The main disadvantage of the oscillating boom is the loss of spray due to tree run-off. Other disadvantages, and the advantages, are summarised in table 26.1.

Table 26.1 Oscillating booms: advantages and disadvantages

Advantages	Disadvantages
• 90% spray coverage	• use high volumes of water and usually require an attendant water tank
• wet the whole tree, and both sedentary and mobile pests	• usually use more pesticide per ha than low-volume machines
• give good coverage of the top and centre of large trees	• up to 40% of the spray runs off onto the ground (high off-target losses)
• low spray drift (because of large droplet sizes)	• cumbersome for spraying trees less than 5 years old
• low power requirements (less than 45 kW)	• slow for spraying large orchards
	• not very adaptable for spraying lower volumes against mobile insects and diseases, and for applying nutrient sprays

Set-up and maintenance of the oscillating boom is important: coverage can be reduced by as much as 50% when there is poor maintenance. See table 26.2 for set-up recommendations. Boom sprayers without oscillation have been used by some growers, but coverage is generally not as good as with oscillating booms and is inadequate for scale and mealybug control.

Plate 26.1 Oscillating boom sprayer.

Table 26.2 Set-up of oscillating booms

Component	Recommendations
spray volume	3000–15 000 L/ha
groundspeed*	2.3–2.75 km/h
pump pressure	3000–5000 kPa
oscillation rate* (preferably two-way)	90–110 oscillations per minute
spray pattern	adjacent cones marrying at about 1.75 m (i.e. narrow cone angle, but the maximum angle consistent with good tree penetration)
agitation	1 large paddle per 500 L, i.e. a 2500 L vat should have 5 paddles; paddles should clear the bottom of the vat by about 10 mm and should have the same orientation
boom set	set angle of top boom to match tree size

*If the oscillation rate is lower than 90–110 and/or the oscillations are only one-way, then groundspeed should not exceed 2.5 km/h.

Low-profile air-blast sprayers

Low-profile air-blast sprayers are unsuitable for spraying sedentary pests, particularly on citrus trees over 2 m in height. They can be adapted for high-volume application by the addition of a full tree-height tower with hydraulic nozzles, or a full tree-height air tower. They normally apply between 200 and 3000 L/ha. Advantages and disadvantages are summarised in table 26.3, and recommendations for set-up are given in table 26.4.

Table 26.3 Low-profile air-blast sprayers: advantages and disadvantages.

Advantages	Disadvantages
• less expensive to buy than most other machines • low maintenance • relatively quick and mobile, especially for young trees • usually use less pesticide per ha than oscillating boom • much less wastage due to run-off than oscillating boom • useful for applying nutrient sprays	• give insufficient spray coverage for sedentary pests (scales, mealybugs, mites) in trees over 2 m high • spray coverage is uneven in higher trees, with excessive deposition low on the tree and under-dosing in the top • high power requirement (total of about 50 kW (70 hp)) • can produce excessive drift of small droplets • require careful air calibration to ensure airflow matched to tree size

Table 26.4 Set-up of low-profile air-blast sprayers

Component	Recommendations
groundspeed	• 2.5–3 km/h
airflow	• to achieve maximum airflow, use the greatest fan blade pitch possible in relation to the available tractor power; tractors with power ratings of 50 kW or higher, and fans capable of delivering around 20 000 cubic metres of air per hour, or greater calibrated output (not fan rating), are required for good results in citrus • use only one side; modify airflow with an air cowling for one-sided spraying (when spraying scale insects and mealybugs, 20 000 cubic metres of air per hour is not enough for two-sided spraying) • adjust fan cowling to widest setting to maximise area of air outlet and thus air volume produced • use upper and lower deflectors to ensure air is directed at the tree canopy • use sprayers with straightening vanes; set left and right side of sprayer independently if not using air straightening vanes
nozzles	• use large numbers of fine, abrasion-resistant or wear-resistant, ceramic hollow-cone nozzles of similar sizes, operated at optimum pressure; most air-blast sprayers will give improved coverage if more nozzles are added (at least 15 per side are recommended) • Spraying Systems TX or Delevan HC nozzles (or equivalent) use a different swirl system and give fine droplets at lower pressures (400–1000 kPa compared to around 2500 kPa for standard disc-core hollow-cone nozzles) • when adding additional nozzles, ensure that the air duct is wide enough to carry all the droplets, otherwise the additional nozzles will not improve spray coverage, and may decrease it • with conventional nozzles and spacings, direct large nozzles towards the top of the trees • rotatable nozzle assemblies enable rapid switching from fine to coarser nozzle sizes, or the assembly to be turned off (e.g. for small trees, upper nozzles can be turned off)

Plate 26.2 Low-profile air-blast sprayer.

Structurally modified low-profile air-blast sprayers

Much-improved results can be achieved with low-profile air-blast sprayers by adding an overhead hydraulic boom to the front of the unit, before the region of air-blast from the fan. The boom should be as high, or higher than, the trees.

Some growers with skirted trees also add a low horizontal boom to pass underneath the trees. A cluster of three or four jets on the end of the boom, passing within about 500 mm of the trunk, is used to spray the inside canopy of the tree. With these modifications, volumes of around 10 000 L/ha can be obtained, and the coverage approaches that achieved by the oscillating boom.

Table 26.5 Set-up of structurally modified air-blast sprayers

Component	Recommendations
Airflow	• for a one-sided unit, add a cowling (short air tower) with nozzles to convert the air-blast component to one side; if the fan rotates clockwise, block off the air on the left side • set other parameters as for the low-profile air-blast sprayer
Nozzles	• use fine, low-output hollow nozzles with air assistance • use high-output, solid-cone nozzles on overhead and under-tree booms with similar spray pattern and nozzle spacings to those used on the oscillating boom • the capacity of standard air-blast pumps is often too low to cope with the flow rate of the additional high-volume nozzles, especially with two-sided units; if this is the case, add a larger pump

Air-blast sprayers with tower air conveyors

Air-blast sprayers fitted with air towers to the full height of the trees can achieve spray coverage approaching that achieved by the oscillating boom. However, poor design in

Plate 26.3 Structurally modified low-profile air-blast sprayer with overhead hydraulic boom.

the tower or insufficient airflow often drastically reduces the efficacy of these sprayers against scales and mites. They normally apply between 1000 and 5000 L/ha, but spray volume should be increased to 10 000 or more L/ha for sedentary pests.

Advantages and disadvantages of air-blast sprayers fitted with air towers are summarised in table 26.6. Their main advantage over oscillating booms is their adaptability: they can be used for high-volume or low-volume application.

In Australia, most units are for one-sided application only. Two-sided units are available from overseas manufacturers, but capital cost and power requirements are extremely high.

Table 26.6 Air-blast sprayers with air tower: advantages and disadvantages

Advantages	Disadvantages
• low maintenance • relatively quick and mobile • much better coverage than low-profile air-blast sprayers • ideal for mobile pests, disease and nutrient sprays • usually use less pesticide per hectare than oscillating boom	• not quite as effective as oscillating boom on sedentary pests • high power requirements (50–75 kW) • poorly designed towers or insufficient airflows result in much-reduced coverage

Table 26.7 Set-up of air-blast sprayers with air tower

Component	Recommendations
airflow	correctly designed internal baffles ensure relatively uniform airflow along the whole tower
nozzles	seek professional help to select and set up

High-velocity sprayers with air-shear nozzles

High-velocity sprayers with air-shear nozzles can be used to apply volumes ranging from 100 L/ha to 1000 L/ha. They normally use an enclosed centrifugal fan with air-ducting to spray-heads containing constricted air outlets with air-shear nozzles of various types. See table 26.8 for advantages and disadvantages of these sprayers.

Air-shear nozzles use air movement to atomise a stream of liquid. Air velocities over the atomisers normally range from 360–650 km/h to shear off very fine droplets, which may be as fine as 50 micrometres vmd (volume median diameter). In comparison, the typical air velocity of a standard air-blast sprayer is about 100 km/h. Pressures required for spray liquid movement to the air-shear nozzles are low, and low-pressure centrifugal pumps and hoses are used. See table 26.9 for set-up recommendations.

Table 26.8 High-velocity sprayers with air-shear nozzles: advantages and disadvantages

Advantages	Disadvantages
• effective coverage for control of mobile pests	• high power requirement
• good for nutrient sprays	• unsatisfactory against sedentary pests
• reliable	• can produce excessive drift
• mobile and fast, with high work rate	

Plate 26.4 Air-blast sprayers with tower.

Plate 26.5 High-velocity air-shear sprayer with tower attachment.

Table 26.9 Set-up of high-velocity sprayers with air-shear nozzles

Component	Recommendations
liquid flow rate and chemical concentration	match to rates and volumes required
fan speeds	high; flooding of nozzles, with consequently larger droplet sizes and poorer coverage, can occur at lower fan speeds, especially at higher flow rates
towers	use full-height towered units, with two, or preferably three, heads per side

Rotary atomisers

Rotary atomisers have multiple spray-heads, each containing an axial fan and high-speed (6000–10 000 rpm) rotary cage, drum or disc atomiser. These atomisers produce fine droplets, generally around 100–150 micrometres vmd (volume median diameter). The spray-heads are normally powered by hydraulic motors. Current sprayers of this type can be used to apply spray volumes from 200 L/ha to 4000 L/ha. See table 26.10 for the advantages and disadvantages of these sprayers.

The heads on later models have electric drive, requiring much less power from the tractor. They are designed for a much-increased airflow and higher volume application against sedentary pests. See table 26.11 for set-up recommendations.

Table 26.10 Rotary atomisers: advantages and disadvantages

Advantages	Disadvantages
• effective coverage for control of mobile pests • reliable • mobile and fast with high work rate • ideal for nutrient spraying	• not effective against sedentary pests at volumes under 4000 L/ha • can produce excessive spray drift • high level of operator expertise required • high maintenance • some types can be unreliable

Table 26.11 Set-up of rotary atomiser sprayers

Component	Recommendations
heads	• 4–6 heads per side are normally needed to spray large citrus trees • converging airstreams give additional turbulence and coverage • direct airflows to converge into the tree canopy • have the top head slightly lower than the tops of trees, and a slight upward (not downward) incline for upper heads and slightly more upward incline on lower heads • airflows should converge laterally at 10°–20° • liquid flow can be turned off top upper heads when spraying small trees, or the heads repositioned to suit where all trees are small
liquid flow rate and chemical concentration	• for best results, match to rates and volumes required (i.e. calibrate accurately), and to the speed and capacity of discs, drums or cages
speed of cages, drums or discs	• maintain at specified revolutions per minute to produce droplets of the required size
groundspeed	• operate at correct groundspeed for the available airspeed and coverage requirements, i.e. accurately calibrate for airflow

Multi-head sprayer

A multi-head sprayer with a large (500 mm diameter) 3000 rpm fan, and three-phase AC electric drive is also available. Air capacity of a 4–6 head sprayer is very high, and power requirement is low, compared to air-blast sprayers. Hollow-cone nozzles spraying through the fan are used instead of a rotary atomiser, to apply high (10 000 L/ha) or low volumes.

Further information

Further information on nozzles, spray tanks, agitation, pumps, pressure gauges, filters, calibration, coverage, testing etc. can be found in other books, e.g. *Pesticide application manual* (2nd edition, Department of Primary Industries, Queensland).

Plate 26.6 One type of rotary atomiser.

Plate 26.7 Prototype of a multi-head sprayer.

27 Petroleum spray oils (PSOs)

Petroleum spray oils (PSOs) (also called white, summer or narrow-range oils) are IPM-compatible. These oils act by suffocating susceptible pests or altering their behaviour. They can also be used as fungicides.

There has been a major resurgence of interest in the use of PSOs over the last ten years. This has occurred because of the need to address issues such as widespread resistance to synthetic pesticides, and environmental and health problems associated with pesticide use.

While PSOs are ideally suited for use in sustainable IPM programs, they should be used to control pests only when natural enemies fail to keep pest populations below action levels.

The advantages of using PSOs instead of broad-spectrum pesticides are:

- only minimum protective clothing (such as overalls, goggles and simple dust masks) needs to be worn when handling PSOs
- their toxicity to vertebrate animals is low—they are almost as pure as the products used for baby and hair oils, skin lotions and creams
- they have fewer detrimental effects on beneficial insects and mites than most synthetic pesticides
- they do not stimulate pest outbreaks
- pests are not known to develop resistance to PSOs
- oil molecules are similar to natural plant waxes, and deposits are broken down within weeks by microbes, oxidation and ultraviolet light to form simple, harmless molecules.

PSOs are composed of molecules of many different shapes and sizes. Current classification of PSOs is based on the temperature at which 50% of molecules boil, and the relationship between the boiling points and the number of carbon (C) atoms in molecules with known structures (hence the terms 'C21', 'C23' and 'C24' oils, used to describe PSOs).

Pesticidal activity and the risk of PSOs damaging plants (phytotoxicity) both increase as the 50% boiling points increase. PSOs can damage plants if applied in hot weather, and can affect colouring of early mandarins if applied after December in Queensland.

PSOs made up of molecules with a limited range of boiling points (narrow-range oils) were developed in the 1960s to maximise pesticidal activity while minimising the risk of phytotoxicity. The most recent development in PSO technology has led to the manufacture of C24 oils with additives that reduce the risk of phytotoxicity while increasing pesticidal activity.

PSOs are applied at rates of 0.25%–2% (250 mL – 2 L oil per 100 L water). The 1% rate is preferred for most scales. The 0.25%–0.5% rates are recommended for citrus leafminer. The 2% rate, once widely used to control scales, is now rarely used.

To achieve effective control of pests, PSOs must be applied properly. They often need to be applied at higher volumes than most other types of pesticides. This is because they have limited residual activity.

PSOs must be applied so that effective films of oil are formed by the oil droplets joining together as the oil and water components of sprays separate when they hit twigs, branches, fruit and leaves.

Equipment and coverage

Some sprayers are far more effective than others for applying PSOs. An oscillating boom with a horizontal outrigger is the best type. Other sprayers which can achieve effective control of pests with PSOs are air-blast sprayers with towers, and rotary atomisers, providing they apply high volumes of spray. Low-profile air-blast sprayers are generally not suitable because they do not deposit sufficient spray inside trees or in the tops of trees taller than 2 m. For more detailed information on the different types of sprayers, see chapter 26.

Figures 27.1, 27.2 and 27.3 compare the effects on different pests of applying PSOs with oscillating boom, air-blast and rotary atomiser sprayers. Against medium to heavy infestations of red scale on trees 3–4 m high, PSO volumes less than 10 000 L/ha give inadequate control. The air-blast and rotary atomiser sprayers are unable to deliver this volume and are virtually useless (figure 27.1). Similar results are obtained when using PSOs against pink wax scale on trees 2–3 m high (figure 27.2). The figures indicate that slightly higher volumes of 1% sprays are required to give the same level of control as 2% sprays, but at a lower cost (almost half). Against citrus leafminer on trees 3–4 m high, control becomes optimum at about 4000 L/ha (figure 27.3).

PSO volumes required for effective control of pests depend on a number of factors. These include target pest, severity of infestation, tree height, planting density and canopy density.

On trees 4 m high, single, annual 10 000–14 000 L/ha 1% sprays are recommended for medium to heavy densities of red scale and other armoured scales. Lower volumes are

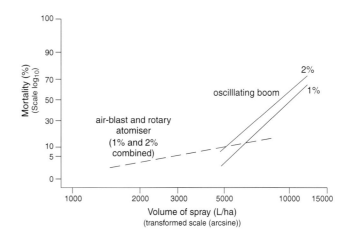

Figure 27.1 Effect on red scale mortality of applying 1% and 2% PSOs with oscillating boom, air-blast and rotary atomiser sprayers. (Based on unpublished work by G.A.C. Beattie et al.)

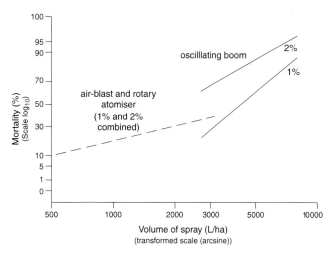

Figure 27.2 Effect on pink wax scale mortality of applying 1% and 2% PSOs with oscillating boom, air-blast and rotary atomiser sprayers. (Based on unpublished work by G. A. C. Beattie et al.)

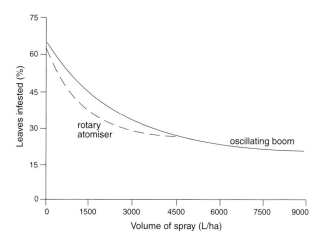

Figure 27.3 Effect on citrus leafminer infestation of applying 1% and 2% PSOs with oscillating boom, air-blast and rotary atomiser sprayers. (Based on unpublished work by Z. M. Liu et al.)

suitable for light infestations only. Two annual 1% sprays, the first in late spring – early summer, and the second in late summer – early autumn, may be required for very heavy infestations of red scale.

On trees 4 m high, 10 000–14 000 L/ha 1% sprays are recommended for mealybug, and for greenhouse thrips and rust thrips. Two sprays may be required annually. PSO sprays by themselves are usually not able to provide the required levels of control of citrophilous mealybug on export fruit.

On trees 4 m high, 8000–10 000 L/ha 1% sprays control young soft scales, and 5000–8000 L/ha 0.7%–1% sprays control mites. To control citrus leafminer in trees 3–4 m high, 0.25%–0.5% sprays should be applied at 4000 L/ha.

For trees shorter or taller than 3–4 m, volumes of spray should be reduced or increased accordingly. Sprayers need to be adjusted carefully.

To control soft scales and mealybugs, PSOs must be applied when the youngest stages of the pests predominate. Applications for greenhouse thrips, rust thrips and mealybugs should be made after harvest and before new season fruit start to touch.

Multiple 0.5% PSO sprays (4–6 per season) applied to control citrus leafminer in summer and autumn will also control armoured scales, soft scales, whiteflies and mites to varying degrees.

Professional advice should be sought to ensure that sprayers are operated correctly (see chapter 26). Professional advisers should include pest scouts who can assess the outcome of spray applications, independent pesticide application specialists and spray machine manufacturers. The pest scouts will also be able to recommend the most appropriate time(s) for spraying, and whether spraying is needed.

Timing of PSO sprays

Table 27.1 is a guide for timing of PSO sprays. Note that PSOs applied for the control of one specific pest will also simultaneously control a range of other pests. Some pests not listed in the table may also be controlled by PSOs at the times indicated, e.g. first instar bronze orange bugs will be susceptible to 1% PSO sprays applied at 6000–12 000 L/ha in February–March.

Table 27.1 Pests controlled by 1% PSOs (1 L of oil per 100 L of water) applied at 10 000–12 000 L/ha to mature trees 3–4 m high

PEST	MONTH											
	Jul	Aug	Sep	Oct	Nov	Dec	Jan	Feb	Mar	Apr	May	Jun
Armoured scales												
chaff scale					■							
circular black scale					■							
Glover's scale					■			■				
purple scale (medium to heavy infestations)					■			■				
red scale (medium to heavy infestations)					■			■				
yellow scale					■			■				
Soft scales												
black scale						■		■				
citricola scale								■				
Florida wax scale						■						
green coffee scale						■						
hard (Chinese) wax scale								■				
hemispherical scale					■							
long soft scale					■							
nigra scale					■							
pink wax scale					■	■						
cottony citrus scale					■							
soft brown scale					■			■				
white wax scale				■		■						
Mealybugs												
citrophilous mealybug *					■			■				
longtailed mealybug *					■			■				
rastrococcus mealybug								■				
Whiteflies												
citrus whitefly					■			■				
Thrips												
greenhouse thrips					■			■				
Mites												
brown citrus rust mite					■			■				
citrus red mite		■	■	■					■	■	■	
citrus rust mite					■			■				
flat mites					■			■				

KEY

▨ Queensland ▨ all areas ▨ all areas, except early varieties in Qld ▨ central coast of NSW

* 10 000–14 000 L/ha

28 Pesticide toxicity to natural enemies

The following table rates the relative toxicity of commonly used pesticides to natural enemies. It shows which pesticides are least compatible with IPM (those of high toxicity) and those that are most compatible (those of low toxicity).

The table is a guide only. Check registrations for your crops. When in doubt, ask your supplier of beneficial insects and mites, and/or another professional adviser, for more specific information.

Table 28.1 Chemical toxicity to natural enemies

CHEMICAL NAME *	TRADE NAMES	*Phytoseiulus persimilis*		*Typhlodromus occidentalis*		*Chilocorus* spp. *Rhyzobius* spp.	
		APPLIED	RESIDUAL	APPLIED	RESIDUAL	APPLIED	RESIDUAL
Insecticides							
Azinphos methyl	Gusathion	M	2	L	0	(H)	2-3
B. thuringiensis	Dipel, Biobit etc.	L	0	L	0	L	0
Buprofezin (IGR)	Applaud	L	0	L	0	H	3
Carbaryl	Various	L	0	M	1	(H)	4
Chlorpyrifos	Lorsban	M	2	M	0	(H)	2-3
Diazinon	Gesapon	M	2	L	0	M	1
Dimethoate	Rogor	H	2	(H)	1	(H)	4
Endosulfan	Various	M	2	L	1	(M)	1
Fenoxycarb (IGR)	Insegar	L	0	L	0	H	2-3
Fenthion	Lebaycid	H	3	(H)	2-3	H	4
Imidicloprid	Confidor	(L)	0	(L)	0	–	–
Maldison	Malathion	M	1	M	0	H	4
Methamidophos	Nitofol	H	3	(H)	1	H	4
Methidathion	Supracide	H	3	H	1	H	4
Methomyl	Lannate, Nudrin etc.	H	1	M	1	(H)	1
Natural pyrethrum	Various	H	1	H	1	(H)	1
Neem	–	L	0	(L)	0	–	–
Parathion	Parathion	L–M	1	L	0	H	4
Pirimicarb	Pirimor	L	1	L	0	(L)	1
Soap sprays	Various	M	0	M	0	L	0
Petroleum spray oils	Various	L–M	1	L–M	1	(L)	0
Sulprofos	Helothion	(M)	1	(M)	1	H	2-3
Synthetic pyrethroids	Various	H	4	H	4	H	4
Tebufenozide (MAC)	Mimic	L	0	L	0	L	0
Thiodicarb	Larvin	(H)	1	(H)	1	H	4
Wettable sulphur	Various	L	0	(M)	1	M	2-3
Miticides							
Avermectin	Vertimec	(L)	0	(L)	0	L	0
Clofentezine	Apollo	L	0	L	0	–	1
Dicofol	Kelthane	H	2	H	1	L	1
Fenbutatin oxide	Torque	L	0	L	0	L	1
Propargite	Omite	L–M	1	L–M	0	L	1
Tebufanpyrad	Pyranica	H	2	M	2	–	–
Tetradifon	Tedion	L	1	L	0	L	1
Fungicides							
Benomyl	Benlate	H	4	H	2-3	(L)	1
Carbendazim	Spin	L–M	1	(L–M)	1	–	–
Iprodione	Rovral	L	0	L	0	L	0
Mancozeb	Dithane	L–M	1	L	1	L	1
Metalaxyl	Ridomil	(L)	1	(L)	1	–	–
Oxythioquinox	Morestan	H	4	(M–H)	2-3	(L)	–
Procymidone	Sumisclex	(L)	1	(L)	1	–	–
Pyrazophos	Afugan	M–H	1	(M)	1	–	–
Triforine	Saprol	M	1	L–M	1	(L)	–

* Not all trade names are given for each chemical.

Table 28.1 Chemical toxicity to natural enemies *(continued)*

Key to toxicity ratings

Applied = toxicity of chemicals when sprayed on the beneficials

    L = low

    M = moderate

    H = high

    ( ) = estimated toxicity

    – = toxicity unknown

Residual = suggested waiting time (in weeks) before introducing beneficials

    – = waiting time unknown

CHEMICAL NAME *	TRADE NAMES	*Cryptolaemus montrouzieri* APPLIED	RESIDUAL	*Aphytis* spp. APPLIED	RESIDUAL	*Encarsia* spp. APPLIED	RESIDUAL
Insecticides							
Azinphos methyl	Gusathion	(H)	2–3	H	4	H	4
B. thuringiensis	Dipel, Biobit, etc.	(L)	0	L	0	L	0
Buprofezin (IGR)	Applaud	H	3	L	0	L	0
Carbaryl	Various	(H)	4	H	4	H	4
Chlorpyrifos	Lorsban	(H)	2–3	H	4	H	4
Diazinon	Gesapon	M	1	H	2–3	H	4
Dimethoate	Rogor	(H)	4	H	4	H	4
Endosulfan	Various	(M)	1	M	1	M	1
Fenoxycarb (IGR)	Insegar	H	2–3	(L)	1	(L)	1
Fenthion	Lebaycid	H	4	H	4	H	4
Imidicloprid	Confidor	–	–	–	–	–	–
Maldison	Malathion	H	4	H	2–3	H	2–3
Methamidophos	Nitofol	H	4	H	4	H	4
Methidathion	Supracide	H	4	H	4	H	4
Methomyl	Lannate, Nudrin etc.	(H)	1	H	1	H	4
Natural pyrethrum	Various	(H)	1	(H)	2	H	1
Neem	–	–	–	M	1	M	1
Parathion	Parathion	H	4	H	4	H	4
Pirimicarb	Pirimor	L	1	L	1	M	1
Soap sprays	Various	L	0	(L)	1	H	1
Petroleum spray oils	Various	(L)	0	L	1	M	1
Sulprofos	Helothion	H	2–3	H	4	H	4
Synthetic pyrethroids	Various	H	4	H	4	H	4
Tebufenozide (MAC)	Mimic	L	0	L	0	L	0
Thiodicarb	Larvin	H	4	H	4	H	4
Wettable sulphur	Various	M	2–3	H	4	M	1
Miticides							
Avermectin	Vertimec	L	0	(L)	0	(L)	0
Clofentezine	Apollo	–	1	–	–	(L)	1
Dicofol	Kelthane	L	1	L	1	L–M	1
Fenbutatin oxide	Torque	L	1	L	1	L	1
Propargite	Omite	L	1	(L)	1	(L)	1
Tebufanpyrad	Pyranica	–	–	–	–	–	–
Tetradifon	Tedion	L	1	(L)	1	(L)	1
Fungicides							
Benomyl	Benlate	L	1	(L)	1	L	0
Carbendazim	Spin	–	–	–	1	(L)	1
Iprodione	Rovral	L	0	L	0	(L)	0
Mancozeb	Dithane	L	1	L	1	L	0
Metalaxyl	Ridomil	–	–	–	1	(L)	0
Oxythioquinox	Morestan	(L)	–	(L)	1	(L)	1
Procymidone	Sumisclex	–	–	–	1	(L)	1
Pyrazophos	Afugan	–	1	(H)	1	–	–
Triforine	Saprol	–	–	(L)	1	(L)	1

* Not all trade names are given for each chemical.

(continued)

Table 28.1 Chemical toxicity to natural enemies (continued)

Key to toxicity ratings

Applied = toxicity of chemicals when sprayed on the beneficials

    L = low

    M = moderate

    H = high

    ( ) = estimated toxicity

    – = toxicity unknown

Residual = suggested waiting time (in weeks) before introducing beneficials

    – = waiting time unknown

CHEMICAL NAME *	TRADE NAMES	*Leptomastix dactylopii* APPLIED	RESIDUAL	*Trichogramma* spp. APPLIED	RESIDUAL	*Mallada signata* APPLIED	RESIDUAL
Insecticides							
Azinphos methyl	Gusathion	H	4	(H)	3	(H)	2
B. thuringiensis	Dipel, Biobit, etc.	L	0	L	0	(L)	0
Buprofezin (IGR)	Applaud	L	0	(L)	0	(L)	0
Carbaryl	Various	H	4	H	3	(M-H)	2
Chlorpyrifos	Lorsban	H	2-3	(H)	3	(H)	2
Diazinon	Gesapon	(M)	1	(H)	3	H	1
Dimethoate	Rogor	H	4	(H)	3	(H)	2
Endosulfan	Various	M	1	H	2-3	(M)	1
Fenoxycarb (IGR)	Insegar	(L)	1	–	–	(L)	1
Fenthion	Lebaycid	H	4	(H)	4	(H)	2
Imidicloprid	Confidor	–	–	–	–	(H)	4
Maldison	Malathion	H	4	(H)	3	(H)	1
Methamidophos	Nitofol	H	2-3	(H)	4	(H)	3
Methidathion	Supracide	H	4	(H)	4	(H)	3
Methomyl	Lannate, Nudrin etc.	H	1	H	2	(H)	1
Natural pyrethrum	Various	(H)	2-3	H	2	(H)	1
Neem	–	–	1	(L)	0	–	1
Parathion	Parathion	H	4	H	3	(H)	3
Pirimicarb	Pirimor	(L)	1	(M)	1	L	1
Soap sprays	Various	L	0	–	1	(L)	0
Petroleum spray oils	Various	(L)	0	–	1	(L)	0
Sulprofos	Helothion	–	4	(H)	3	(H)	2
Synthetic pyrethroids	Various	H	4	H	4	(H)	3-4
Tebufenozide (MAC)	Mimic	L	0	L	0	L	0
Thiodicarb	Larvin	H	4	H	2	(H)	2
Wettable sulphur	Various	H	4	–	–	(M)	2
Miticides							
Avermectin	Vertimec	(L)	0	(L)	0	(L)	0
Clofentezine	Apollo	–	1	(L)	1	(L)	1
Dicofol	Kelthane	L	1	(L)	1	(L)	1
Fenbutatin oxide	Torque	L	1	(L)	1	(L)	0
Propargite	Omite	L	1	(L)	1	(L)	1
Tebufanpyrad	Confidor	–	–	–	–	–	–
Tetradifon	Tedion	L	1	–	1	(L)	1
Fungicides							
Benomyl	Benlate	L	1	M	2-3	(L)	1
Carbendazim	Spin	–	–	–	1	(L)	–
Iprodione	Rovral	L	0	L	0	(L)	0
Mancozeb	Dithane	L	1	–	1	(L)	1
Metalaxyl	Ridomil	–	–	–	0	(L)	1
Oxythioquinox	Morestan	(L)	–	–	1	–	–
Procymidone	Sumisclex	–	–	–	0	–	–
Pyrazophos	Afugan	(H)	1	(H)	2-3	M-H	2
Triforine	Saprol	(L)	–	–	1	L	0

* Not all trade names are given for each chemical.

Appendixes

Appendix 1 Exotic citrus pests

There are many insects and mites which are major citrus pests, but not present in Australia. Some of these are listed below.

Citrus fruit and plant material must not be brought into Australia without official permission from the Australian Quarantine and Inspection Service (AQIS). The whole Australian citrus industry could be put at risk if exotic pests were introduced, and there is an even greater threat from exotic diseases, such as citrus canker.

Table A1.1 Exotic citrus pests and places where they commonly occur

GROUPS	SPECIES	COMMON NAME	PLACES WHERE PESTS COMMONLY OCCUR
Mites	*Aculus pelekassi*	pink citrus rust mite	Japan, Florida
	Calacarus citrifolii	citrus grey mite	South Africa
	Eutetranychus banksi	Texas citrus mite	USA
	Eutetranychus anneckei	Lowveldt citrus mite	South Africa
	Tetranychus cinnabarinus	common red spider mite	South Africa
Thrips	*Scirtothrips citri*	citrus thrips	California
	Scirtothrips aurantii	citrus thrips	South Africa
Psyllids	*Diaphorina citri*	citrus psylla	South-East Asia
	Trioza erythreae	African citrus psylla	South Africa
Fruit flies	*Bactrocera dorsalis*	oriental fruit fly	South-East Asia
	Bactrocera philippinensis	Philippine fruit fly	South-East Asia
	Bactrocera carambolae	carambola fruit fly	South-East Asia
	Ceratitis rosae	Natal fruit fly	South Africa
	Anastrepha ludens	Mexican fruit fly	Mexico
	Anastrepha suspensa	Caribbean fruit fly	Florida, Caribbean
	Anastrepha fraterculus	South American fruit fly	South America
Whiteflies	*Dialeurodes citri*	citrus whitefly	South-East Asia, USA
	Dialeurodes citrifolii	cloudy wing whitefly	South-East Asia, Florida
	Aleurocanthus woglumi	citrus blackfly	South-East Asia, USA, Central America
	Aleurocanthus spiniferus	citrus spiny whitefly	South-East Asia, Japan
	Aleurothrixcis floccosus	woolly whitefly	Mediterranean
	Bemisia citricola	whitefly	South-East Asia
Mealybugs	*Paracoccus burnerae*	oleander mealybug	South Africa
	Pseudococcus maritimus	Baker's mealybug	USA
	Pseudococcus comstocki	Comstock's mealybug	USA
	Pseudococcus cryptus	citrus mealybug	Mediterranean
Armoured scales	*Parlatoria ziziphis*	black parlatoria scale	South-East Asia, Mediterranean, Florida
	Ischnaspis longirostris	black thread scale	South Africa
	Unaspis yannonensis	yanone scale	China, Japan
	Selenaspidus articulatus	West Indian red scale	Caribbean
	Pinnaspis aspidistrae	aspidistra scale	South America, Japan, Florida

GROUPS	SPECIES	COMMON NAME	PLACES WHERE PESTS COMMONLY OCCUR
Soft scales	*Pulvinaria psidii*	green shield scale	South-East Asia
	Pulvinaria aurantii	orange pulvinaria	Japan, China
	Protopulvinaria pyriformis	heart-shaped scale	Mediterranean, South Africa, Florida
	Protopulvinaria floccifera	camellia scale	Mediterranean
	Ceroplastes brevicauda	citrus wax scale	South Africa
	Ceroplastes cirripediformis	barnacle scale	Florida
	Ceroplastes rusci	fig wax scale	Mediterranean
Fruitborers	*Citripestis sagittiferella*	fruitborer	South-East Asia
	Argyrotaenia citrana	orange tortrix	California
	Cryptophlebia leucotreta	false codling moth	South Africa
	Cryptoblabes gnidiella	rindboring orange moth	Mediterranean
Beetles	*Hypomyces squamosus*	green snout weevil	South-East Asia
	Anoplophora chinensis	citrus trunk borer	China
	Diaprepes abbreviatus	sugarcane rootstalk borer	Florida, Caribbean
	Agrilus occipitalis	citrus bark borer	Philippines
	Podagricomela nigricollis	Hong Kong beetle	Hong Kong
Bugs	*Rhynchocoris humeralis*	citrus stink bug	South-East Asia
	Leptoglossus phyllopus	leaf-footed bug	USA
	Empoasca citrusa	green citrus leafhopper	South Africa
Ants	*Solenopsis invicta*	red imported fire ant	USA, Central and South America
	Atta spp.	leafcutting ant	USA, Central and South America

Appendix 2 Keys for identifying common wasp parasites of scales and mealybugs

Over 40 species of wasp parasites have been recorded from scales and mealybugs on citrus in Australia. Some of these parasites are uncommon and are not significant in controlling the populations of hosts. Among the species included in the key, there are 42 primary parasites (which parasitise a scale or mealybug host) and 2 hyperparasites (which parasitise other parasites).

The keys presented below are intended to allow identification of the wasp parasites commonly collected from scales (both soft and armoured scales) and mealybugs associated with citrus in Australia, particularly in south-eastern regions. For user convenience, separate keys have been provided for parasites of soft scales, armoured scales and mealybugs. Note that the information in square brackets describes a complementary character that should be checked as well as the primary character (which is described in unbracketed text).

The keys have been based on specimens, most collected from the Sunraysia and Riverland citrus growing regions during 1992–95, and some received from Dan Smith of Maroochy Horticultural Research Station, Department of Primary Industries, Queensland. The specimens are deposited in the Victorian Agricultural Insect Collection (VAIC) at the Institute for Horticultural Development, Agriculture Victoria, Knoxfield.

Preparation of specimens

To identify a specimen, you will need to kill it, preferably in a stoppered glass tube containing tissue paper to which a couple of drops of ethyl acetate have been added. Examine the specimens under a stereo-microscope (up to 40 × magnification) with a good light source. Since the specimens are small (most are less than 2 mm in body length), a large proportion of the characters included in the key may not be visible using a stereo-microscope. For this reason, they have to be mounted on microscope slides for examination using a compound microscope at high magnification (up to 400 ×).

A simplified method of preparing specimens and mounting them on slides is as follows.

Permanent mounts

1. Soak specimens (which have been freshly killed or dry-preserved) in 10% KOH at room temperature, for several hours to overnight, for clearing and softening. To accelerate clearing, specimens can be placed in 10% KOH and warmed on a heating plate at about 30–40°C for about 1 hour, or they can be warmed in a test tube in a hot-water bath.
2. Rinse off KOH, using distilled water.
3. To dehydrate specimens, transfer to 70% ethyl alcohol for 10–30 minutes, then to absolute ethyl alcohol for 10–15 minutes.
4. Clear the specimens in clove oil for 10–30 minutes (or longer, if the specimen is still not cleared).
5. Transfer specimens to a drop of Canada balsam on a microscope slide. Spread the wings, antennae, mouthparts, etc. The head may be separated from the body if this is required to enable mouthparts to be seen easily. Euparal can be used in place of Canada balsam, but do not clear specimens in clove oil beforehand.
6. Place cover slip over drop and specimen.
7. Before storage, dry the slide in an oven at 40–50° C for a couple of weeks.

Temporary mounts

For quicker identification, particularly when you have a large series of specimens, the specimens can be mounted in a water-based mountant, such as Hoyer's, after step 2 above. However the quality of the preparation may not always be satisfactory for identification. Furthermore, water-based mountants are not recommended for specimens intended for permanent storage.

Morphology

The terminology used in the keys and in associated illustrations presented here have been adopted from various sources, particularly Annecke (1964), Rosen (1966), Annecke and Insley (1974), Hayat (1983), Prinsloo (1984) and Noyes (1988).

A generalised outline drawing of a wasp, with major parts/structures labelled, is provided to assist the reader not familiar with wasp morphology. The term 'thorax' as used here excludes propodeum, and 'abdomen' refers to gaster + propodeum. Although some characters are easier to see than others, certain characters may not be visible clearly if the specimens are not well prepared. Therefore it is essential to have a series of well-prepared specimens, both mounted and unmounted, preferably of both sexes, before attempting to use these keys.

Parasites included in the keys

The soft scale parasites *Encyrtus lecaniorum* (Mayr), *Microterys triguttatus* (Girault) (Encyrtidae) and *Coccophagus ceroplastae* (Howard) (Aphelinidae) (see, for example, pages 46, 55, 193) are not included in key 1, as no specimens were available for examination.

Table A2.1 Soft scale parasites included in key 1

Family	Species
Aphelinidae	*Coccophagus lycimnia* (Walker)
	Coccophagus semicircularis (Foerster)
	Euryischomyia flavithorax Girault & Dodd
	Myiocnema near *comperei* Ashmead (hyperparasite)
Encyrtidae	*Anicetus beneficus* Ishii & Yasumatsu
	Anicetus communis Annecke
	Cheiloneurus near *gonatopodis* Perkins (hyperparasite)
	Diversinervus elegans Silvestri
	Encyrtus infelix Embleton
	Metaphycus bartletti Annecke & Mynhardt+
	Metaphycus helvolus (Compere)
	Metaphycus near *inviscus* Compere
	Metaphycus lounsburyi (Howard)
	Metaphycus varius (Girault)
	Microterys flavus (Howard)
	Paraceraptrocerus nyasicus (Compere)
	Paraceraptrocerus sp.
Eulophidae	*Tetrastichus ceroplastae* (Girault)*
Pteromalidae	*Scutellista caerulea* (Fonscolombe)
	Moranila californica (Howard)

* Not held in Victorian Agricultural Insect Collection (VAIC), Agriculture Victoria, Knoxfield
+ Specimens from California held in VAIC; those from Israel held in collection of D. Smith (Department of Primary Industries, Queensland)
(The above notes apply to tables A2.1, A2.2 and A2.3)

Table A2.2 Armoured scale parasites included in key 2

Family	Species
Aphelinidae	*Aphytis chrysomphali* (Mercet)
	Aphytis columbi (Girault)*
	Aphytis holoxanthus DeBach*
	Aphytis lepidosaphes Compere*
	Aphytis lingnanensis Compere*
	Aphytis melinus DeBach
	Encarsia citrina (Craw)
	Encarsia perniciosi (Tower)
Encyrtidae	*Compyeriella bifasciata* Howard
	Compyeriella lemniscata Compere & Annecke
	Epitetracnemus sp.

Table A2.3 Mealybug parasites included in key 3

Family	Species
Aphelinidae	*Coccophagus gurneyi* Compere
	Coccophagus sp.
Encyrtidae	*Anagyrus agraensis* Saraswat
	Anagyrus fusciventris (Girault)
	Anagyrus pseudococci (Girault)
	Coccidoxenoides peregrinus (Timberlake)
	Leptomastidea abnormis (Girault)*
	Leptomastix dactylopii Howard
	Tetracnemoidea brevicornis (Girault)
	Tetracnemoidea peregrina (Compere)
	Tetracnemoidea sydneyensis (Timberlake)
Platygasteridae	*Allotropa* sp. possibly *citri*
Pteromalidae	*Ophelosia* sp.

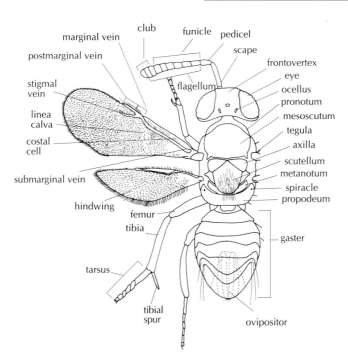

Figure A2.1 Wasp morphology.

Key 1: Common parasites of soft scales on citrus in Australia

1. Tarsi 4-segmented (fig. A2.2); front basitarsus simple (fig. A2.2); spur of fore tibia short and straight (fig. A2.2) .... (Eulophidae) *Tetrastichus ceroplastae* (Girault)

   Tarsi 5-segmented (fig. A2.3); front basitarsus modified to form a strigil (fig. A2.3); spur of fore tibia long, curved (fig. A2.3) and bifid at apex 2

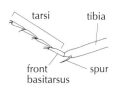

Figure A2.2

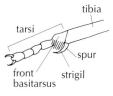

Figure A2.3

2. Inner angles of axillae widely separated from dorsal thoracic midline (fig. A2.4) 3

   Inner angles of axillae meeting or almost meeting on dorsal thoracic midline (fig. A2.5); [mesopleuron greatly enlarged, convex, transformed into a large, undivided shield (fig. A2.6)] .... (Encyrtidae) 8

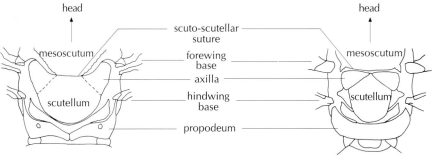

Figure A2.4

Figure A2.5

3. Anterior margins of axillae not advanced in front of scuto-scutellar suture (fig. A2.7) ... . (Pteromalidae) 4

   Anterior margins of axillae strongly advanced in front of level of scuto-scutellar suture (fig. A2.4) .... (Aphelinidae) 5

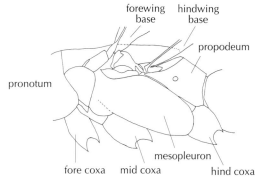

Figure A2.6

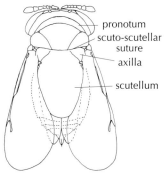

Figure A2.7

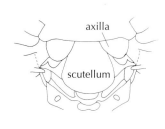

Figure A2.8

4. Scutellum elytriform, covering basal half of gaster and inner margins of wings at rest (fig. A2.7) .... *Scutellista caerulea* (Fonscolombe)

   Scutellum not elytriform, about as long as broad (fig. A2.8) .... *Moranila californica* (Howard)

5. Forewing with postmarginal and stigmal veins well developed (fig. A2.9); proximal downward curvature of marginal vein (= parastigma) enlarged and bearing two conspicuously long bristles (fig. A2.9) 6

   Forewing with postmarginal and stigmal veins not well developed (fig. A2.10); parastigma not conspicuously enlarged and not bearing conspicuously long bristles (fig. A2.10) .... *Coccophagus* 7

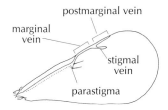

Figure A2.9

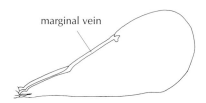

Figure A2.10

6 Forewing narrow, about 3 times as long as wide, costal cell and disc in basal third almost bare (fig. A2.11); maxillary and labial palps each 2-segmented (fig. A2.13) ....
.... *Euryischomyia flavithorax* Girault & Dodd

Forewing broad, about 2.5 times as long as wide, costal cell and disc in basal third with setae (fig. A2.12); maxillary palp 3-segmented, labial palp 2-segmented (fig. A2.14) ......
.... *Myiocnema* near *comperei* Ashmead

Figure A2.11

Figure A2.13

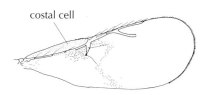

Figure A2.12

Figure A2.14

7 All coxae and femora mostly black; scutellum with only 6 setae (fig. A2.15) ....
.... *C. lycimnia* (Walker)

Mid and hind coxae and hind femora mostly black, fore coxa and other femora yellow/white; scutellum about as densely setose as mesoscutum (fig. A2.16) ....
.... *C. semicircularis* (Foerster)

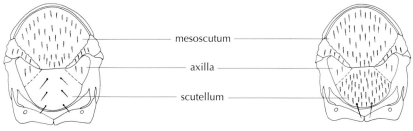

Figure A2.15   Figure A2.16

8 FEMALES (figs A2.17, A2.18) 9
   MALES (fig. A2.19) 21

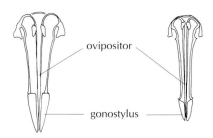

Figure A2.17   Figure A2.18

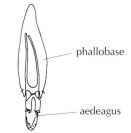

Figure A2.19

9 Flagellum broad and flat (fig. A2.20) 10
   Flagellum not flat, more or less cylindrical to broadly oval in cross-section (fig. A2.22) ..
   13

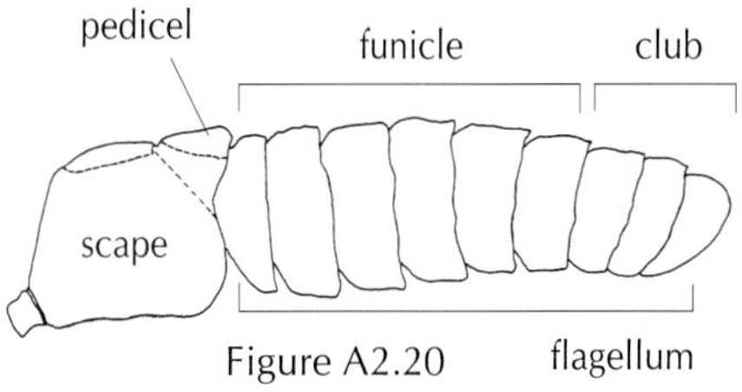

Figure A2.20

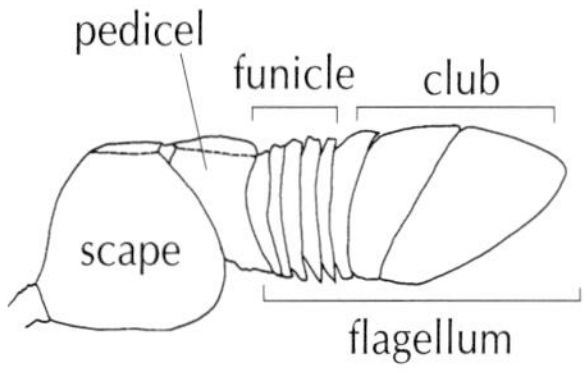

Figure A2.21

10 Greatest (oblique) length of antennal club longer than upper edge of funicle, usually longer than that of funicle and pedicel together; antennal scape tending to be subtriangular (fig. A2.21); body colour largely yellow or yellow-brown; [basal triangle of forewing with a large area devoid of setae (fig. A2.24)] .... *Anicetus* 11
   Greatest (oblique) length of antennal club at most as long as, usually shorter than, funicle measured along upper edge; antennal scape tending to be subrectangular (fig. A2.20); body black and with metallic reflections .... *Paraceraptrocerus* 12

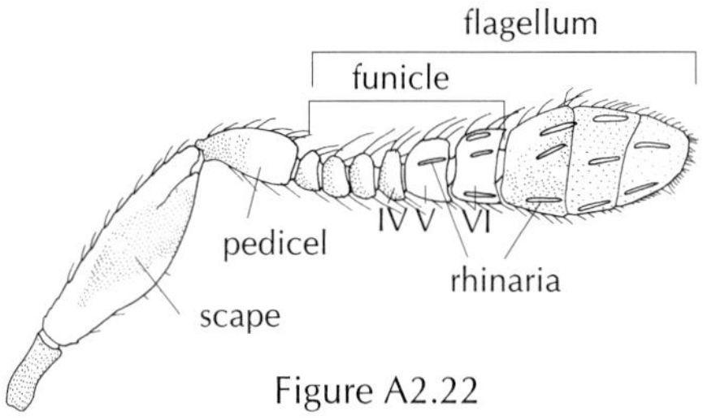

Figure A2.22

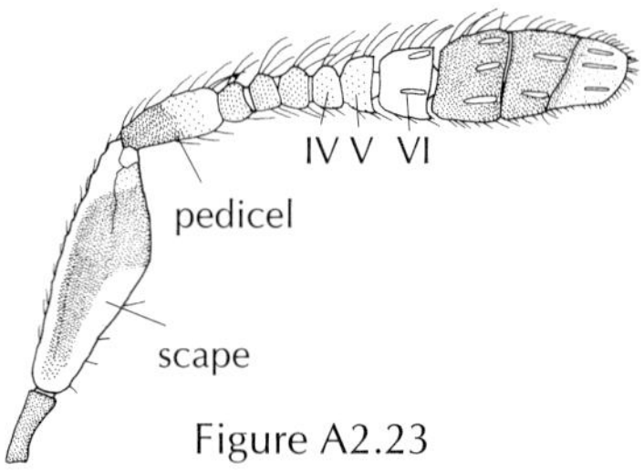

Figure A2.23

11 Propodeum with one seta, rarely two, at each spiracle (fig. A2.26) .... *communis* Annecke
   Propodeum with 15–18 setae at each spiracle (fig. A2.27) .... *beneficus* Ishii & Yasumatsu

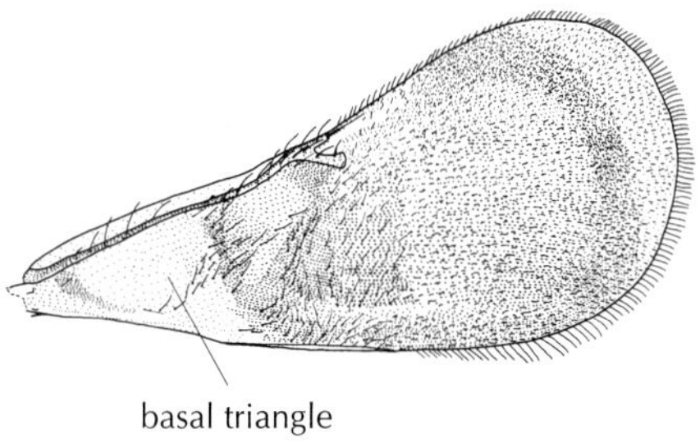

Figure A2.24

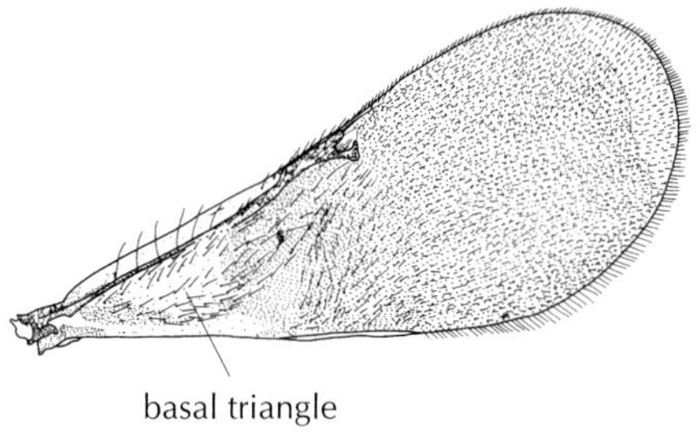

Figure A2.25

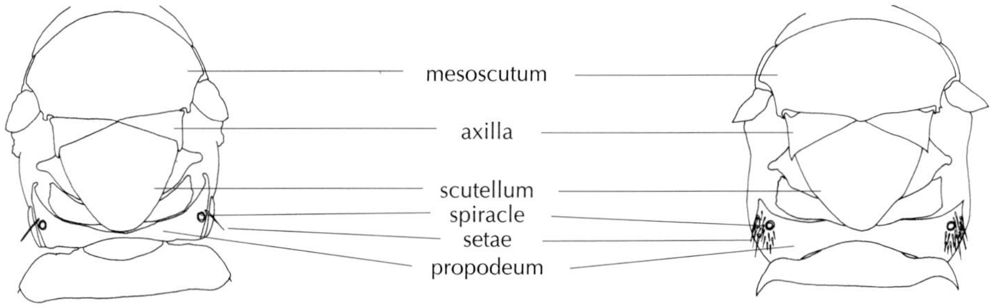

Figure A2.26

Figure A2.27

12 Ovipositor long, extending to almost entire length of gaster; forewing infuscate except for small basal and apical area, densely setose from base to linea calva (fig. A2.25) ....
   .... *nyasicus* (Compere)
   Ovipositor short, extending to only about $1/3$ length of gaster; forewing infuscate only in middle $1/3$, basal triangle devoid of setae .... *Paraceraptrocerus* sp.

13 Forewing hyaline, occasionally with a very small infuscate area below marginal vein and near base of wing (fig. A2.28) .... *Metaphycus* 14
   Forewing with a distinct infuscate pattern (fig. A2.29) 18

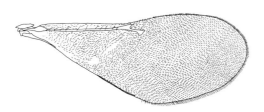

Figure A2.28

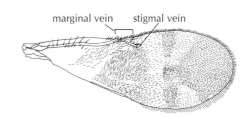

Figure A2.29

14 Tibiae uniformly pale, without any fuscous bands (fig. A2.30) 15
   Tibiae with fuscous bands (fig. A2.31) 16

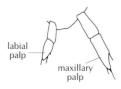

Figure A2.30

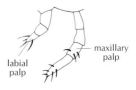

Figure A2.31

15 Maxillary and labial palps each 2-segmented (fig. A2.32); scape narrow; female with gonostylus short, about ¼ as long as ovipositor (fig. A2.18). .... *helvolus* (Compere)
   Maxillary and labial palps each 3-segmented (fig. A2.33); scape broad; female with gonostylus long, about ½ as long as ovipositor (fig. A2.17) .... *varius* (Girault)

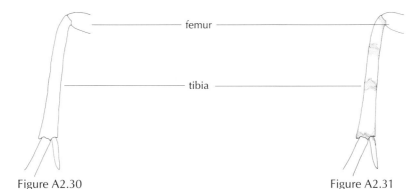

Figure A2.32 Figure A2.33 Figure A2.34

16 Maxillary palp 4-segmented, labial segment 3-segmented (fig. A2.34) ....
   .... *lounsburyi* (Howard)
   Maxillary and labial palps each 3-segmented (fig. A2.33) 17

17 Funicle segments V and VI of comparable size, considerably larger than preceding segments, both with rhinaria (fig. A2.22) .... *bartletti* Annecke & Mynhardt
   Funicle segment V considerably smaller than VI, closer to IV in size, V usually without rhinaria, VI with rhinaria (fig. A2.23) .... near *inviscus* Compere

18 Apex of scutellum without tuft of setae (fig. A2.35); forewing with a pair of interrupted hyaline fasciae distad of apex of venation, marginal vein not longer than stigmal (fig. A2.29)
   .... *Microterys flavus* (Howard)
   Apex of scutellum with a tuft of setae (fig. A2.36) 19

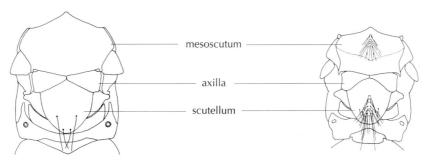

Figure A2.35 Figure A2.36

19 Head with frontovertex expanded into a frontal ledge and face meeting it in a sharp angle (fig. A2.37); [mesoscutum and scutellum each with tuft of setae (fig. A2.36)] .... *Diversinervus elegans* Silvestri

Head not modified as above 20

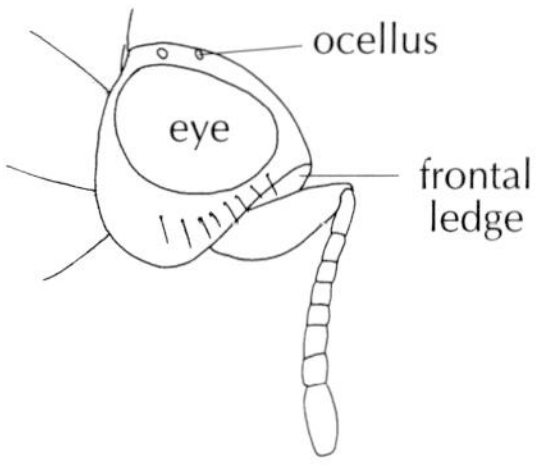

Figure A2.37

20 Marginal vein longer than stigmal vein (fig. A2.38); mandible tridentate (fig. A2.40) .... *Cheiloneurus* near *gonatopodis* Perkins

Marginal vein about as long as stigmal vein (fig. A2.39); mandible edentate (fig. A2.41) .... *Encyrtus infelix* Embleton

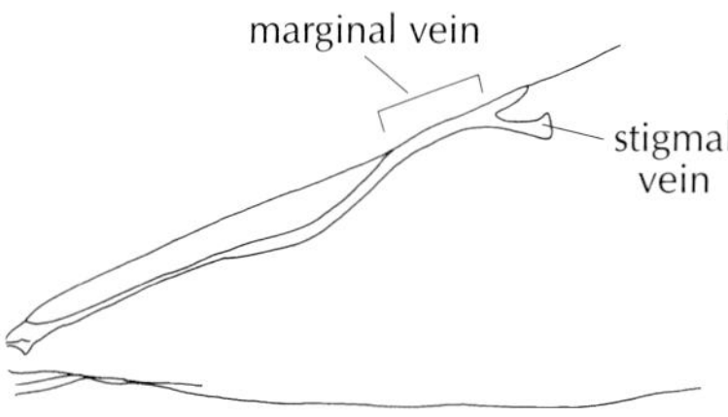

Figure A2.38

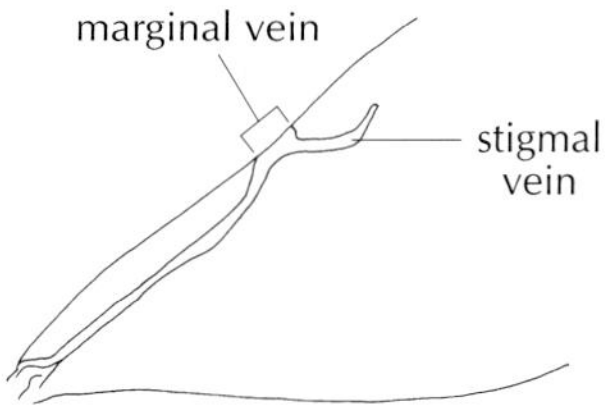

Figure A2.39

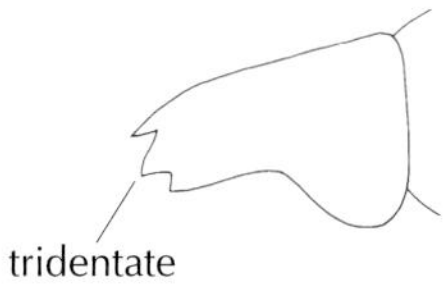

Figure A2.40

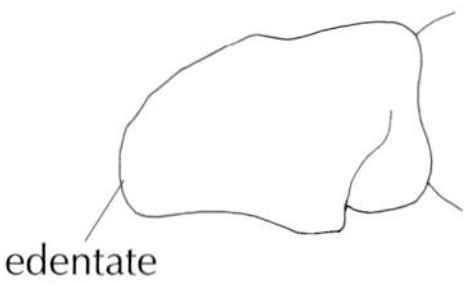

Figure A2.41

21 Funicle segments at most slightly longer than broad (fig. A2.22) .... *Metaphycus* 14

Funicle segments longer than broad (fig. A2.42) 22

22 Funicle segments cylindrical, with setae at most as long as diameter of segment (fig. A2.42) .... *Microterys flavus* (Howard)

Funicle segments not cylindrical, with setae at least 1.5 times as long as diameter of segment (fig. A2.43) 23

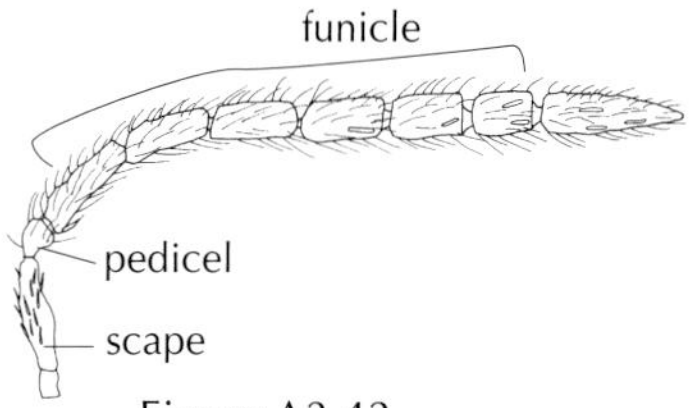

Figure A2.42

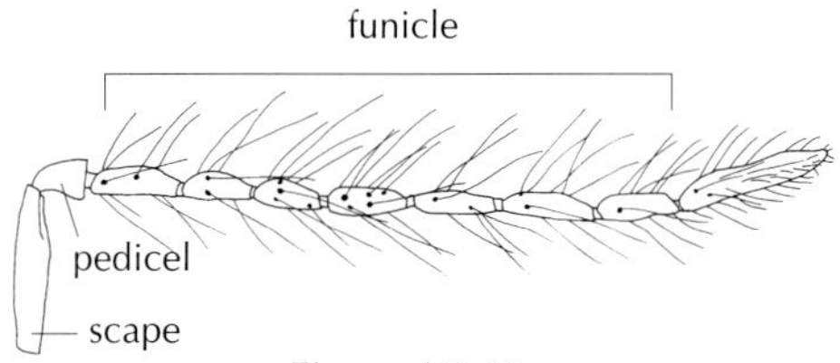

Figure A2.43

23 Head, thorax and gaster predominantly black with green lustre in parts, legs yellowish white .... *Diversinervus elegans* Silvestri

Head, thorax and gaster predominantly yellowish or brownish 24

24 Scutellum without apical tuft of setae (fig. A2.35) .... *Cheiloneurus* near *gonatopodis* Perkins

Scutellum with apical tuft of setae (fig. A2.36) .... *Encyrtus infelix* Embleton

Key 2: Common parasites of armoured scales on citrus in Australia

1. Inner angles of axillae meeting or almost meeting on dorsal thoracic midline (fig. A2.45); mesopleuron greatly enlarged, convex, transformed into a large, undivided shield (fig. A2.46) .... (Encyrtidae) 2

   Inner angles of axillae widely separated from dorsal thoracic midline (fig. A2.44); mesopleuron normal, not transformed into a large shield .... (Aphelinidae) 4

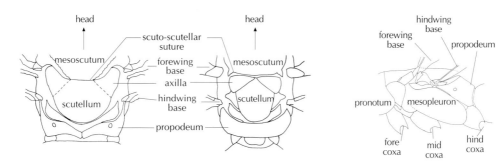

Figure A2.44   Figure A2.45   Figure A2.46

2. Female forewing with 1 or 2 longitudinal infuscate bands (fig. A2.47) .... *Comperiella* 3

   Female forewing without longitudinal infuscate bands, but with infuscation with several hyaline spots (fig. A2.48); [male funicle 2-segmented, club long, unsegmented] .... *Epitetracnemus* sp.

Figure A2.47   Figure A2.48

3. Head with broad dorsal median longitudinal band around ocelli (fig. A2.49); mesoscutum with medial longitudinal greenish band broadly interrupted by cupreous band along midline (fig. A2.49); tibial spur of middle leg longer than basitarsus (fig. A2.51) .... *bifasciata* Howard

   Head with narrow dorsal median longitudinal band around ocelli (fig. A2.50); mesoscutum with median longitudinal bluish green band not interrupted by cupreous band along midline (fig. A2.50); tibial spur of middle leg subequal to basitarsus (fig. A2.52) .... *lemniscata* Compere & Annecke

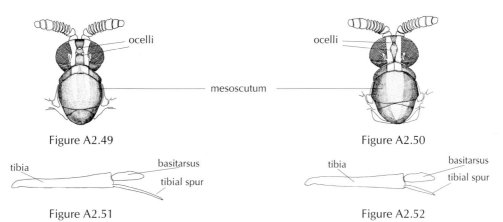

Figure A2.49   Figure A2.50

Figure A2.51   Figure A2.52

4. Antennae at most with 6 segments (fig. A2.53); forewing with linea calva (fig. A2.55) .. .. *Aphytis* 5

   Antennae with at least 7 segments (fig. A2.54); forewing without linea calva (fig. A2.56) .... *Encarsia* 10

5   Female antennae 6-segmented, first funicle segment anneliform, small with an elongate club (fig. A2.53); male antenna 4-segmented .... *columbi* (Girault)

    Both female and male antennae 6-segmented 6

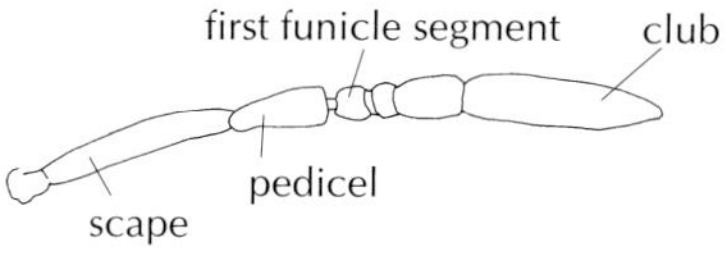

Figure A2.53

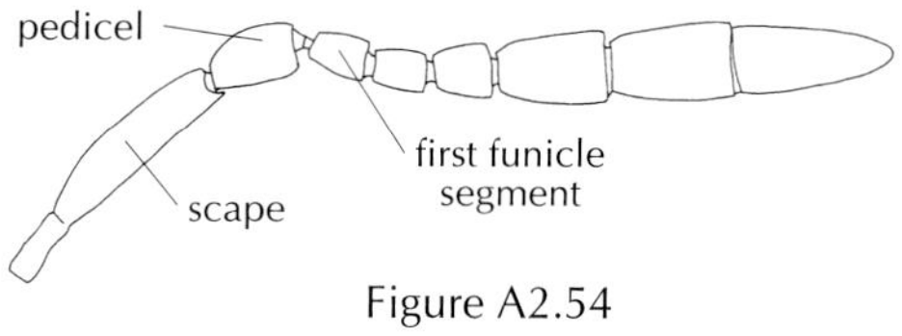

Figure A2.54

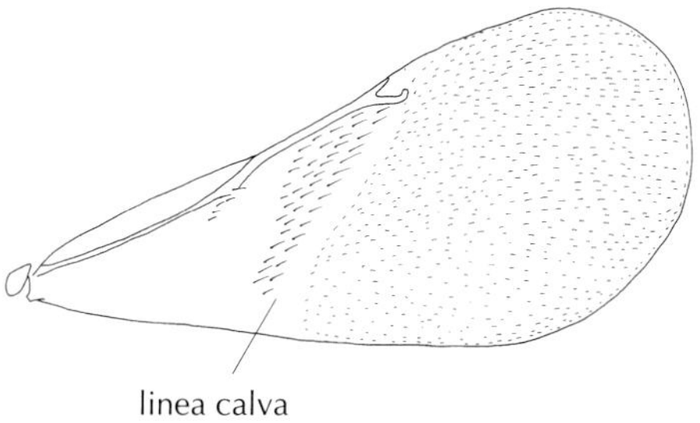

Figure A2.55

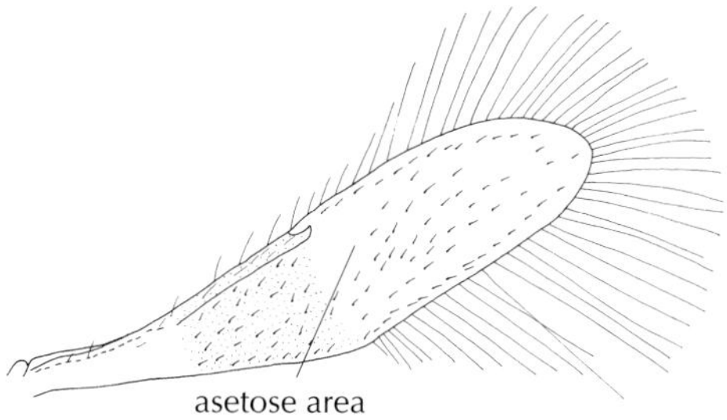

Figure A2.56

6   Crenulae on posterior margin of propodeum large, overlapping (fig. A2.57) 7

    Crenulae on posterior margin of propodeum minute, not overlapping (fig. A2.58); [thoracic sterna dusky] 9

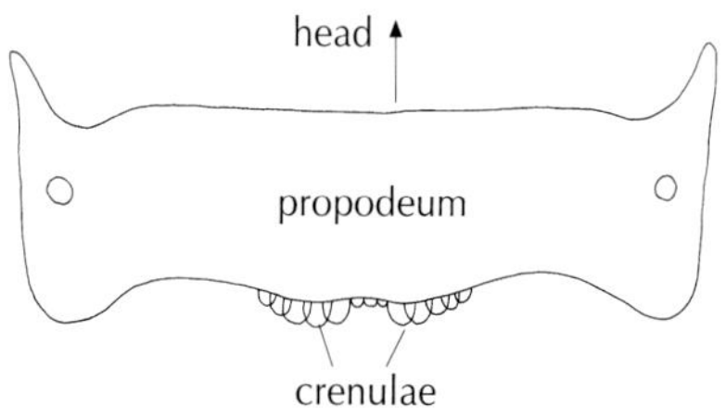

Figure A2.57

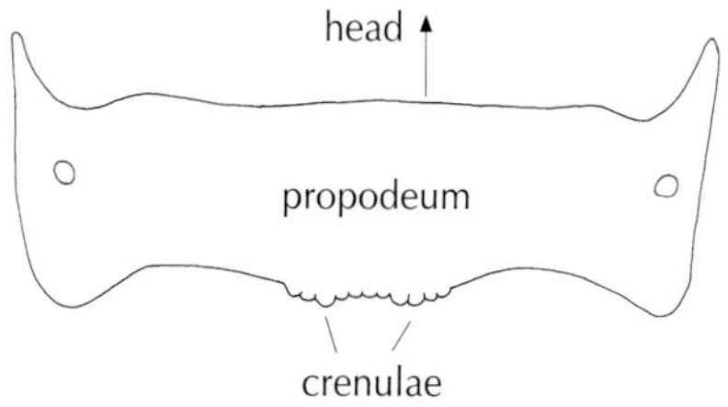

Figure A2.58

7   Thoracic sterna dusky in contrast to rest of body which is pale .... *lingnanensis* Compere

    Thoracic sterna immaculate, yellow as rest of body 8

8   Thoracic setae dark and coarse; antennae dusky, club up to 3.5 times as long as wide (fig. A2.59) .... *holoxanthus* DeBach

    Thoracic setae pale and slender; antennae pale or slightly dusky, club nearly 4 times as long as wide (fig. A2.60) .... *melinus* DeBach

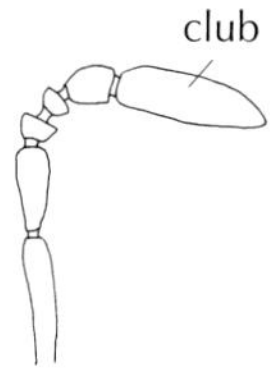

Figure A2.59

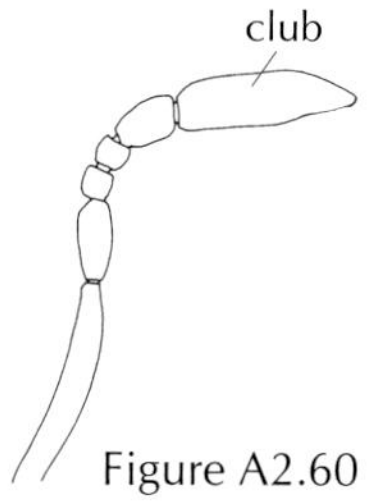

Figure A2.60

9   Mesoscutum usually with 10 (rarely 11 or 13) setae (fig. A2.61), propodeum as long as scutellum .... *chrysomphali* (Mercet)

    Mesoscutum with 10–15 (usually 12) setae (fig. A2.62); propodeum about ¾ as long as scutellum .... *lepidosaphes* Compere

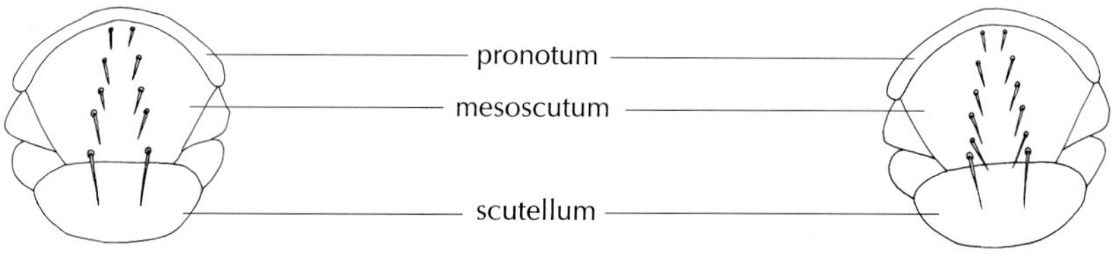

Figure A2.61

Figure A2.62

10 Forewing long and narrow, with an asetose area around stigmal vein (fig. A2.56); disc sparsely setose with long marginal fringe .... *citrina* (Craw)

Forewing broad, without an asetose area around stigmal vein; disc densely setose with short marginal fringe (fig. A2.63) .... *perniciosi* (Tower)

Figure A2.63

Key 3: Common parasites of mealybugs on citrus in Australia

1 Forewing venation characteristic, terminating apically in a club-like shape (fig. A2.64) [head and body completely blackish, shiny] ... (Platygasteridae) *Allotropa* sp. possibly *citri*

Forewing venation otherwise (fig. A2.65) 2

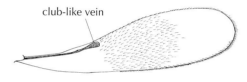

Figure A2.64

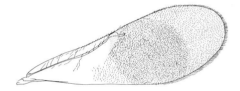

Figure A2.65

2 Inner angles of axillae widely separated from dorsal thoracic midline (fig. A2.66) 3

Inner angles of axillae meeting or almost meeting on dorsal thoracic midline (fig. A2.67); [mesopleuron greatly enlarged, convex, transformed into a large, undivided shield (fig. A2.68)] .... (Encyrtidae) 5

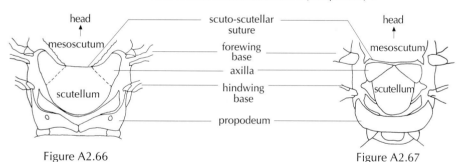

Figure A2.66

Figure A2.67

3 Anterior margin of axillae not advanced in front of scutoscutellar suture (fig. A2.69) [propodeum with a conspicuous horizontal neck, petiole short so that gaster rather loosely attached to thorax] .... (Pteromalidae) *Ophelosia* sp.

Anterior margins of axillae strongly advanced in front of level of scutoscutellar suture (fig. A2.66) .... (Aphelinidae) 4

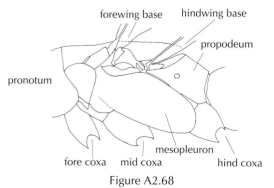

Figure A2.68

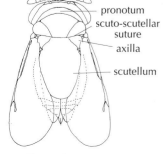

Figure A2.69

4 Female with band across basal part of abdomen yellow; legs yellow ....
   ....*Coccophagus gurneyi* Compere
   Not as above [most of side lobes of mesoscutum (parapsides) and tegula paler than rest of thorax] .... *Coccophagus* sp.

5 Female gaster with last tergite more or less shield-shaped or triangular, its anterior margin almost straight or only slightly curved (fig. A2.70); anterior margin of all other tergites straight or only slightly curved (fig. A2.70) 6
   Female gaster with last tergite U-shaped, its anterior margin strongly curved (fig. A2.71); anterior margin of all other tergites also strongly curved (fig. A2.71) 9

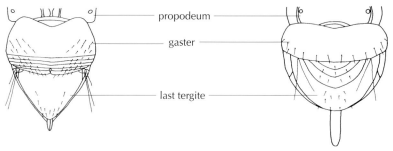

Figure A2.70   Figure A2.71

6 Funicle of female with 5 segments (fig. A2.72); male funicle branched (fig. A2.74) ....
   .... *Tetracnemoidea* 7
   Funicle of female with 6 segments (fig. A2.73); male funicle not branched (fig. A2.73) [Forewing with postmarginal vein not longer than stigmal (fig. A2.75); ovipositor not exserted] .... *Coccidoxenoides peregrinus* (Timberlake)

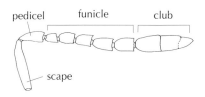 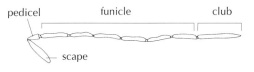

Figure A2.72   Figure A2.73

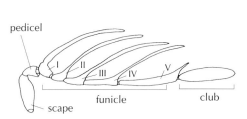 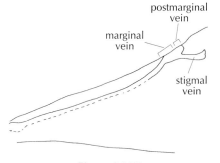

Figure A2.74   Figure A2.75

7 Forewing of female with a large fuscous area in disc extending across wing from marginal vein (fig. A2.65); fifth funicle segment of male with a short but distinct branch (fig. A2.74)
   .... *sydneyensis* (Timberlake)
   Forewing of female more or less hyaline or hardly infuscate; fifth funicle segment of male simple, not branched (fig. A2.76) 8

8 Female with antennal first funicle segment smaller than second segment; male with antennal second funicle segment with at least 1 longitudinal sensillum (fig. A2.77) ....
   .... *peregrina* (Compere)
   Female with antennal first funicle segment larger and longer than second segment; male with antennal second funicle segment without any longitudinal sensilla (fig. A2.76) ......
   .... *brevicornis* (Girault)

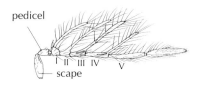

Figure A2.76

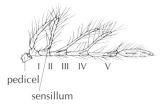

Figure A2.77

9 Female with scape (fig. A2.78) cylindrical; [forewing with postmarginal vein clearly longer than stigmal vein (fig. A2.80)] 10

Female with scape (fig. A2.79) broadly flattened .... *Anagyrus* 11

Figure A2.78

Figure A2.79

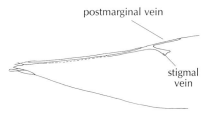

Figure A2.80

10 Forewing with distinct dark cross band (fig. A2.81); generally smaller species (< 1.3 mm); [mainly yellowish brown species with long slender antennae] .... *Leptomastidea abnormis* (Girault)

Forewing without dark pattern (fig. A2.82); generally larger species (>1.4 mm); [funicle and club ferruginous] .... *Leptomastix dactylopii* Howard

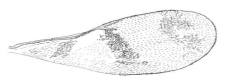

Figure A2.81

Figure A2.82

11 Female with club brown (fig. A2.83) .... *fusciventris* (Girault)

Female with club white (fig. A2.84) 12

12 Female with only first funicle segment partially dark brown (fig. A2.84) .... *pseudococci* (Girault)

Female with first and second funicle segments partially dark brown (fig. A2.85) .... *agraensis* Saraswat

Figure A2.83

Figure A2.84

Figure A2.85

Appendix 3 A brief history of IPM in Australian citrus

Australia has developed one of the most successful IPM programs for pest control in citrus in the world. Orchards in all major producing areas have adopted the program with some variations, depending on the occurrence and abundance of pests. The history of biological control of some of the major pests is described below.

Biological control of some major pests: a brief history

Red scale

Red scale (*Aonidiella aurantii*) was probably introduced to Australia from Asia before 1830. Since the 1880s, it has been the most serious pest of citrus in most regions. The scale was originally described on lemons imported into New Zealand from Sydney, and entomologists originally thought that the scale was native to Australia. The scale was also introduced to California in plant material from Australia.

These events led to attempts in the late 1890s and early 1900s to locate natural enemies of the scale in Australia, and to introductions of some natural enemies, e.g. ladybirds, to North America. By the 1930s it was clearly established that the scale was not native to Australia but to Asia, particularly east Asia.

Until the mid-1980s, most research into citrus pests and most parasite introductions concentrated on red scale. Early attempts at biological control were closely associated with work in California. From 1943 to 1979 a total of eight species of wasp parasites were imported and released against red scale.

Aphytis chrysomphali

Between 1902 and 1909, the Californian government and the Department of Agriculture of Western Australia cooperated in financing overseas exploration for beneficial insects. A parasitic wasp, thought to be *Aphytis chrysomphali*, was introduced into Western Australia from China by G. Compere in 1905, and was successfully established in the Chittering area. Circumstantial evidence suggests that the species was probably *Aphytis lingnanensis*.

A. chrysomphali was introduced into New South Wales from Western Australia by P. C. Hely in 1925–26, and became established in the coastal areas near Gosford. It was released soon afterwards into inland citrus areas of New South Wales, including the Murrumbidgee Irrigation Area (MIA). It was still common in the Upper Murray region in the early 1980s.

A. chrysomphali was recorded as parasitising red scale in Adelaide in 1943, and in Mypolonga in 1949. The parasite has never been actively introduced into South Australia, and how it became established is not known. Collections of red scale in the early 1970s showed that *A. chrysomphali* was well established in the Riverland at Loxton and Waikerie.

A. chrysomphali was reintroduced into Victoria from California by C. J. R. Johnston in 1954. He also obtained stocks of the parasite from P. C. Hely in New South Wales. It was successfully established in the Sunraysia, and by 1958 several growers were using biological control for the control of citrus scale insects.

Aphytis melinus and *Aphytis lingnanensis*

C. J. R. Johnston imported the wasps *Aphytis melinus* (originally from Pakistan) in 1961 and *Aphytis lingnanensis* (originally from Guangzhou, China) in 1962 from Riverside in California. *A. melinus* was established in Victoria in 1963, but *A. lingnanensis* did not become established.

Regional laboratories were established at Mildura in 1967 and Loxton in 1968 with facilities for mass rearing *A. melinus* and *A. lingnanensis*. In the Sunraysia district, widespread establishment of *A. melinus* was achieved, but *A. lingnanensis* was active only on one property for a brief period.

Successful biological control of red scale resulted from releases of *A. melinus*, particularly in the Boundary Bend area of Victoria. During the 1970s, *A. melinus* displaced *A. chrysomphali* as the dominant parasite of red scale on unsprayed citrus in the Sunraysia.

Surveys in 1967–68 in Riverland orchards did not recover any *Aphytis* spp. Releases of *A. melinus* from Boundary Bend by N. L. Richardson in the Riverland resulted in some establishment. M. Campbell and D. Maelzer in South Australia, and I. McLaren and G. Buchanan in Victoria, did extensive research on the use of *A. melinus* during the 1970s.

In 1971, Australia's first commercial insectary, Biological Services Inc., operated by R. George, was established at Loxton. The operations of this insectary resulted in the establishment of *A. melinus* in all Riverland orchards by 1975.

Attempts were made to establish *A. melinus* in Queensland in 1974. The attempts were not successful at that time, but significant numbers of the wasps have been recovered at Emerald after releases in 1990–93. *A. melinus* is now also established in Alice Springs and Western Australia.

By 1978, D. Smith showed that the main parasite in red scale in Queensland was *A. lingnanensis*. In 1978 a commercial insectary for mass-producing *A. lingnanensis* was started up at Mundubbera by D. Papacek.

Comperiella bifasciata

Two strains of the wasp parasite *Comperiella bifasciata* probably occur in Australia: the red scale strain and the yellow scale strain. At least one of these strains was recorded in the 1930s in association with yellow scale in Queensland.

C. bifasciata, possibly the red scale strain, was first imported into Australia from California in 1942. It was cultured by Commonwealth Scientific and Industrial Research (CSIR), and released in all mainland states from 1943 to 1947.

Further introductions from California of the red scale strain were made by CSIR in 1947–49. Releases were made in Queensland, Victoria and New South Wales. This parasite is now well established in all mainland states. G. A. C. Beattie has studied the role of this parasite extensively in New South Wales.

Encarsia perniciosi, red scale strain

The wasp *Encarsia perniciosi* was imported in 1970 from California by I. McLaren. It was bred at Mildura and Loxton. *E. perniciosi* was established in the Sunraysia in 1973, but no wasps were recovered at that time from releases made in the Riverland. *E. perniciosi* is now a useful parasite of red scale in the Sunraysia and the Riverland, particularly in spring when *Aphytis melinus* numbers are low.

E. perniciosi also occurs on the central coast of New South Wales, and has probably been present there since before 1900.

Encarsia citrina

The wasp *Encarsia citrina* occurs on the central coast of New South Wales, and has probably been present there since before 1900, along with *E. perniciosi*.

Status of red scale wasp parasites

Aphytis melinus and *Comperiella bifasciata* are established throughout the Murray–Darling basin in all citrus growing areas. *A. melinus* is the most numerous parasite from about February to late autumn, while *C. bifasciata* is the most numerous parasite from about late autumn to January.

Red scale in these areas is under effective biological control. Spraying has been reduced, or eliminated, with associated cost benefits.

In the inland citrus areas, *A. melinus* is generally thought to kill more red scale than *C. bifasciata* because of its shorter generation time, and because it also kills scale by host feeding. *C. bifasciata* is thought to be more important in some parts of the MIA, and is more important in coastal New South Wales. Further work is required to determine the full effect of each on populations of red scale.

A. melinus has displaced *A. chrysomphali* from many inland districts of south-eastern Australia.

Initial introductions of *C. bifasciata* resulted in the complete biological control of yellow scale (*Aonidiella citrina*), but had little effect on red scale due to wasp egg encapsulation. In 1971 a strain successfully parasitising about 80% of adult female red scale was found on citrus fruit at Barmera and Kingston in the Riverland.

The red scale strain was mass-reared at Loxton. In contrast to previous attempts, this attempt resulted in the wasps breeding very successfully on red scale in the laboratory. Parasites from this mass-rearing were released throughout the Riverland.

In Queensland, *A. lingnanensis* gives good biological control of red scale and is ably assisted by *C. bifasciata*. *A. chrysomphali* also occurs, particularly in coastal areas, and *A. melinus* occurs in inland areas. *Encarsia citrina* is common in second instar and male scales.

Circular black scale

Aphytis holoxanthus

Aphytis holoxanthus, originally from China, was imported from Israel in 1974 by G. Snowball and G. Lukins to control circular black scale (*Chrysomphalus aonidum*). Wasps were reared and released throughout Queensland by D. Smith in 1975–76. *Aphytis holoxanthus* became established and has been extremely successful in controlling circular black scale.

Citrus snow scale (white louse scale)

Four species of parasitic wasps were introduced between 1978 and 1988 by D. Smith and D. Papacek to improve biological control of citrus snow scale (*Unaspis citri*). The species were: *Aphytis lingnanensis* (HK1 strain) from Florida, *Aphytis lingnanensis* (HJK strain) from Japan, *Aphytis* sp. from Thailand, and *Aphytis gordoni* from Southern China. *A. lingnanensis* (HJK strain) and the *Aphytis* sp. from Thailand became established, but they affect populations of scale only on the fruit and foliage, not the limbs.

The predatory ladybird *Chilocorus circumdatus* became established on citrus snow scale in Queensland and northern New South Wales orchards in the early 1990s and has significantly reduced the scale's importance. The species is available from commercial insectaries.

Attempts have been made to establish the small parasite *Encarsia inquerenda* from southern China, so far without success.

Mussel scale (purple scale)

Mussel scale (*Lepidosaphes beckii*) is usually satisfactorily controlled in Queensland and New South Wales by the wasp *Aphytis lepidosaphes*. There is no record of this oriental wasp species being introduced to Australia.

Longtailed mealybug

When efforts were underway in the 1960s and 1970s to introduce wasp parasites along the Murray River for the control of red scale, major secondary pests that sometimes required control were the longtailed mealybug (*Pseudococcus longispinus*) and soft scales.

The soft scales were readily controlled by petroleum spray oils, but no satisfactory control measures were available for longtailed mealybug. A good number of natural enemies were found to be present (mainly *Anagyrus fusciventris*) but two parasites—*Tetracnemoidea sydneyensis* and *Tetracnemoidea peregrina* (subsequently used in California for the biological control of this pest)—were not recorded.

T. sydneyensis was introduced from the Sydney area in 1972, and *T. peregrina* from cultures in Israel by G. Furness. Both were cultured at Loxton, and released widely throughout the Riverland, but no recoveries of either parasite were made. They are now sometimes found in small numbers.

Citrophilous mealybug

The citrophilous mealybug (*Pseudococcus calceolariae*), originally from the east coast of Australia, was first discovered in the Riverland at Loxton in October 1986. Since that time its incidence has continued to increase, and by 1994 it had almost completely displaced longtailed mealybug on citrus.

Citrophilous mealybug is now regarded as the most serious pest of citrus in the Riverland and in some orchards in the Sunraysia. Attempts are underway to introduce natural enemies, including *Coccophagus guerneyi* from near Sydney. Another species, *Tetracnemoidea brevicornis*, was established in the Riverland by J. Altmann.

Citrus mealybug

Citrus mealybug (*Planococcus citri*) is a serious pest in Queensland. The Brazilian wasp *Leptomastix dactylopii*, a parasite of citrus mealybug, was imported from California in 1980 and established throughout Queensland by D. Smith, D. Murray and D. Papacek. Since 1988 it has been mass-reared for augmentative release in Queensland.

Black scale

Black scale (*Saissetia oleae*) is a pest mainly of inland citrus. In 1960, F. Wilson listed the beneficials imported for this pest as including *Scutellista caerulea* (= *cyanea*) (from California in 1904), and *Metaphycus lounsburyi* (from South Africa in 1902).

Metaphycus helvolus was imported from California in 1942 by the CSIR. It occurs in South Australia and Victoria, but mainly in soft brown scale (*Coccus hesperidum*).

Another species, *Metaphycus bartletti*, is soon to be released.

Soft brown scale

The main parasites that were imported to control soft brown scale (*Coccus hesperidum*) were *Coccophagus lycimnia* (from California in 1907), *Metaphycus helvolus*, *Euryischomyia flavithorax*, and *Coccophagus semicircularis*. *Microterys flavus* and *Diversinervus elegans* are currently important control agents, especially in Queensland.

White wax scale

White wax scale (*Ceroplastes destructor*) occurs mainly in coastal Queensland and New South Wales. Researchers in New South Wales imported parasites of white wax scale from Kenya and Uganda in 1935–36, but the parasites were not successfully established. G. Snowball, G. Lukins and D. Sands imported three wasp parasites from South Africa in 1968–70. These were *Paraceraptrocerus nyasicus*, *Anicetus communis* and *Tetrastichus ceroplastae*.

P. nyasicus was particularly successful in Queensland and northern New South Wales, and white wax scale is now scarce in those places. All three species of wasp parasite appear to have combined to suppress the scale in the central coast area of New South Wales, with *A. communis* being the most important.

Pink wax scale

Pink wax scale (*Ceroplastes rubens*) occurs mainly in coastal Queensland and New South Wales. The parasite *Anicetus communis* was introduced from Japan by D. Smith in 1977. It was established throughout Queensland and northern New South Wales, and effectively controls pink wax scale.

Other pests

The beneficial organisms imported have mainly been natural enemies of scales and mealybugs. More recently, wasp parasites of citrus leafminer (*Phyllocnistis citrella*) have been introduced from Thailand and southern China by D. Smith and G. A. C. Beattie. These have included *Ageniaspis citricola*, *Cirrospilus quadristriatus* and *Citrostichus phyllocnistoides*. The first two have become established in Queensland.

Native beneficials

With the increased adoption of IPM over the last 20 years in Australian citrus, the importance of existing native beneficial organisms in controlling native pests is being increasingly realised. For example, native beneficials are known to be important in controlling spined citrus bug (*Biprorulus bibax*) (as shown by D. James), citrus gall wasp (*Bruchophagus fellis*) and citrus planthoppers. Native predatory phytoseiid mites have been shown to be important in controlling pest mites, such as brown citrus rust mite (*Tegolophus australis*). Aphids have a large range of native natural enemies.

Integrated pest management (IPM) in different regions

Along the Murray River

Before the 1950s, fumigation of citrus trees with hydrogen cyanide was commonly used in the Murray River area to control citrus pests, principally red scale. With the advent of modern insecticides, fumigation was replaced by the use of organophosphate insecticides and petroleum spray oils (PSOs). Insecticides continued to be the main method of control until the 1970s.

Chemical control of red scale in the 1960s was achieved by two sprays per year of either oil, oil plus insecticide, or insecticide. Extra insecticide sprays were occasionally used, especially for controlling outbreaks of secondary pests, such as mealybugs and soft scales.

Biological control of red scale began along the Murray River with the introduction of the *Aphytis* parasites in the 1960s and 1970s as described above. The widespread establishment of *Aphytis melinus* was the catalyst for fundamental changes in the control of red scale.

During the 1970s, growers changed from chemical methods of insect control, based on organophosphates, to increased use of oil and biological control (with no spraying at all). A number of surveys have documented this change (see table A3.1). The savings resulting from biological control along the Murray River were from $5 million to $15 million. By the 1980s over 70% of growers found that in most years they could rely completely on biological control, and not spray for any citrus pests.

During the early 1970s, it was found that occasional outbreaks of secondary pests, such as longtailed mealybug, lightbrown apple moth (*Epiphyas postvittana*) and black citrus aphid (*Toxoptera citricida*) could not be controlled by PSOs. In general, growers tolerated losses caused by these pests because they did not want to disrupt the biological control of red scale by using organophosphate insecticides. It was later found that lightbrown apple moth could be controlled by non-disruptive sprays containing the bacterium *Bacillus thuringiensis*.

Other secondary scale insects on citrus, such as soft brown scale and black scale, could, like red scale, be controlled with PSOs. G. Furness in South Australia and G. A. C. Beattie in New South Wales have researched the use of oil sprays and associated spray technology.

From the 1980s, IPM programs began to be developed in the area, notably by Biological Services Inc. (which had helped

Table A3.1 Change in pesticide use in the Sunraysia and Riverland from 1966 to 1978

Location	Year	Percentage orchards unsprayed (biological control)	Percentage orchards using oil sprays only (integrated control)	Percentage orchards using organophosphate insecticides
Sunraysia	1966	19	–	81
	1971	33	39	28
	1974	49	51	0
Riverland	before 1970	10 (approx.)	–	90
	1974	45	19	36
	1978	74	18	8

to achieve rapid establishment of A. melinus in the Riverland in the 1970s). The company continues to operate and offers a commercial pest management service. Several other pest management services have also developed, including Yandilla Park, and HPMS.

Changed markets and new pests

During the 1990s, pest control along the Murray changed for the following reasons:

- The price of juice fruit fell dramatically to well below the cost of production. This meant that the industry had to increase the amount of fresh fruit packed from 20–40% to 60–80%, and concentrate on new, high-priced export markets such as the USA and Japan, as well as maintain other export and local markets.

- Several new pests appeared, increasing the complexity of pest control. These pests include citricola scale, citrophilous mealybug, spined citrus bug, citrus leafminer and megalurothrips (*Megalurothrips kellyanus*).

Queensland

Early foundations for citrus entomology in Queensland were laid by the basic biology studies of W. A. T. Summerville before World War II. Following a series of intensive biological control studies by D. Smith in the 1970s, IPM was first adopted by one large grower in 1978. Initially the organophosphate pesticides were withdrawn and a procedure for monitoring using a scout or consultant was established. As better understanding of the relationships between natural enemies and their hosts developed, the monitoring system was continually fine-tuned.

In association with this first IPM program, an insectary was developed by D. Papacek of Mundubbera (Integrated Pest Management Pty Ltd (Bugs for Bugs)). The insectary provided beneficial insects for augmentative release to control two key pests: red scale and citrus mealybug. As interest in the success of the program grew, the service was made commercially available to other growers. A manual on IPM called *Protect your citrus* was also produced by the Department of Primary Industries, Queensland.

IPM is now used on approximately 80% of citrus trees in production in Queensland. Participating orchards have been able to reduce insecticide and miticide use by approximately 75%. There are pest scouts at Gayndah, Mundubbera, Bundaberg, Beerwah, Mareeba and Emerald.

The most significant outcomes from the use of IPM have been reductions in pesticide usage, overall better pest management and large savings to growers. IPM costs range from 37% to 53% of the costs of chemical control.

New South Wales

P. C. Hely and then J. G. Gellatley expanded knowledge of citrus entomology in New South Wales from 1940 to 1975. IPM techniques developed along the Murray River spread into the MIA during the 1970s and 1980s. In recent years, PSOs have formed the basis of scale control in New South Wales, when natural enemies fail.

Western Australia

In Western Australia, the early work on parasite introductions for red scale by G. Compere in 1905 was continued by later researchers, e.g R. Jenkins. A wide range of *Aphytis* spp., encyrtid wasps and ladybirds were released. Releases of sterile male flies have been used successfully to eradicate Queensland fruit fly.

Australia in general

Throughout Australia today, many growers recognise the importance of biological control and IPM. There are many professional pest scouts and several commercial insectaries, as well as effective research, development and extension programs focusing on IPM.

Bibliography

Abdelrahman, I. (1973). Toxicity of malathion to California red scale (*Aonidiella aurantii*) (Mask.) (Hemiptera: Diaspididae). *Australian Journal of Agricultural Research* 24:111–118.

Abdelrahman, I. (1974). Growth, development and innate capacity for increase in *Aphytis chrysomphali* Mercet and *A. melinus* DeBach, parasites of California red scale, *Aonidiella aurantii* (Mask.), in relation to temperature. *Australian Journal of Zoology* 22:213–230.

Allender, W. J. and Beattie, G. A. C. (1991). Determination of petroleum oil deposits on *Citrus sinensis* (L.) Osbeck. *Pesticide Science* 31:133–139.

Andersen, A. N. (1991). *Ants of southern Australia: a guide to the Bassian fauna*. CSIRO, Melbourne.

Annecke, D. P. (1964). *The encyrtid and aphelinid parasites (Hymenoptera: Chalcidoidea) of soft brown scale, Coccus hesperidum Linnaeus (Hemiptera: Coccidae), in South Africa*. Entomology Memoirs, vol. 7. Department of Agricultural Technical Services, Republic of South Africa.

Annecke, D. P. and Insley, H. P. (1974). *The species of Coccophagus Westwood, 1833 from the Ethiopian region (Hymenoptera Aphelinidae)*. Entomology Memoirs, no. 37. Department of Agricultural Technical Services, Republic of South Africa.

Bailey, P. T. and Furness, G. O. (1981). A survey of insecticide use on fruit trees and grapevines in the Riverland of South Australia. *Agricultural Record* 8 and 9:13–19.

Batchelor, Leon and Webber, Herbert (eds) (1948). *The citrus industry*, vol. 11. University of California Press.

Beattie, B. B. and Revelant, L. J. (1995). *Product description language: oranges*. Australian Horticultural Corporation.

Beattie, G. A. C. (1978a). Biological control of citrus mites in coastal NSW. In *1978 Proceedings of the International Society of Citriculture*, pp. 156–158.

Beattie, G. A. C. (1978b). Citrus leafminer control with systemic insecticides. *Insecticide and Acaricide Tests* 3:57.

Beattie, G. A. C. (1980). Control of Chinese wax scale. *Insecticide and Acaricide Tests* 5:47.

Beattie, G. A. C. (1985). *Ecology of citrus red scale Aonidiella aurantii (Maskell): report on a study tour to China 6 June 1983 – 22 May 1984*. Miscellaneous Bulletin no. 2. Biological and Chemical Research Institute, NSW Department of Agriculture.

Beattie, G. A. C. (1989). *Citrus leafminer*. Agfact H2.AE.4. NSW Agriculture & Fisheries.

Beattie, G. A. C. (1990a). *Citrus petroleum spray oils*. Agfact H2.AE.5. NSW Agriculture & Fisheries.

Beattie, G. A. C. (1990b). *Pest management guide: coastal citrus*. NSW Agriculture & Fisheries.

Beattie, G. A. C. (1992). The use of petroleum spray oils in citrus and other horticultural crops. In *Proceedings of the First National Conference of the Australian Society of Horticultural Science, Sydney, September–October 1991*, pp. 351–362.

Beattie, G. A. C. (1993). *Integrated control of citrus leafminer*. NSW AgFact H2.AE.4. NSW Agriculture.

Beattie, G. A. C. (1995). *Reducing pesticide use in citrus: an update on petroleum spray oils*. Agnote DPI/118. NSW Agriculture.

Beattie, G. A. C. and Gellatley, J. G. (1983a). *Citrus scale insects*. Agfact H2.AE.2. NSW Department of Agriculture.

Beattie, G. A. C. and Gellatley, J. G. (1983b). *Mite pests of citrus*. Agfact H2.AE.3. NSW Department of Agriculture.

Beattie, G. A. C., and Jiang, L. (1990). Greenhouse thrips and its parasitoids in coastal New South Wales. *General and Applied Entomology* 22:21–24.

Beattie, G. A. C., and Kaldor, C. J. (1990). Comparison of high-volume oscillating boom and low-volume fan-assisted rotary atomiser sprayers for the control of Chinese wax scale, *Ceroplastes sinensis* del Guercio (Hemiptera: Coccidae), on Valencia orange, *Citrus sinensis* (L.) Osbeck. *General and Applied Entomology* 22:49–53.

Beattie, G. A. C. and Liu, Z. M. (1995). *Reducing pesticide use in citrus: citrus leafminer: an update on the use of petroleum spray oils*. Agnote DPI/106. NSW Agriculture.

Beattie, G. A. C and Rippon, L. E. (1978). Phytotoxicity and scalicide efficacy of citrus spray oils. In *1978 Proceedings of the International Society of Citriculture*, pp. 171–174.

Beattie, G. A. C. and Smith, D. (1993). *Citrus leafminer*. Agfact H2.AE.4, 2nd edn. NSW Agriculture.

Beattie, G. A. C., Barkley, P., Weir, R. and Gellatley, J. G. (1984). *Citrus in the home garden: pests, diseases and nutrient disorders*. Department of Agriculture NSW.

Beattie, G. A. C., Roberts, E. A., Rippon, L. E., and Vanhoff, C. L. (1989). Phytotoxicity of petroleum spray oils to Valencia orange, *Citrus sinensis* (L.) Osbeck, in New South Wales. *Australian Journal of Experimental Agriculture* 29:273–283.

Beattie, G. A. C., Roberts, E. A., Vanhoff, C. L., and Flack, L. K. (1991). The effect of climate, natural enemies and biocides on three citrus mites in coastal New South Wales. *Experimental and Applied Acarology* 11: 271–295.

Beattie, G. A. C., Weir, R. G., Clift, A. D., and Jiang, L. (1990). Effect of nutrients on the growth and phenology of *Gascardia destructor* (Newstead) and *Ceroplastes sinensis* del Guercio (Hemiptera: Coccidae) infesting citrus. *Journal of the Australian Entomological Society* 29:199-203.

Beattie, G. A. C., Broadbent, P., Baker, H., Gollnow, B., and Kaldor, C. J. (1989). Comparison of conventional medium to high-volume and high-volume sprayers with a low-volume sprayer for the control of black spot, *Guignardia citricarpa* Keily, on Valencia orange. *Plant Protection Quarterly* 4:146-148.

Beattie, G. A. C., Clift, A. D., Allender, W. J., Jiang, L., and Wang Y. A. (1991). Efficacies of low- to high-volume (960–10 700 litre ha^{-1}) orchard sprayers for applying petroleum spray oil to control Chinese wax scale. *Pesticide Science* 32:47-56.

Beattie, G. A. C., Liu, Z. M., Watson, D. M., Clift, A. D. and Jiang, L. (1995). Evaluation of petroleum spray oils and polysaccharides for control of citrus leafminer, *Phyllocnistis citrella* Stainton (Lepidoptera: Gracillaridae). *Journal of the Australian Entomological Society* 34:349-353.

Beattie, G. A. C., Somsook, V., Watson, D. M., Clift, A. D. and Jiang, L. (1995). Field evaluation of *Steinernema carpocapsae* (Weiser) (Rhabditida: Steinernematidae) and selected pesticides and enhancers for control of citrus leafminer, *Phyllocnistis citrella* Stainton (Lepidoptera: Gracillaridae). *Journal of the Australian Entomological Society* 34:335-342.

Bedding, R., Akhurst, R. and Kaya, H. (1993). *Nematodes and the biological control of insect pests*. CSIRO Publications, East Melbourne.

Bedford, E. C. G. (ed.) (1971). *Citrus pests in the Republic of South Africa*. Science Bulletin no. 391. Department of Agricultural Technical Services, Republic of South Africa.

Behncken, G. B. and Broadley, R. H. (eds) (1990). *Pesticide application manual*, 2nd edn. Department of Primary Industries, Queensland.

Ben-Dov, Y. (1993). *A systematic catalogue of soft scale insects of the world*. Sandhill Crane Press, Gainesville, Florida.

Boucek, Z. (1988). *Australasian Chalcidoidea (Hymenoptera). A biosystematic revision of genera of fourteen families, with a reclassification of species*. C.A.B. International, Wallingford Oxon UK.

Brewer, R. H. (1971). The influence of the parasite *Comperiella bifasciata* How. on populations of two species of armoured scale insects, *Aonidiella aurantii* (Mask) and *A. citrina* (Coq.), in South Australia. *Australian Journal of Zoology* 19:53–63.

Broadley, Roger and Thomas, Michael (eds) (1995). *The good bug book*. Australasian Biological Control Inc., with Department of Primary Industries, Queensland, and Rural Industries Research and Development Corporation.

Broadley, R. H., Smith, D., Papacek, D., Owen-Turner, J. C., Chapman, J. C., Banks, A. G. and Mayers, P. (1987). *Protect your citrus*. Department of Primary Industries, Queensland.

Brough, Elaine; Elder, Rod; Beavis, Colin (eds) (1994). *Managing insect and mites in horticultural crops*. Department Primary Industries, Queensland.

Browning, H. W., McGovern, R. J., Jackson, L. K., Calvert, D. V., and Wardowski, W. F. (1995). *Florida citrus diagnostic guide*. Florida Science Source, Lake Alfred, Florida.

Browning, T. O. (1959). The longtailed mealybug, *Pseudococcus adonidum* (L.) in South Australia. *Australian Journal of Agricultural Research* 10:322–329.

Buchanan, G. A. (1977). The seasonal abundance and control of lightbrown apple moth, *Epiphyas postvittana* (Walker) (Lepidoptera: Tortricidae), on grapevines in Victoria. *Australian Journal of Agricultural Research* 28:125–132.

Campbell, M. M. (1972). Counts of adults of *A. melinus* on citrus trees before and after spraying with an emulsion containing 1% of pre-emulsified oil in water. *Experimental Record* 7:44.

Campbell, M. M. (1975). Establishing *Aphytis melinus* in citrus orchards by a new simple method. *Journal of the Australian Institute of Agricultural Science* 41:62–63.

Campbell, M. M. (1976). Colonisation of *Aphytis melinus* DeBach (Hymenoptera: Aphelinidae) in *Aonidiella aurantii* (Mask). (Hemiptera: Coccidae) on citrus in South Australia. *Bulletin of Entomological Research* 65:659–668.

Campbell, M. M. (1987a). Evaluation of equipment and methods for applying agricultural chemicals. In T. H. Lee (ed.), *Proceedings Sixth Australian Wine Industry Technical Conference, Adelaide, 14–17 July, 1986*, pp. 180–183.

Campbell, M. M. (1987b). *Orchard sprayers*, parts 1–4. Agnotes Nos 3757/87, 3762/87, 3769/87, 3770/87. Victorian Department of Agriculture and Rural Affairs.

Cant, R. G., Spooner-Hart, R. N., Beattie, G. A. C. and Meats. A. (in press). The biology of the bronze orange bug, *Musgraveia sulciventris* (Stål). A literature review. Part I: description, biology, host species and distribution. *General and Applied Entomology*.

Cant, R. G., Spooner-Hart, R. N., Beattie, G. A. C. and Meats, A. (in press). The biology of the bronze orange bug, *Musgraveia sulciventris* (Stål). A literature review. Part II: feeding, control, defensive secretions, pheromones, reproduction and aggregation. *General and Applied Entomology*.

Carman, G. E. (1975). Spraying procedures for pest control in citrus. In *Citrus*, Ciba Geigy Agrochemicals Technical Monograph no. 4, pp. 28–34.

Carman, G. E. (1977). Evaluation of citrus sprayer units with air towers. *Citrograph* 62:134–139.

Carman, G. E. and Jeppson, L. R. (1974). Low-volume applications to citrus trees: method for evaluation of spray droplet distributions. *Journal of Economic Entomology* 67:397–402.

Citrus pests and diseases: quality management guide for field operations (1994). Australian Horticultural Corporation.

Citrus pest control 1968-69 for Northern Victorian districts (1968). Leaflet no. H57. Dept Agriculture, Victoria.

Clift, A. D. and Beattie, G. A. C. (1989a). A computer model to simulate changes in numbers of citrus red scale, *Aonidiella aurantii* (Maskell) (Hemiptera: Diaspididae) in NSW. In *Abstracts of the 20th Annual General Meeting and Scientific Conference of the Australian Entomological Society, Sydney*, p. 42.

Clift, A. D. and Beattie, G. A. C. (1989b). Use of a database and spreadsheet to develop simulation models based on field data. In *Proceedings of the Biennial Conference of the Simulation Society of Australia, Canberra 1989*, pp. 71–76.

Clift, A. D. and Beattie, G. A. C. (1993). SCALEMAN: a computer program to help citrus growers manage red scale, *Aonidiella aurantii* (Hemiptera: Diaspididae). In S. A. Corey, D. J. Dall, and W. M. Milne (eds), *Pest control and sustainable agriculture*, pp. 470–472. CSIRO Division of Entomology, Canberra.

Common, I. F. B. (1990). *Moths of Australia*. Melbourne University Press, Melbourne.

Compere, H. (1961). The red scale and its insect enemies. *Hilgardia* 31(7):173–278.

Compere, H., and Smith, H. S. (1932). The control of the citrophilous mealybug *Pseudococcus gahani* by Australian parasites. *Hilgardia* 6(7):585–618.

Cunningham, G., Hughes, P. and Harden, J. (1994). *Efficient pesticide use in citrus*. University of Queensland, Gatton College, and Department of Primary Industries, Queensland.

Dahms, E. C. and Smith, D. (1994). The *Aphytis* fauna of Australia. In D. Rosen (ed.), *Advances in the study of Aphytis (Hymenoptera: Aphelinidae)*, pp. 245–255. Intercept Ltd, Andover.

Davies, R. A. H. and McLaren, I. W. (1977). Tolerance of *Aphytis melinus* DeBach (Hymenoptera: Aphelinidae) to 20 orchard chemical treatments in relation to integrated control of red scale, *Aonidiella aurantii* (Mask) (Homoptera: Diaspididae). *Journal of Experimental Agriculture and Animal Husbandry* 17:323–328.

DeBach, P. and Sundby, R. A. (1963). Competitive displacement between ecological homologues. *Hilgardia* 34:105–166.

Drew, R. A. I. and Hancock, D. L. The *Bactrocera dorsalis* complex of fruit flies (Diptera: Tephritidae: Dacinae) in Asia. *Bulletin of Entomological Research*, supplement no. 2.

Drew, R. A. I., Hooper, G. H. S. and Bateman, M. S. (1982). *Economic fruit flies of the south Pacific region*, 2nd edn. Dept of Primary Industries, Queensland.

Fisher, R. W., Menzies, D. R. and Hickichi, A. (1976). *Orchard sprayers*. Publication No. 373. Ministry of Agriculture and Food, Ontario.

Forsyth, J. B. and Barkley, P. (1989). *Citrus rootstocks*. Agfact H2.2.2. NSW Agriculture and Fisheries.

Furness, G. O. (1975). Developing the integrated control of red scale for grower use. In *Australian Applied Entomological Research Conference, Proceedings*, Mildura, Victoria 1975.

Furness, G. O. (1976). The dispersal, age-structure and natural enemies of the longtailed mealybug, *Pseudococcus longispinus* (Targioni-Tozzetti), in relation to sampling and control. *Australian Journal of Zoology* 24:237–247.

Furness, G. O. (1977a). Apparent failure of two parasites, *Anarhopus sydneyensis* (Hymenoptera: Encyrtidae) and *Hungariella peregrina* (Hymenoptera: Pteromalidae) to establish on field populations of *Pseudococcus longispinus* (Hemiptera: Coccidae) in South Australia. *Australian Journal of Agricultural Research* 28:319–332.

Furness, G. O. (1977b). Chemical and integrated control of the longtailed mealybug, *Pseudococcus longispinus* (Targioni-Tozzetti) (Hemiptera: Coccidae) in the Riverland of South Australia. *Australian Journal of Agricultural Research* 28:319–332.

Furness, G. O. (1981a). The phytotoxicity of narrow distillation range petroleum spraying oils to Valencia orange trees in South Australia. Part II: The influence of distillation temperature and spray timing on fruit quality. *Pesticide Science* 12:603–608.

Furness, G. O. (1981b). The phytotoxicity of narrow distillation range petroleum spraying oils to Valencia orange trees in South Australia. Part III: The influence of distillation temperature and spray timing on leaf and fruit drop. *Pesticide Science* 12:609–613.

Furness, G. O. (1982). Lessons learnt from a successful IPM program. In *Proceedings of an Australian workshop on development and implementation of IPM*. DSIR Auckland, New Zealand.

Furness, G. O. (1987). High resolution g.l.c. specification for plant spray oils. *Pesticide Science* 18:113–128.

Furness, G. O. and Maelzer, D. A. (1981). The phytotoxicity of narrow distillation range petroleum spraying oils to Valencia orange trees in South Australia. Part I: The influence of distillation temperature and spray timing on yield and alternate cropping. *Pesticide Science* 12:573–602.

Furness, G. O. and Pinczewski, V. W. (1985). A comparison of the spray distribution obtained from sprayers with converging and diverging airjets with low-volume, air-assisted spraying on citrus and grapevines. *Journal of Agricultural Engineering Research* 32:291–310.

Furness, G. O., Buchanan, G. A., George, R. S. and Richardson, N. L. (1983). A history of the biological and integrated control of red scale *Aonidiella aurantii* on citrus in the Lower Murray Valley of Australia. *Entomophaga* 28(3):199–212.

Gerson, U. and Smiley, R. L. (1990). *Acarine biocontrol agents: an illustrated key and manual*. Chapman and Hall, London.

Green, J. (1996). *Field guide to spiders*. Cooperative Research Centre for Tropical Pest Management and Dept of Entomology, University of Queensland.

Greenslade, P. J. M. (1979). *A guide to ants of South Australia*. South Australian Museum, Adelaide.

Gurney, N. B. (1936). Parasites of white wax scales collected in East Africa. *Agriculture Gazette of NSW* 47:453–56, 464, 472.

Hardman, J. R., Papacek, D. and Smith, D. (1993). The use of integrated pest management in citrus orchards in Queensland: an economic perspective. In *Aggregated papers of the 37th Annual Conference of the Australian Agricultural Research and Economics Society, University of Sydney, 9–11 February 1993*.

Hardy, S., Barkley, P., Beattie, A., Forsyth, J., Sarooshi, R. and Wild, B. (1995). *Pest and disease management guide for coastal districts in New South Wales*. NSW Agriculture.

Hayat, M. (1983). The genera of Aphelinidae (Hymenoptera) of the world. *Systematic Entomology* 8:63–102.

Hely, P. C. (1968). *The entomology of citrus in New South Wales: W. W. Froggatt Memorial Lecture*. Miscellaneous Publication no. 1. Australian Entomological Society.

Hely, P. C., Pasfield, G. and Gellatley, J. G. (1982). *Insect pests of fruit and vegetables in NSW*. Inkata Press, Melbourne.

Herron, G. A., Beattie, G. A. C, Parkes, R. A. and Barchia, I. (1995). Potter spray tower bioassay of selected citrus pests to petroleum spray oil. *Journal of the Australian Entomological Society* 34:255–263.

Insects of Australia, vols 1 & 2 (1991). CSIRO. Melbourne University Press, Melbourne.

Integrated pest management for citrus (1984). University of California Publication 3303.

James, D. G. (1988a). Fecundity, longevity and overwintering of *Trissolcus biproruli* Girault (Hymenoptera: Scelionidae) a parasitoid of *Biprorulus bibax* Breddin (Hemiptera: Pentatomidae). *Journal of the Australian Entomological Society* 27:297–301.

James, D. G. (1988b). Induction of pupal diapause in *Papilio aegeus aegeus* Donovan and *Graphium sarpedon choredon* Felder (Lepidoptera: Papilionidae). *Australian Entomological Magazine* 15:39–44.

James, D. G. (1989a). Effect of pesticides on survival of *Amblyseius victoriensis*, an important predatory mite in southern New South Wales peach orchards. *Plant Protection Quarterly* 4:141–143.

James, D. G. (1989b). Influence of diet on development, survival and oviposition in an Australian phytoseiid, *Amblyseius victoriensis* (Acari: Phytoseiidae). *Experimental and Applied Acarology* 6:1–10.

James, D. G. (1989c). Population biology of *Biprorulus bibax* Breddin (Hemiptera: Pentatomidae) in a southern New South Wales citrus orchard. *Journal of the Australian Entomological Society* 28:279–286.

James, D. G. (1990a). Energy reserves, reproductive status and population biology of *Biprorulus bibax* Breddin (Hemiptera: Pentatomidae) in southern New South Wales citrus groves. *Australian Journal of Zoology* 38:415–422.

James, D. G. (1990b). Incidence of egg parasitism of *Biprorulus bibax* Breddin (Hemiptera: Pentatomidae) in southern New South Wales. *General and Applied Entomology* 22:55–60.

James, D. G. (1990c). Seasonality and population development of *Biprorulus bibax* Breddin (Hemiptera: Pentatomidae) in south-western New South Wales. *General and Applied Entomology* 22:61–66.

James, D. G. (1990d). The effect of temperature on development and survivorship of the spined citrus bug *Biprorulus bibax* (Hemiptera: Pentatomidae). *Environmental Entomology* 19:874–877.

James, D. G. (1991a). An evaluation of chemical and physical treatments to prevent Fuller's rose weevil oviposition on citrus fruit. *Plant Protection Quarterly* 6:79–81.

James, D. G. (1991b). Maintenance and termination of reproductive dormancy in an Australian stink bug, *Biprorulus bibax*. *Entomologia Experimentalis et Applicata* 60:1–5.

James, D. G. (1992a). Effect of citrus host variety on oviposition, fecundity and longevity in *Biprorulus bibax* Breddin (Heteroptera). *Acta Entomologia Bohemoslovica* 89:65–67.

James, D. G. (1992b). Effect of temperature on development and survival of *Pristhesancus plagipennis* (Hemiptera: Reduviidae). *Entomophaga* 37:259–264.

James, D. G. (1992c). Summer reproductive dormancy in *Biprorulus bibax* (Hemiptera: Pentatomidae). *Australian Entomological Magazine* 19:65–68.

James, D. G. (1993a). Apparent overwintering of *Biprorulus bibax* (Hemiptera: Pentatomidae) on *Eremocitrus glauca* (Rutaceae). *Australian Entomologist* 20:129–132.

James, D. G. (1993b). Biology of *Anastatus biproruli* (Hymenoptera: Eupelmidae), a parasitoid of *Biprorulus bibax* (Hemiptera: Pentatomidae). *Entomophaga* 38:155–161.

James, D. G. (1993c). New egg parasitoid records for *Biprorulus bibax* (Breddin) (Hemiptera: Pentatomidae). *Journal of the Australian Entomological Society* 32:67–68.

James, D. G. (1993d). Toxicity and use of endosulfan against spined citrus bug, *Biprorulus bibax*, and some of its egg parasitoids. *Plant Protection Quarterly* 8:54–56.

James, D. G. (1993e). Toxicity of endosulfan to adult *Aphytis melinus* (Hymenoptera: Aphelinidae). *Pakistan Journal of Scientific and Industrial Research* 36:44.

James D. G. (1994a). Effect of citrus host variety on lipid reserves in overwintering *Biprorulus bibax* (Breddin) (Heteroptera). *Victorian Entomologist* 24:42–45.

James D. G. (1994b). Prey consumption by *Pristhesancus plagipennis* Walker (Hemiptera: Reduviidae) during development. *Australian Entomologist* 21:43–47.

James, D. G. (1994c). The development of suppression tactics for *Biprorulus bibax* (Heteroptera: Pentatomidae) as part of an integrated pest management program in inland citrus of south-eastern Australia. *Bulletin of Entomological Research* 84:31–38.

James, D. G. and Mori, K. (1995). Spined citrus bugs, *Biprorulus bibax* Breddin (Hemiptera: Pentatomidae) do not discriminate between enantiomers in their aggregation pheromone. *Journal of Chemical Ecology* 21:403–406.

James, D. G. and Taylor, A. (1992). Effect of temperature on development and survival of *Amblyseius victoriensis* (Womersley) (Acari: Phytoseiidae). *International Journal of Acarology* 18:93–96.

James, D. G. and Warren, G. N. (1989). Sexual dimorphism of dorsal abdominal glands in *Biprorulus bibax* Breddin (Hemiptera: Pentatomidae). *Journal of the Australian Entomological Society* 28:75–76.

James, D. G. and Warren G. N. (1991). Effect of temperature on development, survival, longevity and fecundity of *Trissolcus oenone* Dodd (Hymenoptera: Scelionidae). *Journal of the Australian Entomological Society* 30:303–306.

James, D. G., Faulder, R. J. and Warren, G. N. (1990). Phenology of reproductive status, weight and lipid reserves of *Biprorulus bibax* (Hemiptera: Pentatomidae). *Environmental Entomology* 19:1710–1715.

James, D. G., Moore, C. and Aldrich, J. R. (1994). Identification, synthesis and bioactivity of male-produced aggregation pheromone in the assassin bug, *Pristhesancus plagipennis* (Hemiptera: Reduviidae). *Journal of Chemical Ecology* 21:3281–3295.

James, D. G., Mori, K., Aldrich, J. R. and Oliver, J. E. (1994). Flight-mediated attraction of *Biprorulus bibax* Breddin (Hemiptera: Pentatomidae) to natural and synthetic aggregation pheromone. *Journal of Chemical Ecology* 20:71–80.

James, D. G., Stevens, M. M., O'Malley, K. and Heffer, R. (1996). Activity of a hydramethylnon-based bait against the citrus ant pests *Iridomyrmex rufoniger* gp. spp. and *I. purpureus*. *Plant Protection Quarterly*.

Jenkins, C. F. H. (1946). Biological control in Western Australia. Presidential address. *Journal of the Royal Society of Western Australia* 32:35–40.

Kirejtshuk, A. G., James, D. G. and Heffer, R. (1996). Description and biology of a new species of *Cybocephalus* (Coleoptera: Nitidulidae), a predator of Australian citrus whitefly. *Journal of the Australian Entomological Society*.

Knapp, Joseph L. (ed.) (1991). *Florida citrus integrated pest and crop management handbook*. Institute of Food and Agricultural Sciences, University of Florida.

Mascord, R. (1979). *Australian spiders in colour*. Reed, Sydney.

Matthews, G. A. (1979). *Pesticide application methods*. Longman, London.

Matthews, G. A. and Hislop, E. C. (eds) (1993). *Application technology for crop protection*. CAB International, Wallingford Oxon UK.

McLaren, I. W. (1971). A comparison of the population growth potential in California red scale, *Aonidiella aurantii* (Mask), and yellow scale, *A. citrina* (Coquillet), on citrus. *Australian Journal of Zoology* 19:89–204.

McLaren, I. W. and Buchanan, G. A. (1973). Parasitism by *Aphytis chrysomphali* Mercet and *A. melinus* DeBach of California red scale, *Aonidiella aurantii* (Mask), in relation to season availability of suitable stages of the scale. *Australian Journal of Zoology* 21:111–117.

Moulds, M. S. (1990). *Australian cicadas*. New South Wales University Press, Sydney.

Neale, C., Smith, D., Beattie, G. A. C. and Miles, M. (1995). The importation, host specificity testing, rearing and releasing of three parasitoids of citrus leafminer in citrus in eastern Australia. *Journal of the Australian Entomological Society* 34(4):343–348.

Nicholas, P. R., Magarey, P. A. and Wachtel, M. F. (eds) (1994). *Diseases and pests*. Grape Production Series. Winetitles, Adelaide.

Noble, N. S. (1936). *The citrus gall wasp, Eurytoma fellis Girault*. NSW Department of Agriculture Scientific Bulletin 53.

Noble, N. S. (1938). *Epimegastigmus brevivalvus Girault, a parasite of the citrus gall wasp, Eurytoma fellis Girault, with notes on several other species of hymenopterous gall inhabitants*. NSW Department of Agriculture Scientific Bulletin 65.

Noyes, J. S. (1988). *Encyrtidae (Insecta: Hymenoptera)*. Fauna of New Zealand, no. 13.

Noyes, J. S. and Hayat, M. (1984). A review of the genera of Indo-Pacific Encyrtidae (Hymenoptera: Chalcidoidea). *Bulletin of the British Museum (Natural History) Entomology Series* 48(3):131–395.

Noyes, J. S. and Hayat, M. (1994). *Oriental mealybug parasitoids of the Anagyrini (Hymenoptera: Encyrtidae)*. CAB International, Wallingford Oxon UK.

Oliver, J. E., Aldrich, J. R., Lusby, W. R., Waters, R. M. and James, D. G (1992). A male-produced pheromone of the spined citrus bug. *Tetrahedron Letters* 33:891–894.

Owen-Turner, J. (1995). *Growing citrus in Queensland*. Department of Primary Industries, Queensland.

Papacek, D. F. and Smith, D. (1985). *Aphytis lingnanensis*. In P. Singh and R. F. Moore (eds), *Handbook of insect rearing*, vol. 1. Elsevier, Amsterdam.

Papacek, D. F. and Smith, D. (1989). Insecticidal control of citrus gall wasp in Queensland. *General and Applied Entomology* 21:2–4.

Parkes, R. A., Herron, G. A. and Beattie, G. A. C. (1992). A confinement technique for arthropods allowing bioassay of highly motile forms with a Potter spray tower. In *Proceedings of the First National Conference of the Australian Society of Horticultural Science, Sydney, September–October 1991*, pp 497–502.

Persley, D. (ed.) (1993). *Diseases of fruit crops*. Department of Primary Industries, Queensland.

Prinsloo, G. L. (1984). *An illustrated guide to the parasitic wasps associated with citrus pests in the Republic of South Africa*. Science Bulletin no. 402. Department of Agriculture.

Qin, T. K., Fletcher, M. J., Beattie, G. A. C. and Gullan, P. J. (1992). Where does the Chinese wax scale (Insecta: Coccidae) come from? Implications for the control of pests of agricultural importance. In *Proceedings of the First National Conference of the Australian Society of Horticultural Science, Sydney, September–October 1991*, pp. 451–454.

Qin, T. K., Gullan, P. J., Beattie, G. A. C., Trueman, J. W. H., Cranston, P. S., Fletcher, M. J. and Sands, D. P. A. (1994). The current distribution and geographical origin of the scale insect pest *Ceroplastes sinensis* (Hemiptera: Coccidae). *Bulletin of Entomological Research* 84:541–549.

Rae, D. J., Beattie, G. A. C., Watson, D. M., Liu, Z. M. and Jiang, L. (1996). Effect of petroleum spray oils without and with copper fungicides on the control of citrus leafminer, *Phyllocnistis citrella* Stainton (Lepidoptera: Gracillaridae). *Australian Journal of Entomology* 35:247–251.

Rae, D. J., Watson, D. M., Liang, W. G., Tan, B. L., Li, M., Huang, M. D., Ding, Y., Xiong, J. J., Du, D. P., Tang, J. and Beattie, G. A. C. (1996). Comparison of petroleum spray oils, abamectin, cartap and methomyl for citrus leafminer (Lepidoptera: Gracillariidae) control in southern China. *Journal of Economic Entomology* 89(2):493–500.

Raisgl, U. and Felber, H. (1991). Comparison of different mistblowers and volume rates for orchard spraying. In *Air-assisted spraying in crop protection*, pp. 185–196. Monograph no. 46, British Crop Protection Council.

Randall, J. M. (1971). The relationships between air volume and pressure on spray distribution in fruit trees. *Journal of Agricultural Engineering Research* 16:1–31.

Rentz, D. (1996). *Grasshopper country: the abundant orthopteroid insects of Australia—grasshoppers, katydids, crickets, cockroaches, mantids, stick insects*. University of New South Wales Press, Sydney.

Revelant, Lou and Beattie, Brian (eds) (1993). *Guide to quality management in the citrus industry*. Australian Horticultural Corporation.

Richardson, N. L. (1978). Biological aspects of coexistence between *Comperiella bifasciata* Howard (Hymenoptera: Chalcidoidea: Encyrtidae) and *Aphytis* spp. Howard (Hymenoptera: Chalcidoidea: Aphelinidae). In *Proceedings of the International Society for Citriculture 1978*.

Rosen, D. (1966). Key for the identification of the hymenopterous parasites on scale insects, aphids and aleyrodids on citrus in Israel. *Scripta Hierosolymitana*. Publications of the Hebrew University, Jerusalem. 18:43–79, figures 1–267.

Rosen, D. (ed.) (1990). *Armoured scale insects: their biology, natural enemies and control*, vol. B. Elsevier.

Rosen, D., and DeBach, P. (1979). *Species of Aphytis of the world (Hymenoptera: Aphelinidae)*. Israel Universities Press, Jerusalem, and W. Junk, The Hague.

Sands, D. P. A., Lukins, R. G. and Snowball, G. J. (1986). Agents introduced into Australia for the biological control of *Gascardia destructor* (Newstead) (Hemiptera: Coccidae). *Journal of the Australian Entomological Society* 25:51–59.

Smith, A. D. M. and Maelzer, D. A. (1986). Aggregation of parasitoids and density-independence of parasitism in field populations of the wasp *Aphytis melinus* and its host the red scale *Aonidiella aurantii*. *Ecological Entomology* 11(4):425–434.

Smith, D. (1978a). Biological control of scale insects in south-eastern Queensland. 1. Control of red scale *Aonidiella aurantii* (Maskell). *Journal of the Australian Entomological Society* 17:367–371.

Smith, D. (1978b). Biological control of scale insects in south-eastern Queensland. 2. Control of circular black scale, *Chrysomphalus ficus*. Ashmead, by the introduced parasite, *Aphytis holoxanthus* DeBach. *Journal of the Australian Entomological Society* 17:373–377.

Smith, D. (1986). Biological control of *Ceroplastes rubens* Maskell by the introduced parasitoid *Anicetus beneficus* Ishii and Yasmatsu. *Queensland Journal of Agricultural and Animal Sciences* 43(2):101–105.

Smith, D. (1990). Integrated pest management in Queensland citrus. *Australian Citrus News* December 1990:6–12.

Smith, D. (1995). Effect of the insect growth regulator bupofezin against citrus pests *Coccus viridis* (Green), *Polyphagotarsonemus latus* (Banks) and *Aonidiella aurantii* (Maskell) and the predatory coccinellid *Chilocorus circumdatus* Gyllenhal. *Plant Protection Quarterly* 10(3):112–115.

Smith, D. and Liu Nannan (1988). Yeast autolysate bait sprays for control of Queensland fruit fly on passionfruit in Queensland. *Queensland Journal of Agricultural and Animal Sciences* 45(2): 169–177.

Smith, D. and Papacek, D. F. (1985). Integrated pest management in Queensland citrus. *Queensland Agricultural Journal* 44:404–408.

Smith, D. and Papacek, D. F. (1991a). Recent advances in integrated pest management in citrus in Queensland. In *Proceedings 1st National Conference Australian Society of Horticultural Science, Macquarie University, Sept. 30 – Oct. 3 1991*, pp. 375–383.

Smith, D. and Papacek, D. F. (1991b). Studies of the predatory mite *Amblyseius victoriensis* (Womersley) (Acarina: Phytoseiidae) in citrus orchards in south-east Queensland: control of *Tegolophus australis* Kiefer and *Phyllocoptruta oleivora* (Ashmead) (Acarina: Eriophyidae), effect of pesticides, alternate host plants and augmentative release. *Experimental and Applied Acarology* 12:195–217.

Smith, D. and Papacek, D. F. (1992). Integrated pest management in Queensland citrus: recent developments. In *Pest control and sustainable agriculture: proceedings of the Fifth Australian Applied Entomological Research Conference, Canberra 1992*.

Smith, D. and Papacek, D. F. (1995). Biological control of citrus snow scale in south-east Queensland. *Israel Journal of Entomology* 29:253–260.

Smith, D., Papacek, D. F. and Murray, D. A. H. (1989). The use of *Leptomastix dactylopii* Howard (Hymenoptera: Encyrtidae) to control *Planococcus citri* (Risso) (Hemiptera: Pseudococcidae) in Queensland citrus orchards. *Queensland Journal of Agricultural and Animal Sciences* 45(2):157–164.

Snowball G. J. (1969). Prospects for biological control of white wax scale (*Gascardia destructor*) in Australia by South African natural enemies. *Journal of the Entomological Society of Australia* 5:23–33.

Sproul, A. N. (1981). Citrus red scale, from sprays to parasites. *Journal of Agriculture of Western Australia* 2:50–51.

Stevens, M. M., James, D. G. and O'Malley, K. T. (1995) Evaluation of alphacypermethrin-treated proprietary trunk barriers for the exclusion of *Iridomyrmex* spp. (Hymenoptera: Formicidae) from young citrus trees. *International Journal of Pest Management* 41:22–26.

Summerville, W. A. T. (1931). *The larger horned citrus bug Biprorulus bibax Breddin*. Entomology Bulletin. Queensland Department of Agriculture and Stock.

Summerville, W. A. T. (1934). Queensland citrus scale insects and their control. *Queensland Agricultural Journal* 41:568–591.

Sutton, T. B. and Unrath, C. R. (1984). Evaluation of the tree-row-volume concept with density adjustments in relation to spray deposits in apple orchards. *Plant Disease* 68:480–484.

Swaine, G., Ironside D. A. and Yarrow, W. H. T. (eds) (1985). *Insect pests of fruit and vegetables in colour*. Department of Primary Industries, Queensland.

Thornton, I. R. and El-Zeftawi, M. M. (1983). *Culture of irrigated citrus fruits*. Department of Agriculture, Victoria.

Walter, D. E., Halliday, R. B. and Smith, D. The oriental red mite, *Eutetranychus orientalis* (Klein) (Acarina: Tetranychidae), in Australia. *Journal of the Australian Entomological Society* 34(4):307–308.

Waterhouse, D. F. (1993). *Biological control: Pacific prospects: supplement 2*. ACIAR Monograph no. 20. ACIAR, Canberra.

Waterhouse, D. F. and Norris, K. R. (1987). *Biological control: Pacific prospects*. Inkata Press, Melbourne.

Watson, D. M. and Beattie, G. A. C. (1995). The effect of weather station logging interval on the precision of degree-day estimates. *Australian Journal of Experimental Agriculture* 35:795–805.

Willard, J. R. (1971). The rhythm of emergence of crawlers of California red scale *Aonidiella aurantii* (Mask.) (Hemiptera: Diaspididae). *Australian Journal of Zoology* 20:49–65.

Willard, J. R. (1974). Horizontal and vertical dispersal of California red scale in the field. *Australian Journal of Zoology* 22:531–548.

Wilson, F. (1960). *A review of the biological control of insects and weeds in Australian and New Guinea*. Commonwealth Agricultural Bureau Technical Communication no. 1. Commonwealth Institute of Biological Control, Ottawa, Canada.

Index

This index contains references to scientific and common names of citrus pests and their natural enemies, as well as references to citrus diseases, citrus pathogens, citrus disorders and damage symptoms.

As the text on each pest includes the same kind of information under the same subheadings (e.g. 'Distinguishing features', 'Habits', 'Hosts', 'Origin and distribution', 'Monitoring', 'Action level'), these topics are not included in the index. They are easily found by referring to the main section on each pest.

In the index, the reference to the beginning of the main section on each pest is given in **bold** type. References to colour plates are in ***bold italic*** type, and those to life cycle illustrations and graphs are in *italic* type. (Note that, in almost every case, the first page of the main section on each pest contains a colour plate of that pest.)

The book contains six chapters and one appendix on topics other than pests and natural enemies (the Australian citrus industry, concepts of integrated pest management (IPM), practical IPM, pesticide application, petroleum spray oils, pesticide toxicity to natural enemies, and a brief history of IPM). These topics are not indexed, and can easily be found in the contents list.

A

Ablerus 99
Acalolepta vastator **149**
Acanthiza chrysorrhoa 198
Acanthorhynchus tenuirostris 198
Acaridae ***36***, 184
Acarus 184
Achalcerinys 97, 193
Acroclisoides tectacorisi 110, 193
Aculus pelekassi 238
Adoxophyes **125**
African citrus psylla 238
Ageniaspis citricola 120, ***121***, 193, 254
Agistemus 184, 197
Agrilus occipitalis 239
Aleurocanthus **106**
    spiniferus 238
    woglumi 238
aleurocanthus whitefly **106**, ***107***, **217**
Aleurodothrips fasciapennis 197
Aleurothrixcis floccosus 238
Allotropa 79, 194, 241, 249
    citri 241, 249
Alternaria
    alternata 12
    citri 12
alternaria brown spot **12**, ***220***
Amblypelta lutescens lutescens 183
Amblyseius 184
    herbicolus 19, 22, 27, 33, 197
    lentiginosus 19, 22, 33, 197
Amyotea hamatus 113, 196
Anagyrus 83, 88, 193, 251
    agraensis 86, ***87***, 193, 241, 251
    fusciventris 79, 81, 193, 241, 251, 253
    pseudoc occi 83, 193, 241, 251
Anastatus 194
    biproruli 110, 112, ***113***, 194

Anastrepha
    fraterculus 238
    ludens 238
    suspensa 238
Anicetus 244
    beneficus ***53***, 55, 192, 193, 241, 244
    communis 58, ***59***, 193, 241, 244, 254
Anoplophora chinensis 239
ant 5, **37**, 42, 45, 51, 110, ***221***, ***222***, ***224***, ***225***, 239
    Argentine 38
    leafcutting 239
    meat 37, 38
    red imported fire 239
anthracnose 5, **13**, ***223***
Anystis 197
Aonidiella
    aurantii **64**
    citrina **68**
    orientalis 183
Apanteles 194
Aphanomerus 97, 194
Aphelinus 90, 94
aphid 10, 25, 29, 37, **89**, ***91***, 192, 194, 195, 196, 197, 199, 254
    black citrus 11, **89**, ***90***, ***91***, **92**, 93, 195, 196, 197, ***216***, ***221***, ***222***, ***224***, ***225***, 254
    brown citrus **89**
    citrus **89**, ***90***, 195
    cotton 93
    melon 89, **92**, ***93***
    spiraea 89, **92**, ***93***, ***216***
Aphidius 90, 94, 194
Aphis
    gossypii **92**
    spiraecola **92**
Aphytis 68, 71, 75, 192, 202, 235, 247, 253, 254, 255
    chrysomphali 65, **66**, 69, 192, 241, 248, 252
    columbi 70, 192, 241, 248
    gordoni 253
    holoxanthus **69**, 70, 192, 241, 248, **253**
    lepidosaphes **73**, 74, 192, 241, 248, 253
    lingnanensis 65, **66**, 71, 192, 241, 248, 252, 253
    melinus 65, **66**, 192, 241, 248, 252, 254, 255
aphytis wasp 66, 76, ***77***, ***204***
apple weevil **147**
arachnids **185**
Argentine ant 38
Argyrophylax proclinata 128, 194
Argyrotaenia citrana 239
Armillaria luteobubalina 9
armillaria root rot **9**
armoured scale **64**, 82, 183, 231, 232, 233, 238, 240, 241, 247
armyworm 196
Arthrocnodax 196
Asilus ferrugineiventris 196
aspidistra scale 238
assassin bug 97, 110, ***111***, 113, 115, 117, 128, 134, 137, 139, 142, 157, 162, 195, 196
Asynonychus cervinus **141**
Atta 239
Austacris 184
Australian citrus whitefly **104**, ***105***, 192, 196, 197, ***217***, ***222***, ***224***, ***225***
Austropeplus **115**

B

Bacillus thuringiensis 123, 126, 136, 199, 234, 235, 236, 254
Bactrocera 160
    aquilonis 184
    cacuminatus 165
    carambolae 238

dorsalis 238
endriandrae 165
jarvisi 184
neohumeralis 160
papayae 160, **165**
philippinensis 238
tryoni **160**
Badumna longinqua **185**
Baker's mealybug 238
banana fruit caterpillar **137**, 193, 194, ***216***
bark damage 10, 11, 152
    chewed 144, ***145***
    cracking 73
    holes (in green bark) 183
    splitting 71, ***71***
barnacle scale 239
Batrachedra arenosella 72, 197
Bdella 184, 197
bdellid mite 26, ***26***
Bdellidae 184
Beauveria 142, 173, 175, 176, 180
    bassiana 72, 199
bee 59
beetle 31, 36, 72, 136, **141**, 151, 183, 195, 196, 197, 199, 226, 239
    apple weevil **147**
    citrus bark borer 239
    citrus branch borer **148**, ***149***, *219*
    citrus fruit weevil **145**
    citrus leafeating weevil **142**, ***143***, 146, ***216***
    citrus longicorn **149**, ***150***
    citrus trunk borer 239
    dicky rice weevil **143**, ***146***
    dried-fruit beetle 183
    elephant weevil **144**, ***145***
    fig longicorn **149**, ***150***, ***151***, *219*
    fruit-tree borer 183
    Fuller's rose weevil 5, **141**, ***142***, 194, 196, ***221***, ***222***, ***224***, ***225***
    green snout weevil 239
    Hong Kong beetle 239
    large auger beetle 183
    leaf beetle 141
    monolepta beetle **153**, ***154***, 194, ***216***
    mottled flower scarab beetle 183
    pitted apple beetle 183
    pittosporum longicorn **149**, ***150***
    redshouldered leaf beetle 153
    rhyparida beetle **153**, ***154***, ***216***
    rove beetles 196
    soft-winged flower beetles 196
    soldier beetle 196
    speckled longicorn **151**, ***152***, *219*
    spider longicorn 150
    spinelegged citrus weevil 143, **146**, ***147***
    sugarcane rootstalk borer 239
    tiger beetle 196
    whitestriped weevil 183
beetle mite **36**
Bemisia citricola 238
big-eyed bug 196
Biprorulus bibax **108**, 254
bird 13, ***13***, 58, 59, 99, 103, 113, 117, 139, 157, 162, 191, 198
black citrus aphid 11, **89**, *90*, **91**, **92**, 93, 195, 196, 197, ***216***, ***221***, ***222***, ***224***, ***225***, 254
black parlatoria scale 238
black scale 5, 39, **47**, 193, 195, 197, ***217***, 219, ***221***, ***222***, ***224***, 233, **254**
black spot 11, **12**, ***220***
black thread scale 238
bladder cicada **102**
blastobasid fruitborers 124, **126**, ***127***, *219*, *222*
Blastobasis **126**, ***127***
Blepharipa 139, 194
    fulviventris 139, ***139***, 194
blind pocket 11
blue mould 5, **12**, ***220***
blue-tongue lizard 191
Bostrychopsis jesuita 183

braconid wasps 162, 166
branch damage
    bark splitting 9
    destruction 182
    girdling 9
    holes 152
    ringbarking 149
    stem pitting ***10***, 11
    tunnelling 144, ***148***, 149, ***149***, 150, 152
brevipalpid mite 19
Brevipalpus
    californicus **34**
    lewisi **34**
broad mite 22, **25**, ***26***, ***27***, 196, 197, ***216***, ***218***, 219, ***224***
bronze orange bug **111**, ***112***, ***113***, 194, 196, ***216***, 232
brown citrus aphid **89**
brown citrus rust mite **17**, *18*, *20*, 21, 196, 197, 201, 202, ***218***, 219, ***222***, ***224***, ***225***, 233, 254
brown house spider **185**
brown lacewing ***41***, 48, ***92***
brown rot 5, 9, **11**, ***220***, ***223***
brown smudge bug 196
Bruchophagus fellis **168**, 254
bud moth ***216***
bug 197, 199, 239
    assassin 97, 110, ***111***, 113, 115, 117, 128, 134, 137, 139, 142, 157, 162, 195, 196
    big-eyed 196
    bronze orange **111**, ***112***, ***113***, 194, 196, ***216***, 232
    brown smudge 196
    citrus blossom **115**
    citrus stink 239
    crusader **116**, ***117***, ***218***, ***221***, ***224***, ***225***
    damsel 196
    flower 196
    fruitspotting 183, 193, 194
    green shield 108
    green vegetable **113**, ***114***, 193
    grey cluster 183
    leaf-footed 239
    passionvine 183
    Rutherglen 183
    spined citrus **108**, ***109***, ***110***, ***111***, 114, 192, 193, 194, 196, 197, ***219***, ***221***, ***222***, 254, 255
butterfly 118, 195, 197, 199
    large citrus **138**, ***139***, ***140***, 194
    northern citrus 183
    small citrus **138**, 194

C

Caedicia **155**
    simplex **155**
Calacarus citrifolii 238
Cales noacki 105, 192
camellia scale 239
Capnodium 10
carambola fruit fly 238
Caribbean fruit fly 238
Carpophilus 183
caterpillar, scale-eating 40, 45, 48, 50, 55, 57
Catoblemma dubia 40, 45, 48, 50, 51, 55, 57, 197
cecid fly 80, 81
Cecidomya **166**
cecidomyiid flies 19, 22
centre rot **12**, ***220***
Centrodora
    darwini 110, 192
    scolypopae 99, 192
Ceranisus 178
Ceratitis
    capitata **163**
    rosae 238
Cermatulus nasalis 139, 196
Ceroplastes
    brevicauda 239
    cirripediformi 239

destructor **57**, **254**
floridensis **56**
rubens **53**, **254**
rusci 239
sinensis **59**
Chaetanaphothrips orchidii **175**
Chaetophthalmus biseriatus 194
chaff scale **76**, **77**, 192, **218**, 219, 233
Chauliognathus pulchellus 196
Cheiloneurus near *gonatopodis* 241, 246
chemical burn 18, 22
cheyletid mite 25, 76, **76**, **77**
Chilean predatory mite **28**, 29, **30**
Chilocorus
    baileyi 195
    circumdatus 67, 71, **71**, 72, **72**, 75, 76, 195, 253
chilocorus ladybird 67, 71, **71**, 72, **72**, 75, 234
chimera 13, **223**
Chinese wax scale **59**, 233
Chrysomphalus aonidum **69**, **253**
Chrysopa 40, 94
cicada **102**, **103**
    bladder **102**
circular black scale **69**, **70**, 192, **218**, 219, 233, **253**
Cirrospilus
    near *ingenuus* 120, 193
    quadristriatus 120, **121**, 193, 254
citricola scale **39**, **40**, 192, 195, 197, **221**, **222**, **224**, 233, 255
Citripestis sagittiferella 239
citrophilous mealybug **78**, 80, 82, 192, 193, 194, 195, 196, 197, **221**, **222**, 232, 233, **253**, 255
Citrostichus phyllocnistoides 193, 254
citrus aphid **89**, **90**, 195
citrus bark borer 239
citrus blackfly 238
citrus blast **13**
citrus blossom bug **115**
citrus blossom midge 166, **167**, **216**
citrus branch borer **148**, **149**, 219
citrus bud mite 24, **25**, 197, **216**, **224**
citrus flat mite **34**, **218**, 219
citrus flower moth **131**, **132**
citrus fruit weevil **145**
citrus gall wasp **168**, **170**, 194, **219**, 254
citrus grey mite 238
citrus jassid **99**
citrus katydid **155**
citrus leafeating cricket 155, **157**, **158**, **218**
citrus leafeating weevil **142**, **143**, 146, **216**
citrus leafhopper **99**, **100**, **219**
citrus leafminer 30, **118**, **119**, **120**, **134**, 193, 194, **219**, **222**, **225**, 231, 232, 254, 255
citrus leafroller **133**, **134**
citrus longicorn **149**, **150**
citrus mealybug 78, **82**, **83**, **84**, 128, 193, 195, **217**, 219, 238, **254**, 255
citrus nematode 4, **186**, **187**
citrus planthopper **95**, **96**, **97**, 193, 194, 219, 254
citrus psylla 238
citrus red mite 9, 31, **32**, **33**, 195, 196, 197, **216**, **218**, 219, 233
citrus rindborer **125**, **126**, **219**
citrus rust mite 17, 19, **21**, **22**, **23**, 24, 26, 196, 197, 201, 202, **218**, 219, 233
citrus rust thrips **175**, **176**, 199, **219**
citrus snow scale **70**, **71**, **72**, 73, 76, 192, 195, 196, 197, **218**, 219, **253**
citrus spiny whitefly 238
citrus stink bug 239
citrus thrips 238
citrus tristeza virus **10**
citrus trunk borer 239
citrus wax scale 239
citrus whitefly 233, 238
Cladosporium 10
Cleridae 151
closterovirus 11
cloudy wing whitefly 238

cluster caterpillar 183
Coccidoxenoides peregrinus 83, 193, 241, 250
Coccinella transversalis 90, **91**, 195
Coccobius atrithorax 55
Coccophagus **40**, 241, 242, 250
    ceroplastae 46, 55, 192, 240
    gurneyi 79, **79**, 192, 241, 250, 253
    lycimnia 40, 45, 192, 241, 243, 254
    near *rusti* 42, **42**, **43**, 192
    semicircularis 40, **40**, 45, 192, 241, 243, 254
Coccus 41, 52
    hesperidum **43**, 254
    longulus **46**
    pseudomagnoliarum **39**
    viridis **41**
Coelophora inaequalis 90, **91**, 195
Coleoptera 195
Colgar peracutum **95**
Colgaroides acuminata **95**
collar rot 4, 5, 9
Colletotrichum gloeosporioides 10, 13
common garden snail **190**, **191**, **221**, **225**
common hoverfly 90, 94
common paper wasp **170**
common spotted ladybird 55, 90, **91**
Comperiella 247
    bifasciata 65, 66, 68, 69, 193, 241, 247, 252
    lemniscata 193, 241, 247
Comstock's mealybug 238
concave gum 11
Conogethes punctiferalis **126**
core rot 12
corn earworm **134**, **135**, **136**, 137, 193, 194, 195, 196, 197, 199, **216**
corrugations, on fruit 13
Corticium salmonicolor 9
cotton aphid 93
cottony citrus scale **52**, **53**, 195, **217**, 219, 233
cottony cushion scale **61**, **62**, 193, 194, 195, 196, **217**, 219, **222**
cricket **155**
    citrus leafeating 155, **157**, **158**, **218**
crotch rot 10
crusader bug **116**, **117**, **218**, **221**, **224**, **225**
Cryptoblabes
    adoceta **126**
    gnidiella 239
cryptochetid fly 62, 196
Cryptochetum 194
    iceryae 62, **62**, 194, 196
Cryptolaemus montrouzieri 40, 42, 45, 46, 48, 50, 51, 52, 53, **53**, 55, 57, 62, **63**, 80, 81, 83, 86, 195, 235
Cryptophlebia leucotreta 239
Cunaxa 184
cutworm 196
Cybocephalus 72, 196
    aleyrodiphagus 105, 196
Cytosoma schmeltzi **102**

D

damsel bug 196
Deraeocoris 136
    signatus 196
Diachasmimorpha 166
    tryoni 162, 194
Diadiplosis koebelei 80, 81, **84**, 196
Diaea 197, **198**
Dialeurodes
    citri 238
    citrifolii 238
Diaphorina citri 238
Diaporthe citri 10, 74
Diaprepes abbreviatus 239
dicky rice weevil 143, **146**
dieback 10, 62, 65, 69, 71, 73, 75, 76, 102, 154, 168
Diomus 48
    notescens 40, 45, 55, **81**, 195
Diptera 194

Dirioxa pornia 184
diseases
- alternaria brown spot **12**, **220**
- anthracnose 5, **13**, **223**
- armillaria root rot **9**
- black spot **11**, **12**, **220**
- blind pocket 11
- blue mould 5, **12**, **220**
- brown rot 5, 9, **11**, **220**, **223**
- centre rot **12**, **220**
- citrus blast **13**
- citrus tristeza virus 4, 9, **10**, 90
- collar rot 4, 5, 9
- crotch rot 10
- exocortis 4, **11**
- greasy spot **10**, **223**
- green mould **12**, 164, **220**
- melanose 5, **10**, 18, 22, **220**
- phytophthora 4, **9**, 153, 188
- pink disease 5, **9**
- psorosis **11**
- root rot 4, 5, 9
- scab **9**, **220**
- scaly butt **11**
- septoria spot 5, **12**, **223**
- shell bark **10**
- sooty blotch **13**, **220**
- sooty mould **10**, 13, 35, 36, 37, **38**, 39, 40, 41, 42, **42**, 43, 44, 46, 48, 49, 51, 52, 54, **55**, 56, 58, 60, 62, 78, 79, 81, 83, 86, 88, 89, 90, 93, 95, **96**, 97, 99, 104, 105, 106, 107, **123**, 204, **220**
- sour rot **12**
- sudden death syndrome **9**
- tristeza virus 4, 9, **10**, 90

Diversinervus
- *elegans* **44**, 45, 56, 193, 241, 246, 254
- *stramineus* 42

Dolichogenidea arisanus 123, 194
dolichopodid flies 197
dried-fruit beetle 183
Drosophila 161, 164, 165
dryinid wasp 97

E

earwig 197
eastern spinebill 198
echidna 182
elephant weevil **144**, **145**
Eleppone anactus **138**
Empoasca
- *citrusa* 239
- *smithi* **99**

Encarsia 42, **42**, 70, 105, 192, 235, 247
- *australiensis* 55
- *citrina* 67, 69, 71, 192, 241, 249, 253
- *inquerenda* 253
- *perniciosi* 67, 192, 241, 249, 253

encyrtid wasps 79, 255
Encyrtus
- *infelix* 49, **50**, 193, 241, 246
- *lecaniorum* 193, 240

Entomophthora 91, 136, 173, 175, 176, 180
Entomophthorales 199
Ephippityha trigintiduoguttata **155**
Epiphyas postvittana **122**, 254
epipyropid moth **95**, 97
Epitetracnemus 241, 247
Eretmocerus 192
Eriophyes sheldoni **24**
Eudocima salaminia **129**
Eupalopsellidae 184
Eupalopsis 184
- *jamesi* 67, 184, 197
Euplectrus kurandaensis 137, 193
Eupodes 184
Eupodidae 184

Euryischomyia flavithorax 192, 241, 243, 254
Euseius
- *elinae* 19, **19**, 22, 23, **23**, 25, 27, 29, 33, 67, 178, 197
- *victoriensis* 19, 22, 23, **23**, 25, 27, 29, 32, 35, 67, 105, 107, 197, 202

Eutetranychus
- *anneckei* 238
- *banksi* 238
- *orientalis* **31**

Eutinophaea bicristata **142**
exocortis 4, **11**
Exoristae sorbillans 131, 194

F

Fabrictilis gonagra 183
false codling moth 239
ferment flies 161, 164, 165
Fidiobia citri 142, 194
fig longicorn **149**, **150**, **151**, **219**
fig wax scale 239
flat mite 197, 233
flatid **95**
Florida red scale **69**
Florida wax scale 54, **56**, 193, 194, **217**, 219, 233
flower bugs 196
flower damage
- blackening 24
- blistering 179, **179**
- bud, swollen **167**
- distortion 24, **24**, 86, 90, 93
- flower drop 83, 86, 90, 93, 112, 123, 136, 167
- flowers eaten **131**, 132, **132**, 137, **153**, 154, **154**, 156
- petal browning 179
- webbing 123, 132
- wilting 115

flower end rot 12
flower moth 197
flower spiders 198, **198**
fly 192, 194, 196
- cryptochetid 62, 196
- ferment 161, 164, 165
- robber 196

Fopius 166
- *arisanus* 162, 194
- *deeralensis* 162, 194

Frankliniella schultzei 184
fruit burn 13
fruit damage 12
- albedo, discoloured 114
- black spots 11, 12, **12**
- blemish, black 22
- blemish, grey 26
- blemish, silver-grey 22
- blemish, spotted, grey 35
- bronzing 21
- brown spots 12, 100, **100**, 109, **109**, **110**
- calyx, dry, brown 73
- chlorosis 28, 31
- disfiguring 69, 71, 75, 76, 156, 159, 174
- distortion 24, 25, **25**, 83, 86, **86**
- feeding in fruit **127**, 128, **160**, 162, 164, 165
- fruit drop 3, 12, 83, 86, 102, 109, **109**, 112, 114, **114**, 123, 124, 126, 128, 130, **131**, 137, 146
- fruit eaten 13, **13**, 132, **134**, 136, 137, **137**, 154, 159, **159**
- grey scarring 177, **178**
- halo damage 123, **123**, 173, **173**, 174
- holes 126, **126**, **127**, **130**, 136, 137, 146, **190**, 191
- 'mudcake' 10
- navel end rot 83
- netting 13, **13**, **220**, **223**
- pitting 12, 65
- pulp sucked out 130
- pustules, raised, black 10, **10**
- rind damage 11, 13, **18**, **22**, 23, **27**, **28**, 146
- rind discolouration **174**

rind holes 124, 126, 128
rind scarring 13, 34, *35*, 114, 146, *147*, 156, *157*, *174*, 177, *177*, 191
rind softening 178
ring scarring 173, 174
rot 79, 130, 162, 164, 165
russeting 17, 18, 22
rust marks, circular 176, *176*
scarring 123
scurfing 142, 173, 174
seed damage 109
'sharkskin' 22, 26
stem-end scarring 174
stippling 18, 33
yellow spots 68
fruit fly 72, **160**, 167, 184, 238
    carambola 238
    Caribbean 238
    island 184
    Jarvis' 184
    lesser Queensland 160
    medfly 163
    Mediterranean 163, *224*, *225*
    Mexican 238
    Natal 238
    Northern Territory 184
    oriental 238
    papaya 160, **165**
    Philippine 238
    Queensland 5, 130, **160**, *161*, 163, 165, 166, 194, 196, **219**, 255
    solanum 165
    South American 238
fruit-tree borer 183
fruitpiercing moth 118, **129**, *130*, *131*, 193, 194, **219**
fruitspotting bug 183, 193, 194
Fuller's rose weevil 5, **141**, *142*, 194, 196, **221**, *222*, *224*, *225*
fungi, nematode-trapping 187, 188, 189
fungi-feeding mites 35, 184
fungus, red-headed 67, 72, 74, 75, 76
Fusarium 188
    coccophilum 67, 72, 74, 75, 76, 199
    henningsii 199
    lavarum 199
    moniliforme var. *subglutinans* 60, 199
    stilboides 59, 199

G

Geloptera porosa 183
Geocoris 136
    lubra 196
giant grasshopper **159**, 193
giant northern termite **181**, *182*, **219**
Gloeodes pomigena 13
Glover's scale 71, 73, **74**, *75*, 192, **218**, 219, 233
Glycyphagidae 184
Glycyphagus 184
Goetheana shakespeari 178, 194
golden mealybug 183
Goniozus 123, 194
granulosis 136, 199
grasshopper **155**, 184, 199
    giant **159**, 193
    spur-throated locust 184
greasy spot **10**, **223**
green citrus leafhopper 239
green coffee scale **37**, **41**, *42*, *43*, 192, 195, **217**, 219, 233
green lacewing **40**, 48, 83, *90*, *92*, 107
green mould **12**, 164, **220**
green planthopper **95**, *96*, *97*, 219
green shield bug 108
green shield scale 239
green snout weevil 239
green vegetable bug **113**, *114*, 193
greenhouse thrips 176, *177*, 194, 197, **219**, *224*, *225*, 232, 233
grey cluster bug 183
ground beetles 196
Gryon meridionis 193
Guignardia citricarpa 11
gummosis 9, 102
Gyranusoidea 88

H

hail damage 13, **220**, **223**
Halmus chalybeus 33, 52, **53**, 59, 60, 67, 70, 74, 195
Haplothrips 173, 175, 176, 197
hard wax scale **59**, **60**, 193, 195, 199, **219**, **222**, **225**, 233
Harmonia
    conformis 40, 45, 55, 90, **91**, 94, 195
    testudinaria 90, **91**, 195
heart-shaped scale 239
Helicoverpa
    armigera **134**
    punctigera **134**
heliothis moth **135**
Heliothrips haemorrhoidalis **176**
Helix aspersa **190**
Hemicheyletia 197
Hemicriconemoides 186
Hemisarcoptes 72, 184
    mali 197
Hemisarcoptidae 184
hemispherical scale **37**, **49**, **50**, 193, **217**, 219, 233
Heterorhabditis 142, 199
Hirsutella 23, 88
    kirchneri 199
    thompsonii 29, 199
Holconia 198
Hong Kong beetle 239
hoverfly 196
    common 90, 94
huntsman spider 198, **198**
Hyalarcta huebneri 183
Hypomyces squamosus 239

I

Icerya 195
    egyptiaca 183
    purchasi **61**
    seychellarum 183
ichneumonid wasp 125, 134, 136, 137
inland katydid **155**, *156*
Iridomyrmex 38
    purpureus 37, 38
    rufoniger 38
Ischnaspis longirostris 238
island fruit fly 184
Isotenes miserana **124**

J

Jarvis' fruit fly 184
jumping spider 198

K

Karnyothrips flavipes 197
katydid **155**, 158, 196, **216**, **221**
    citrus **155**
    inland **155**, *156*
    spotted **155**, *156*, 158

L

Labidura riparia truncata 197
lacewing 29, 33, 45, 48, 55, 59, 62, 67, 69, 80, 81, 83, 84, 90, 94, 99, 105, 110, 120, 123, 136, 195, 197
    brown **41**, 48, **92**
    green **40**, 48, 83, *90*, *92*, 107

ladybird 27, 29, 31, 33, 40, 45, 48, 50, 52, *53*, 62, *62*, 69, 70, 71, 72, *72*, 76, 80, 81, 83, 88, 90, *91*, *92*, 94, 105, 195, 196, 252, 255
    chilocorus 67, 71, *71*, 72, *72*, 75, 234
    common spotted 55, 90, *91*
    mealybug 42, 46, 48, 50, 51, 52, 53, *53*, 55, 57, 62, *63*, 80, 81, 83, *84*, 86, 87, 141, 195
    steel-blue 33, 52, *53*, 59, 60, 67, 70, 74
    stethorus 29, 31, 33, *33*
    transverse 90, *91*
    variable 90, *91*
    vedalia 195
    yellow-shouldered 90, *91*
large auger beetle 183
large citrus butterfly 138, *139*, *140*, 194
leaf beetle 141
leaf case moth 183
leaf damage 12, *12*, *29*, *32*, *33*, 154, 158, 183
    black spots 10
    blistering 10, 18, 179
    brown spots 12, *12*
    chewed 142, 143, *143*, 158, 159, *159*
    chlorosis 28, *29*, 31, 179, *180*
    distortion 24, 26, 79, 81, 86, *86*, 90, 93, *93*, 120, *120*
    grey scarring 26, 177, *178*
    holes 126, *126*, 137, 154, *154*, 156, *157*, 158, *158*, 183
    leaf drop 9, 10, 11, 13, 18, 28, 31, 33, 40, 44, 48, 52, 62, 64, 65, 68, 69, 71, 73, 75, 76, 100, 102, 107, 143, 183
    multiple budding 24
    pustules, raised, black 10
    rolled 134
    russeting 17, 22
    scarred and skeletonised 191, *191*
    scarred and twisted 26, 173
    serrated edges 142, 146, *147*
    spotting *32*, 34, 68, 71
    stippling 18, 31, 33, *33*
    webbing 123, 124
    wilting 9, 149, *149*, 183
    yellowing 9, 65, 73, 177
leaf-footed bug 239
leafcutting ant 239
leafeating cricket *157*
leafhopper **95**, 195, 196, 197
    citrus jassid **99**
    citrus leafhopper 99, *100*, *219*
    green citrus leafhopper 239
lemon bud moth **131**, *132*
Lepidosaphes
    beckii 73, **253**
    gloverii **74**
Leptoglossus phyllopus 239
Leptomastidea abnormis 83, 193, 241, 251
Leptomastix dactylopii 83, 193, 236, 241, 251, 254
leptomastix wasp 83, *84*, 85
lesser Queensland fruit fly 160
lichen 72
lightbrown apple moth 5, 20, 23, **122**, 125, 194, 196, *216*, *221*, *222*, 254
Linepithema humile 38
Lissopimpla
    excelsa 194
    semipunctata 137, 194
lizard, blue-tongue 191
long soft scale **46**, 192, *217*, 219, 233
longicorn 141
    citrus **149**, *150*
    fig **149**, *150*, *151*, *219*
    pittosporum **149**, *150*
    speckled **151**, *152*, *219*
    spider 150
longtailed mealybug 78, **80**, 82, 193, 195, 196, 197, *217*, 219, ***221***, ***222***, ***225***, 233, **253**, 254
Lowveldt citrus mite 238
Lycosa 197
lygaeid bug 131
lynx spider 198, *198*

M

Maleuterpes spinipes **146**
Mallada 40, 48, 55, 59, 67, 80, 81, 197
    signata 83, *90*, 236
mango planthopper **95**, *96*
Maroga 183
Mastotermes darwiniensis **181**
mealybug 10, 29, 30, 35, 37, **78**, 183, 195, 196, 197, 226, 232, 233, 238, 240, 241, 249, 254
    Baker's 238
    citrophilous **78**, 80, 82, 192, 193, 194, 195, 196, 197, ***221***, ***222***, 232, 233, **253**, 255
    citrus 78, **82**, *83*, *84*, 128, 193, 195, *217*, 219, 238, **254**, 255
    Comstock's 238
    golden 183
    longtailed 78, **80**, 82, 193, 195, 196, 197, *217*, 219, ***221***, ***222***, ***225***, 233, **253**, 254
    oleander 238
    rastrococcus **87**, *88*, *217*, 219, 233
    spherical 25, **85**, *86*, 193, 195, *217*, 219
mealybug ladybird 42, 46, 48, 50, 51, 52, 53, *53*, 55, 57, 62, *63*, 80, 81, 83, *84*, 86, 87, 141, 195
meat ant 37, 38
mechanical injury 13
medfly **163**
Mediterranean fruit fly **163**, ***224***, ***225***
megalurothrips **174**, 177, ***221***, ***222***, 255
Megalurothrips kellyanus **174**, 255
Megaphragma mymaripenne 178, 194
Megastigmus
    brevivalvus 169, ***170***, 194
    trisulcus 169, 194
Melanastoma agrolas **92**
Melangyna viridiceps 196
melanose 5, **10**, 18, 22, ***220***
Meliola 10
melon aphid 89, **92**, *93*
Metaphycus **45**, 244, 246
    bartletti 48, **48**, 193, 241, 245, 254
    helvolus **44**, 45, 48, 193, 241, 245, 254
    lounsburyi 48, **48**, 193, 241, 245, 254
    near *inviscus* 241, 245
    near *varius* 193
    varius 55, 241, 245
Metarhizium 142, 173, 175, 176, 180, 199
Mexican fruit fly 238
Micraspis frenata 59, 60, 195
Micromus 83, 123
    tasmaniae 40, *41*, 45, 48, **92**, 197
Microplitis demolitor 136, 194
Microterys
    flavus 45, 56, 193, 241, 245, 246, 254
    triguttatus 193, 240
Mictis profana **116**
midge 84, 160, 196
    citrus blossom **166**, *167*, *216*
mite 17, 29, 72, 184, 195, 196, 197, 199, 226, 228, 232, 233, 238
    acarid 26
    bdellid 26, *26*
    beetle **36**
    brevipalpid 19
    broad 22, **25**, *26*, *27*, 196, 197, *216*, *218*, 219, ***224***
    brown citrus rust 17, *18*, *20*, 21, 196, 197, 201, 202, *218*, 219, ***222***, ***224***, ***225***, 233, 254
    cheyletid 25, 76, *76*, *77*
    Chilean predatory *28*, 29, *30*
    citrus bud **24**, *25*, 197, *216*, ***224***
    citrus flat **34**, *218*, 219
    citrus grey 238
    citrus red 9, 31, *32*, *33*, 195, 196, 197, *216*, *218*, 219, 233
    citrus rust 17, 19, **21**, *22*, *23*, 24, 26, 196, 197, 201, 202, *218*, 219, 233
    common red spider 238
    flat 197, 233
    fungi-feeding 35, 184

Lowveldt citrus 238
mould 35, **36**
oriental spider 28, **31**, *32*, 196, 197, ***218***, 219
phytoseiid 14, 19, 22, 25, 27, 29, 31, 32, 33, 35, 67, 173, 175, 176, 178, 254
phytoseiid, native 29
pink citrus rust 238
scavenging 35, 184
soil 35, **36**
Texas citrus 238
two-spotted **28**, *29*, *30*, 31, 35, 196, 197, ***218***, 219, ***221***
typhlodromid ***29***, *30*
vegetable spider 184
Monolepta australis **153**
monolepta beetle **153**, ***154***, 194, ***216***
Monoleptophaga caldwelli 154, 194
Moranila 56, 193
  californica 55, 193, 241, 242
moss 72
moth 29, 51, **118**, 183, 195, 196, 197, 199
  armyworm 196
  banana fruit caterpillar **137**, 193, 194, ***216***
  blastobasid fruitborers 124, ***126***, ***127***, ***219***, ***222***
  bud moth ***216***
  caterpillar, scale-eating 40, 45, 48, 50, 55, 57
  citrus flower moth **131**, ***132***
  citrus leafminer 30, **118**, ***119***, ***120***, ***134***, 193, 194, ***219***, ***222***, ***225***, 231, 232, 254, 255
  citrus leafroller **133**, ***134***
  citrus rindborer **125**, ***126***, ***219***
  cluster caterpillar 183
  corn earworm **134**, ***135***, ***136***, 137, 193, 194, 195, 196, 197, 199, ***216***
  cutworm 196
  epipyropid moth **95**
  false codling moth 239
  flower moth 197
  fruitpiercing moth 118, **129**, ***130***, ***131***, 193, 194, ***219***
  heliothis moth **135**
  leaf case moth 183
  lemon bud moth **131**, ***132***
  lightbrown apple moth 5, 20, 23, **122**, 125, 194, 196, ***216***, ***221***, ***222***, 254
  native budworm **134**, 137, 193, 194, 196, 197, 199
  orange fruitborer **124**, 125, 130, 194, ***219***
  orange tortrix 239
  rindboring orange moth 239
  sorghum head caterpillar **126**, ***127***, ***128***
  yellow peach moth **126**, ***127***, ***128***, 194
mottled flower scarab beetle 183
mould mite 35, **36**
Musgraveia sulciventris **111**
mussel scale 71, **73**, ***74***, 75, 76, 192, ***218***, 219, ***253***
Mycosphaerella 10
Myiocnema near comperei 241, 243
Myllocerus 183
mymarid wasp 101

N

Nabis 136
  kinbergii 196
Natal fruit fly 238
native budworm **134**, 137, 193, 194, 196, 197, 199
native snails 190
Nectria
  aurantiicola 199
  flammea 199
nematode-trapping fungi 187, 188, 189
nematode 4, 142, **186**
  citrus 4, **186**, ***187***
  parasitic 199
  root lesion **188**
  stubby root **189**
Neomerimnetes sobrinus **145**
Neozygites 29

Nezara viridula **113**
nigra scale **50**, *51*, 193, 197, ***217***, 219, 233
Nipaecoccus
  aurilanatus 183
  viridis **85**
nitidulid beetle 105
Nomadacris 184
Nomuraea rileyi 136, 199
northern citrus butterfly 183
Northern Territory fruit fly 184
Nosema locustae 199
NPV 123
nuclear polyhedrosis virus 123, 136, 199
Nysius
  clevelandensis 183
  vinitor 183

O

Oechalia 136
  schellembergii 123, ***136***, 139, 196
oleander mealybug 238
oleander scale ***204***
oleocellosis 13, ***220***, ***223***
Oligochrysa lutea 83, ***92***, 197
onion thrips 184
Ooencyrtus 97, 131, 193
Ophelosia 62, 79, 193, 241, 249
Opisthoncus 198
Opius perkinsi 162, 194
orange fruitborer **124**, 125, 128, 130, 194, ***219***
orange pulvinaria 239
orange tortrix 239
Orchamoplatus citri **104**
orchid thrips **175**
Oribatida **36**
oriental fruit fly 238
oriental scale 183, 192, 193, 195, 196, 197
oriental spider mite 28, **31**, *32*, 196, 197, ***218***, 219
Oripodidae 184
Orius 136
  armatus 196
  tantillus 196
Orthorhinus cylindrirostris **144**
Othreis
  fullonia **129**, ***130***
  materna **129**
Otiorhynchus cribricollis **147**
Oxyopes 198

P

Pachyneuron kingsleyi 139, 193
Paecilomyces fumosa rosea 173, 175, 176, 180, 199
Palexoristus solemis 137, 194
Paniscus testaceous 137, 194
Panonychus citri **32**
papaya fruit fly 160, **165**
paper wasp **170**
  common **170**
  yellow **170**
papillae 13, ***220***
Paraceraptrocerus 241, 244
  nyasicus 58, 193, 241, 244, 254
Paracoccus burnerae 238
Paradisterna plumifera **151**
Parapriasus australasiae 40, 45, 48, **48**, 195
Parasaissetia nigra **50**
parasitic nematodes 199
Parasitidae 184
Parasitus 184
Paratylenchulus 186
Parlatoria pergandii **76**
Parlatoria ziziphis 238
passionvine bug 183
passionvine hopper **98**, 192

pathogens **199**
Penicillium
    digitatum **12**
    italicum **12**
pentatomid bug 113, ***136***
Perperus lateralis 183
Pheidole 38
Philippine fruit fly 238
Phomopsis citri 10
Phyllocnistis citrella ***118***, 254
Phyllocoptruta oleivora **21**
Phyllostictina citricarpa 11
physiological fruit drop 3
Phytodietus 125, 194
phytophthora 4, **9**, 153, 188
Phytophthora 188
    citrophthora 9, 11
    parasitica 9
phytoseiid mite 14, 19, 22, 25, 27, 31, 32, 33, 35, 67, 173, 175, 176, 178, 254
Phytoseiidae 184
Phytoseiulus persimilis 29, 30, 197, 234
picking damage 13, ***220, 223***
pink citrus rust mite 238
pink disease 5, **9**
pink wax scale **53**, ***54, 55***, 192, 193, 195, 197, 202, ***217***, 219, ***224, 225***, 231, ***232***, 233, **254**
Pinnaspis aspidistrae 238
pitted apple beetle 183
pittosporum longicorn ***149, 150***
plague thrips **179**, ***180***
Planococcus citri **82**, **254**
planthopper 10, 37, **95**, 193, 195, 199, ***217, 222***
    citrus **95**, ***96, 97***, 193, 194, 219, 254
    green **95**, ***96, 97***, 219
    mango **95**, ***96***
platygasterid 79
Plesiochrysa ramburi ***40***, 45, 48, 197
Podagricomela nigricollis 239
Polistes 197
    dominulus **170**
    humilis synoecus **170**
    tepidus **170**
Polyphagotarsonemus latus **25**
Pratylenchus 188
praying mantis 110, 117, 137, 139, 142, 157, 162, 195, 197
Prays
    nephelomima **131**
    parilis **131**
pre-harvest fruit drop 3
predatory beetle 105, ***105***, 196
predatory bug 110, 123, 136, 173, 175, 176
predatory fly 80, 81
predatory ladybird 74, 253
predatory mite 5, 14, 17, ***19***, 25, 29, 30, 33, 67, 105, 107, 184, 195, 202, 203, 254
predatory thrips 29, 105, 173, 175, 176, 178
Princeps
    aegeus **138**
    fuscus canopus 183
Pristhesancus 136
    plagipennis 110, 113, 115, 117, 128, 134, 139, 142, 157, 162, 196
Pronematus 184
Prospaltella 105
Protaetia fusca 183
Protopulvinaria
    floccifera 239
    pyriformis 239
protozoans 199
Pseudococcus
    calceolariae **78**, 253
    comstocki 238
    cryptus 238
    longispinus **80**, **253**
    maritimus 238
Pseudomonas syringae 13

Psilopus 197
psorosis **11**
Psorosticha zizyphi **133**
psyllids 238
pteromalid 79
Pteromalus puparum 139, 193
Pulvinaria
    aurantii 239
    polygonata **52**, 61
    psidii 239
pulvinaria scale **52**
purple scale **73**, 233, **253**

Q

Quadrastichus 193
Queensland fruit fly 5, 130, **160**, *161*, 163, 165, 166, 194, 196, ***219***, 255

R

rastrococcus mealybug **87**, ***88, 217***, 219, 233
Rastrococcus truncatispinus **87**
red imported fire ant 239
red scale 5, 23, 30, 37, ***64***, *65*, ***66, 67***, 68, 69, 76, 192, 193, 195, 196, 197, 199, 202, 204, 205, ***218***, 219, ***222, 224, 225***, 231, 233, **252**, 253, 254, 255
red-headed fungus 67, 72, 74, 75, 76
redshouldered leaf beetle 153
regreening 13, ***223***
Rhynchocoris humeralis 239
Rhyparida 153
rhyparida beetle 153, ***154, 216***
Rhytidoponera 38
Rhyzobius 72, 234
    lophanthae 40, 45, 67, 69, 195
    near *lophanthae* 45, 48, 195
    ruficollis 80, 81, 195
    ventralis 55, 70, 195
rind creasing 13, ***220, 223***
rindboring orange moth 239
robber flies 196
Rodolia 62, 195
    cardinalis 62, 195
    koebelei 62, 195
root damage 142, 187, 188
    destruction of major roots 182
    stunting 189
    tunnelling 144
root lesion nematodes **188**
root rot 4, 5, 9
rove beetles 196
Rutherglen bug 183

S

Saissetia 52
    coffeae **49**
    oleae **47**, **254**
scab **9**, ***220***
scale **39**, 195, 196, 197, 226, 228, 231, 240, 254, 255
    armoured **64**, 82, 183, 231, 232, 233, 238, 240, 241, 247
    aspidistra 238
    barnacle 239
    black 5, 39, **47**, 49, 193, 195, 197, ***217***, 219, ***221, 222, 224***, 233, **254**
    black parlatoria 238
    black thread 238
    camellia 239
    chaff **76**, ***77***, 192, ***218***, 219, 233
    Chinese wax **59**, 233
    circular black **69**, ***70***, 192, ***218***, 219, 233, **253**
    citricola **39**, ***40***, 44, 47, 49, 192, 195, 197, ***221, 222, 224***, 233, 255
    citrus snow **70**, ***71, 72***, 73, 76, 192, 195, 196, 197, ***218***, 219, **253**
    citrus wax 239

cottony citrus 52, *53*, 61, 195, ***217***, 219, 233
cottony cushion 52, **61**, *62*, 193, 194, 195, 196, ***217***, 219, ***222***
fig wax 239
Florida red scale **69**
Florida wax 54, **56**, 193, 194, ***217***, 219, 233
Glover's 71, 73, **74**, *75*, 192, ***218***, 219, 233
green coffee **37**, **41**, ***42***, ***43***, 192, 195, ***217***, 219, 233
green shield 239
hard wax 54, 56, 57, **59**, *60*, 193, 195, 199, ***219***, ***222***, ***225***, 233
heart-shaped 239
hemispherical **37**, 47, **49**, *50*, 193, ***217***, 219, 233
long soft **46**, 192, ***217***, 219, 233
mussel 71, **73**, **74**, 75, 76, 192, ***218***, 219, **253**
nigra 50, *51*, 193, 197, ***217***, 219, 233
oleander ***204***
orange pulvinaria 239
oriental scale 183, 192, 193, 195, 196, 197
pink wax **53**, *54*, **55**, 56, 57, 192, 193, 195, 197, 202, ***217***, 219, ***224***, ***225***, 231, ***232***, 233, **254**
pulvinaria **52**
purple **73**, 233, **253**
red 5, 23, 30, 37, **64**, *65*, **66**, ***67***, 68, 69, 76, 192, 193, 195, 196, 197, 199, 202, 204, 205, ***218***, 219, ***222***, ***224***, ***225***, 231, 233, **252**, 253, 254, 255
Seychelles 183
soft brown 39, 41, 46, 47, 49, 51
West Indian red 238
white louse **70**, **253**
white wax **57**, *58*, *59*, 193, 194, 197, 198, 199, ***217***, 219, ***221***, ***224***, ***225***, 233, **254**
yanone 238
yellow 64, **68**, 192, 193, 195, 233, 252
scaly butt **11**
scavenging mites 35
Scelio flavicornis 159, 193
scelionid 99
sciomyzid fly 191
scirtothrips **172**, 175, ***216***
Scirtothrips
  albomaculatus **172**
  aurantii 238
  citri 238
  dorsalis **172**, *173*
Scolothrips sexmaculatus 29, 197
Scolypopa australis **98**
Scutellista
  caerulea **47**, 48, ***48***, 49, 50, 51, ***51***, 55, 56, 58, 60, 193, 241, 242, 254
  cyanea 254
Scymnodes lividigaster 59, 60, 90, ***91***, 195
Scymnus 27, 80, 81, 88, 196
Selenaspidus articulatus 238
Semielacher petiolatus 120, ***121***, 193
Septoria citri 10, 12
septoria spot 5, **12**, ***223***
Serangium bicolor 33, 59, 60, 196
Seychelles scale 183
shell bark **10**
'shoe-string' fungus 9
shoot damage 12
  blackening 12
  distortion ***25***, 89, 90, 93, ***93***, 100, ***120***, 134, ***134***
  wilting 112, 115, 116, ***117***
silvereye 198
Simosyrphus grandicornis 90, ***92***, 94, 196
Siphanta
  acuta **95**, ***96***
  hebes 95
six-spotted thrips **29**
Skeletodes tetrops **149**
small citrus butterfly **138**, 194
snail 5, **190**
  common garden **190**, ***191***, ***221***, ***225***
  native 190
soft brown scale 39, **43**, ***45***, 104, 192, 193, 195, 197, ***217***, 219, ***221***, ***222***, ***224***, 233, 254

soft scale 5, 10, 35, 37, **39**, 183, 195, 196, 197, 199, 232, 233, 239, 240, 241, 242, 253, 254
soft-winged flower beetles 196
soil mite 35, **36**
solanum fruit fly 165
soldier beetle 196
Solenopsis invicta 239
sooty blotch **13**, ***220***
sooty mould **10**, 13, 35, 36, 37, **38**, 39, 40, 41, 42, ***42***, 43, 44, 46, 48, 49, 51, 52, 54, **55**, 56, 58, 60, 62, 78, 79, 81, 83, 86, 88, 89, 90, 93, 95, ***96***, 97, 99, 104, 105, 106, 107, ***123***, 204, ***220***
sorghum head caterpillar **126**, ***127***, ***128***
sour rot 12
South American fruit fly 238
speckled longicorn **151**, ***152***, **219**
Sphaceloma fawcettii var. *scabiosa* 9
sphecid wasp 157
spherical mealybug 25, **85**, ***86***, 193, 195, ***217***, 219
spider 97, 99, 110, 123, 136, 162, 173, 175, 176, **185**, 195, 197
  brown house **185**
  flower 197, ***198***, ***216***
  huntsman 198, ***198***
  jumping 198, ***198***
  lynx 198, ***198***
  webbing **185**
  wolf 197
spider longicorn 150
spined citrus bug **108**, ***109***, ***110***, ***111***, 114, 192, 193, 194, 196, 197, ***219***, ***221***, ***222***, 254, 255
spinelegged citrus weevil 143, **146**, ***147***
spiraea aphid 89, **92**, ***93***, ***216***
Spodoptera litura 183
spore-forming bacteria 187, 188, 189
spotted katydid **155**, ***156***, 158
spray burn 13, ***220***, ***223***
spur-throated locust 184
steel-blue ladybird 33, 52, **53**, 59, 60, 67, 70, 74
Steinernema carpocapsae 199
stem end rot 5, 10, 12
Stethorus 195
  fenestralis 31, 196
  histrio 31, 196
  nigripes 33, 196
  vagans 196
stethorus ladybird 29, 31, 33, ***33***
stigmaeid mite 19, 22, 25, 35
Stigmaeidae 184
strepsiptera 38, 97
Strongylurus thoracicus **149**
stubby root nematode **189**
stunting of tree 10, 11
Sturmia 137, 194
stylar end rot 12
sugarcane rootstalk borer 239
sulphur-crested cockatoo **13**
sunburn 13, ***220***, ***223***
Sympiesis 120, 193
syrphid fly 90, ***92***, 94, 105

T

tachinid fly 110, 114, 128, 131, 136, 137, 139, 154, 157
Tamborina australis **157**
tardigrade 187, 188, 189
tarsonemid mite 19, 35, ***35***
Tarsonemidae 184
Tarsonemus 184
  waitei **35**, 184
'tear stain' melanose 10
Tegolophus australis **17**, 254
Telenomus 131, 136, 193
Telsimia 72, 196
Termite **181**
  giant northern **181**, ***182***, **219**
Tetracnemoidea
  bicarinatum 38

       brevicornis 79, 193, 241, 250, 253
       peregrina 81, 193, 241, 250, 253
       sydneyensis 81, 193, 241, 250, 253
tetranychid mite 19, 31, 33, 195
Tetranychus
       cinnabarinus 238
       neocaledonicus 184
       urticae **28**
Tetrastichus 88
       ceroplastae 57, 58, 60, **61**, 194, 241, 242, 254
Texas citrus mite 238
Thripobius semiluteus 178, 194
thrips 29, **172**, 184, 195, 196, 197, 233, 238
       citrus 238
       citrus rust **175**, *176*, 199, *219*
       greenhouse **176**, *177*, 194, 197, *219*, *224*, *225*, 232, 233
       megalurothrips **174**, 177, **221**, **222**, 255
       onion 184
       orchid **175**
       plague **179**, *180*
       scirtothrips **172**, 175, *216*
       six-spotted **29**
       tomato 184
Thrips
       imaginis **179**
       tabaci 184
thrush 191
tiger beetle 196
Tiracola plagiata **137**
tomato thrips 184
Toxoptera
       aurantii **89**
       citricida **89**, *90*, 254
transverse ladybird 90, **91**
tree decline 76
Trichodorus **189**
Trichogramma 123, 131, *136*, 139, **139**, 194, 236
       chilonis 194
       funiculatum 194
       near *brassicae* 194
Trichogrammatoidea 136, 194
Trichopoda 114
Trioza erythreae 238
Trissolcus **110**
       basalis 114, 193
       oenone 110, 193
       ogyges 110, 193
tristeza virus 4, 9, 10, **10**, 90
trunk damage **9**, **10**, 11, **11**, 150, **151**, 152, **152**, **181**, 182, **182**
twig damage **158**
       blistering 18
       chlorosis 28, 31
       death 150
       dieback 9, 33, 183
       distortion 24
       gall 168
       mining 120
       pustules, raised, black 10
       russeting 22
       scarred and skeletonised 191
       scarring, grey 26
       splitting 102
       spotting 34
       stippling 18, 31
       twisting 86
two-spotted mite **28**, *29*, 30, 31, 35, 196, 197, *218*, 219, *221*
tydeid mite 19, 26, 35, 36, **36**
Tydeidae 184
Tydeus 184
       californicus 36, **36**, 184
Tylenchulus semipenetrans **186**
typhlodromid predatory mite **29**, *30*
Typhlodromus occidentalis 29, 30, 197, 234

U

Unaspis
       citri **70**, **253**
       yannonensis 238
Uracanthus cryptophagus **148**

V

Valanga irregularis **159**
variable ladybird 90, **91**
vedalia ladybird 195
vegetable spider mite 184
Verticillium lecanii 43, 46, 48, 50, 51, 52, **53**, 55, 57, 60, 105, 199
virus 33, 140, 199
Voriella 123, 194

W

wasp
       citrus gall 168, **169**, **170**, 194, **219**, 254
       paper **170**
       yellow paper **170**
wasp morphology *241*
water burn 13, **220**, **223**
weather staining 18, 22
webbing spider **185**
weevil 141
       apple **147**
       citrus fruit **145**
       citrus leafeating **142**, *143*, 146, *216*
       dicky rice 143, **146**
       elephant **144**, *145*
       Fuller's rose 5, **141**, *142*, 194, 196, **221**, **222**, **224**, **225**
       green snout 239
       spinelegged citrus 143, **146**, *147*
       whitestriped 183
West Indian red scale 238
white louse scale **70**, **253**
white wax scale 57, **58**, **59**, 193, 194, 197, 198, 199, **217**, 219, **221**, **224**, **225**, 233, **254**
whitefly 37, **104**, 192, 197, 232, 233, 238
       aleurocanthus **106**, *107*, **217**
       Australian citrus **104**, *105*, 192, 196, 197, **217**, **222**, **224**, **225**
       citrus 233, 238
       citrus blackfly 238
       citrus spiny 238
       cloudy wing 238
       woolly 238
whitestriped weevil 183
wind rub 13, **13**, 204, **220**, **223**
wolf spider 197
woolly whitefly 238

X

Xanthopimpla 123, 194
Xiphinema 186

Y

yanone scale 238
yellow paper wasp **170**
yellow peach moth **126**, *127*, **128**, 194
yellow scale 64, **68**, 192, 193, 195, 233, 252
yellow-rumped thornbill 198
yellow-shouldered ladybird 90, **91**

Z

Zaommomentedon brevipetiolatus 120, 194
Zosteromyia 123, 194
Zosterops lateralis 198